Life-Cycle Impact Assessment: Striving towards Best Practice

Other titles from the Society of Environmental Toxicology and Chemistry (SETAC):

Fifth LCA Symposium for Case Studies (Presentation summaries)
1997

Sixth LCA Symposium for Case Studies (Presentation summaries)
1998

A Conceptual Framework for Life-Cycle Impact Assessment
Fava, Consoli, Denison, Dickson, Mohin, Vigon, editors
1993

Guidelines for Life-Cycle Assessment: A "Code of Practice"
Consoli, Allen, Boustead, Fava, Franklin, Jensen, De Oude, Parrish, Perriman, Postlethwaite, Quay, Séguin, Vigon, editors
1993

Integrating Impact Assessment into LCA
Udo de Haes, Jensen, Klepffer, Lindfors, editors
1994

Life-Cycle Assessment Data Quality: A Conceptual Framework
Fava, Jensen, Lindfors, Pomper, De Smet, Warren, Vigon, editors
1994

Life-Cycle Impact Assessment: The State of the Art, 2nd edition
Barnthouse, Fava, Humphreys, Hunt, Laibson, Noesen, Norris, Owens, Todd, Vigon, Weitz, Young, editors
1997

Public Policy Applications of Life-Cycle Assessment
Allen, Consoli, Davis, Fava, Warren, editors
1997

Simplifying LCA: Just a Cut?
Christiansen, editor
1997

A Technical Framework for Life-Cycle Assessment
Fava, Denison, Jones, Vigon, Curran, Selke, Barnum, editors
1991

For information about any SETAC publication, including SETAC's international journal, *Environmental Toxicology and Chemistry*, contact the SETAC Office nearest you.

1010 North 12th Avenue	Avenue de la Toison d'Or 67
Pensacola, Florida, USA 32501-3367	B-1060 Brussels, Belgium
T 850 469 1500	T 32 2 772 72 81
F 850 469 9778	F 32 2 770 53 83
E setac@setac.org	E setac@setaceu.org

www.setac.org

Environmental Quality Through Science®

Life-Cycle Impact Assessment: Striving towards Best Practice

Edited by

Helias A. Udo de Haes
Centre of Environmental Science (CML)
Leiden University
Leiden, The Netherlands

Göran Finnveden
Swedish Defence Research Agency (FOI)
Stockholm, Sweden

Mark Goedkoop
PRé Consultants B.V.
Amersfoort, The Netherlands

Michael Hauschild
Technical University of Denmark (DTU)
Lyngby, Denmark

Edgar G. Hertwich
Norwegian University of Science and Technology
Trondheim, Norway

Patrick Hofstetter
U.S. Environmental Protection Agency (USEPA)
Cincinnati, Ohio, USA

Olivier Jolliet
Ecole Polytechnique Fédérale de Lausanne (EPFL)
Lausanne, Switzerland

Walter Klöpffer
C.A.U. GmbH
Dreieich, Germany

Wolfram Krewitt
German Aerospace Research Centre (DLR)
Stuttgart, Germany

Erwin Lindeijer
TNO Industry
Eindhoven, The Netherlands

Ruedi Müller-Wenk
University of St. Gallen
St. Gallen, Switzerland

Stig I. Olsen
Institute for Product Development (IPU)
Lyngby, Denmark

David W. Pennington
Ecole Polytechnique Fédérale de Lausanne (EPFL)
Lausanne, Switzerland

José Potting
National Institute for Public Health and the Environment (RIVM)
Bilthoven, The Netherlands

Bengt Steen
Chalmers University
Göteborg, Sweden

Current Coordinating Editor of SETAC Books
Andrew Green
International Lead Zinc Research Organization

Publication sponsored by the Society of Environmental Toxicology and Chemistry (SETAC)

Cover by Michael Kenney Graphic Design and Advertising
Copyediting and typesetting by Wordsmiths Unlimited
Indexing by IRIS

Library of Congress Cataloging-in-Publication Data

Life-cycle impact assessment : striving towards best practice / edited by Helias A. Udo de Haes.
      p. cm.
    Includes bibliographical references and index.
    ISBN 1-880611-54-6 (alk. paper)
      1. New products--Environmental aspects. 2. Product life cycle--Environmental aspects. I. Udo de Haes, H. A.

TS170 .L564 2002
658.5'75--dc21

2002030432

Information in this book was obtained from individual experts and highly regarded sources. It is the publisher's intent to print accurate and reliable information, and numerous references are cited; however, the authors, editors, and publisher cannot be responsible for the validity of all information presented here or for the consequences of its use. Information contained herein does not necessarily reflect the policy or views of the Society of Environmental Toxicology and Chemistry (SETAC).

No part of this publication may be reproduced, stored in a retrieval system, or transmitted in any form or by any means, electronic, electrostatic, magnetic tape, mechanical, photocopying, recording, or otherwise, without permission in writing from the copyright holder.

All rights reserved. Authorization to photocopy items for internal or personal use, or the personal or internal use of specific clients, may be granted by SETAC, provided that the appropriate fee is paid directly to Copyright Clearance Center, 222 Rosewood Drive, Danvers, MA 01923 USA (telephone 978-750-8400. Before photocopying items for educational classroom use, please contact the Copyright Clearance Center (http://www.copyright.com).

SETAC's consent does not extend to copying for general distribution, for promotion, for creating new works, or for resale. Specific permission must be obtained in writing from SETAC for such copying. Direct inquiries to the Society of Environmental Toxicology and Chemistry (SETAC), 1010 North 12th Avenue, Pensacola, FL 32501-3367, USA.

© 2002 Society of Environmental Toxicology and Chemistry (SETAC)
SETAC Press is an imprint of the Society of Environmental Toxicology and Chemistry.
No claim is made to original U.S. Government works.

International Standard Book Number 1-880611-54-6
Printed in the United States of America
09 08 07 06 05 04 03 02   10 9 8 7 6 5 4 3 2 1

∞ The paper used in this publication meets the minimum requirements of the American National Standard for Information Sciences—Permanence of Paper for Printed Library Materials, ANSI Z39.48-1984.

Reference Listing: Udo de Haes HA, Finnveden G, Goedkoop M, Hauschild M, Hertwich EG, Hofstetter P, Jolliet O, Klöpffer W, Krewitt W, Lindeijer E, Müller-Wenk R, Olsen SI, Pennington DW, Potting J, Steen B. 2002. Life-cycle impact assessment: Striving towards best practice. Published by the Society of Environmental Toxicology and Chemistry (SETAC), Pensacola FL, USA. 272 p.

SETAC Publications

The publication of books by the Society of Environmental Toxicology and Chemistry (SETAC) provides in-depth reviews and critical appraisals on scientific subjects relevant to understanding the impacts of chemicals and technology on the environment. The books explore topics reviewed and recommended by the Publications Advisory Council and approved by the SETAC Board of Directors for their importance, timeliness, and contribution to multidisciplinary approaches to solving environmental problems. The diversity and breadth of subjects covered in the series reflect the wide range of disciplines encompassed by environmental toxicology, environmental chemistry, and hazard and risk assessment. These volumes attempt to present the reader with authoritative coverage of the literature, as well as paradigms, methodologies, and controversies; research needs; and new developments specific to the featured topics. The books are generally peer reviewed for SETAC by acknowledged experts.

SETAC Publications, which include Technical Issue Papers (TIPs), workshop summaries, newsletter *(SETAC Globe)*, and journal *(Environmental Toxicology and Chemistry)*, are useful to environmental scientists in research, research management, chemical manufacturing and regulation, risk assessment, and education, as well as to students considering or preparing for careers in these areas. The publications provide information for keeping abreast of recent developments in familiar subject areas and for rapid introduction to principles and approaches in new subject areas.

SETAC would like to recognize the past SETAC Special Publication Series editors:

C.G. Ingersoll, Midwest Science Center
U.S. Geological Survey, Columbia, Missouri, USA

T.W. La Point, Institute of Applied Sciences
University of North Texas, Denton, Texas, USA

B.T. Walton, U.S. Environmental Protection Agency
Research Triangle Park, North Carolina, USA

C.H. Ward, Department of Environmental Sciences and Engineering
Rice University, Houston, Texas, USA

Contents

List of Figures .... *xii*
List of Tables .... *xiii*
About the Editors .... *xiv*
Contributors .... *xix*
Foreword .... *xxi*

Chapter 1: Introduction 1
Edgar G. Hertwich, David W. Pennington, Jane C. Bare

Life-Cycle Assessment 1
Approaches to Impact Assessment 4
    Top–down and bottom–up approaches 4
    Midpoint and endpoint indicators 5
    Temporal and spatial differentiation 6
Towards Best Available Practice in LCIA 8
Outline of *Life-Cycle Impact Assessment: Striving towards Best Practice* 9
References 10

Chapter 2: Impact Assessment of Resources and Land Use 11
Erwin Lindeijer, Ruedi Müller-Wenk, Bengt Steen

Life-Cycle Impact Assessment Framework for Resources and Land Use 12
    Aim of this chapter 12
    Areas of protection 12
    Interventions and their impacts on AoPs 13
    Environmental mechanisms of resources and land use 13
    Characteristics of impact assessment methods for resources and land use 14
Impact Assessment of Abiotic Resources 15
    Starting point 15
    Definitions and distinctions 17
    Operationalisation of impact assessment of abiotic resources 20
    Conclusions 25
Impact Assessment of Biotic Resources 26
    Starting point 26
    The need for impact assessment of biotic resource extractions 26
    Definitions and distinctions 27
    Guidelines for impact assessment of biotic extractions 30
    Operationalisation 33
    Conclusions 39

Impact Assessment of Land Use 39
    Starting point 39
    Definitions and distinctions 40
    The relation between land use and its impacts on AoPs 48
    Operationalisation of land-use impact assessment for AoPs 50
    Weighting within impact category land use 60
    Conclusions 61
References 62

Chapter 3: Climate Change, Stratospheric Ozone Depletion, Photooxidant Formation, Acidification, and Eutrophication 65

José Potting, Walter Klöpffer, Jyri Seppälä, Greg Norris, Mark Goedkoop

Introduction 65
General Issues 66
    Best available practice and uncertainties 66
    Models and model domains 67
    Definition of the category indicator 67
    Substances covered 68
    Spatial aspects 68
    Temporal resolution 69
    Interrelations between impact categories 70
Climate Change 71
    Description of the impact category 71
    Relevant substances 71
    State-of-the-art science 72
    Category indicator and characterisation factors 73
    Spatial aspects 73
    Temporal aspects 73
    Other aspects 73
    Recommendations 74
Stratospheric Ozone Depletion 74
    Description of the impact category 74
    Relevant substances 75
    State-of-the-art science 75
    Category indicator and characterisation factors 76
    Spatial aspects 76
    Temporal aspects 76
    Other aspects 76
    Recommendations 77
Photooxidant Formation 77
    Description of the impact category 77
    Relevant substances 78
    State-of-the-art science 78
    Category indicator and site-generic characterisation factors 79

    Spatial aspects and site-dependent characterisation factors 80
    Temporal aspects 81
    Other aspects 81
    Recommendations 81
Acidification 82
    Description of the impact category 82
    Relevant substances 82
    State-of-the-art science 82
    Category indicator and site-generic characterisation factors 84
    Spatial aspects and site-dependent characterisation factors 84
    Temporal aspects 86
    Other aspects 87
    Recommendations 87
Terrestrial Eutrophication 87
    Description of the impact category 87
    Relevant substances 88
    State-of-the-art science 88
    Category indicators and site-generic characterisation factors 88
    Spatial aspects and site-dependent characterisation factors 88
    Temporal aspects 89
    Other aspects 90
    Recommendations 90
Aquatic Eutrophication 90
    Description of the impact category 90
    Relevant substances 91
    State-of-the-art science 91
    Category indicators and site-generic characterisation factors 91
    Spatial aspects and site-dependent characterisation factors 92
    Temporal aspects 93
    Other aspects 93
    Recommendations 94
Overall Recommendations 94
    Recommendations on a general level 94
    Recommendations by impact category 95
References 97

Chapter 4: Fate and Exposure Assessment in the Life-Cycle Impact Assessment of Toxic Chemicals 101
Edgar G. Hertwich, Oliver Jolliet, David W. Pennington, Michael Hauschild, Carsten Schulze, Wolfram Krewitt, Mark Huijbregts

Introduction 101
Methodological Choices in the Assessment of Toxicity 103
    Desired level of sophistication 103
    The interface between fate and effect assessment 106

Multimedia Models in LCA	106
Medium-Specific Approaches	109
Application of air quality modelling: The EcoSense Model	110
Application of water quality and aquatic fate modelling	112
Use of the Models in LCIA	112
Evaluation of Fate and Exposure Methods	113
Environmental relevance and comprehensiveness	113
Scientific validity and reliability	114
Transparency and reproducibility	115
Relevance to the decision context	115
Feasibility	116
Recommendations	116
An approach to the location issue	116
Temporal considerations	118
Uncertainty and the standard of evidence	118
Model selection	119
Conclusions	119
References	120

Chapter 5: Indicators for Human Toxicity in Life-Cycle Impact Assessment 123
Wolfram Krewitt, David W. Pennington, Stig I. Olsen, Pierre Crettaz, Olivier Jolliet

Introduction	123
Current Practice in LCIA	125
Toxicological Potency: Dose–Response	127
Noncarcinogens	129
Carcinogens	131
Radiation	132
Severity	133
Qualitative approaches to account for severity	133
Quantitative approaches to account for severity	135
Availability of data to address expected severity	137
Evaluation of Human Toxicity Indicators	139
Scientific validity and reliability	139
Transparency and reproducibility	141
Comprehensiveness and sophistication	142
Relevance to the decision context	143
Feasibility	143
Conclusions and Recommendations	144
References	146

Chapter 6: Indicators for Ecotoxicity in Life-Cycle Impact Assessment 149
Michael Hauschild, David W. Pennington

Introduction 149
Laboratory Test Data for Toxicity to Key Species 152
Predicted No-Effect Concentration 152
    Measuring PNEC in field meso- and microcosm tests 153
    Estimating PNEC on the basis of laboratory tests performed on individual species 153
    Evaluation and comparison of PNEC estimation methods 157
    Estimation of PNEC for terrestrial ecosystems and sediments 158
Marginal Measures, Mixtures, and Endpoint Relevance 159
    Existing marginal ecosystem effect measures 160
    Potentially affected fraction of species in mixtures: Combi-PAF 162
    Ecosystem endpoint measure relevance: PDFs and PAFs 164
Criteria to Evaluate Ecotoxicity Indicators 166
    Pre-selection of feasible indicators 166
Conclusions 169
References 171
Annex: Data Sources 174
    Key sources of measured data 174
    Data prediction: QSARs 174

Chapter 7: Normalisation, Grouping, and Weighting in Life-Cycle Impact Assessment 177
Göran Finnveden, Patrick Hofstetter, Jane C. Bare, Lauren Basson, Andreas Ciroth, Thomas Mettier, Jyri Seppälä, Jessica Johansson, Greg Norris, Stephan Volkwein

Introduction 178
On the Overall Structure and Methodology for Life-Cycle Impact Assessment 179
Normalisation 181
    External normalisation 181
    Case-specific normalisation 183
    Comparison between external and case-specific normalisation 183
    When is normalisation needed? 184
Grouping 184
Overview of Weighting Methods 185
    Proxy weighting methods 185
    Technology abatement weighting methods 185
    Monetization weighting methods 186
    Panel weighting methods 188
    Distance-to-target weighting methods 191
    Other classifications 192

Criteria for the Evaluation of Weighting Methods 192
    Input-related criteria 193
    Procedure-related criteria 193
    Output-related criteria 195
Evaluation of Methods 195
    Evaluation 196
    Concluding remarks 201
Conclusions 202
References 204

Chapter 8: The Conceptual Structure of Life-Cycle Impact Assessment 209
Helias A. Udo de Haes, Erwin Lindeijer

Introduction 209
LCIA Terminology 210
    Environment 210
    Environmental mechanism 210
    Environmental intervention 211
    Midpoint and endpoint 211
    Environmental relevance 212
    Areas of protection 212
Viewpoints and Approaches in LCIA 213
    Decision support versus systems analysis 213
    The scope of LCIA 214
    Midpoint versus endpoint approaches 215
Classification of AoPs in LCIA 216
    Classification of AoPs according to physical characteristics 217
    Classification of AoPs according to societal values 217
    Exploration of manmade environment in relation to system boundary 218
    Splitting up the AoP natural environment 219
    Comparison of two types of classification of AoPs 220
Overview of the AoPs 220
    Visual representation of the classification of AoPs according to physical characteristics 221
    Visual representation of the classification of AoPs according to societal values 222
Conclusions 222
References 225

Abbreviations 227

Index 235

List of Figures

Figure 1-1	Phases of LCA	2
Figure 1-2	Elements of the LCIA phase	3
Figure 1-3	LCIA as a number of matrix multiplications, based on a hypothetical LCI	3
Figure 1-4	Concept of CIs according to ISO 14042	5
Figure 1-5	Structure of the book according to assessment steps and impact categories	5
Figure 1-6	Ozone depletion midpoint or endpoint (damage) modelling	7
Figure 2-1	The cause–impact network, focusing on resources and land use	14
Figure 2-2	Land occupation with renaturalisation	41
Figure 2-3	Transformation	42
Figure 2-4	State of flora, fauna, soil, or soil surface before, during, and after a human activity	42
Figure 2-5	State of flora, fauna, soil, or soil surface before, during, and after human activity, indicating gradual and continuous degradation of renaturalization potential and reference state if human activity were terminated	43
Figure 2-6	State of flora, fauna, soil, or soil surface before, during, and after 2 human activities	44
Figure 4-1	Causality chain from emission to impact	102
Figure 4-2	Limits in the amount of information that can be used in comparing chemicals to each other	104
Figure 4-3	Human toxicity impact chain, with benzene as an example	105
Figure 4-4	The general integration of source, dispersion, and exposure in integrated environmental fate and exposure models	108
Figure 4-5	Flowchart of the EcoSense model	111
Figure 5-1	Outline of stages for calculating characterization factors for human health	124
Figure 5-2	Dose–response measure β_{ED10} for acephate insecticide predicted by fitting the Crouch multistage model to data observed in a mice bioassay	131
Figure 6-1	Example of causal relationships in the environmental mechanism between environmental emissions, midpoints, and (category) endpoints for ecotoxicity	151
Figure 6-2	Determination of the environmental concentration HCp that affects $p\%$ of all species in the ecosystem through extrapolation from a statistical distribution estimated from measured species sensitivities to a substance	154
Figure 6-3	Tangential (marginal) versus secant (average) gradient measures	160
Figure 6-4	Tangential gradient to the SSD curve ($\Delta PAF/\Delta C$) as a function of the PAF of species for 4 types of distribution model	161
Figure 6-5	Comparison of dose-additive and response-additive curves for 2 mixtures of similar chemicals, assuming a straight line and a log-logistic dose–response curve	164
Figure 7-1	Input, procedure, and output as 3 principal elements of a weighting method	192
Figure 7-2	System of criteria for weighting methods	194
Figure 8-1	Product system and economy: Two levels of system definition	214
Figure 8-2	Classification of AoPs according to physical characteristics	221
Figure 8-3	Classification of AoPs according to societal values	223

List of Tables

Table 2-1	ISO requirements and recommendations for selecting impact categories and indicators	16
Table 2-2	General items to address when discussing category indicators in subsequent tables	17
Table 2-3	Present options for operationalisation of deposit abiotic resources	24
Table 2-4	Minimum list of important 'wild' biotic resources threatened by high extraction rates	32
Table 2-5	Proposals for operationalisation of biotic resources	37
Table 2-6	Operationalisation characteristics for competition over land	52
Table 2-7	Operationalisation characteristics for biodiversity	54
Table 2-8	Operationalisation characteristics for life support	59
Table 3-1	Site-generic acidification factors typically used in LCA and new site-generic factors based on sophisticated spatially resolved models	85
Table 3-2	Site-generic factors to assess terrestrial eutrophication based on sophisticated spatially resolved models	89
Table 4-1	Characteristics of a consistent fate and exposure analysis, for different types of effect coefficients	107
Table 4-2	Prominent multimedia models	110
Table 4-3	Two different modelling approaches in comparison	113
Table 5-1	Availability of toxicity data for high production volume substances	128
Table 5-2	Proposal for the definition of human toxicity subcategories according to the severity of effect	134
Table 5-3a	Summary review of different human health indicators with respect to their use in calculating characterisation factors in LCIA: Potency-based indicators	139
Table 5-3b	Summary review of different human health indicators with respect to their use in calculating characterisation factors in LCIA: Severity-based indicators	140
Table 6-1	Assessment factors for determination of $PNEC_{water}$ from ecotoxicological test data	156
Table 6-2	Assessment factors for determination of $PNEC_{soil}$ from ecotoxicological test data	159
Table 6-3	Preliminary summary of the evaluation of the approaches for ecotoxicological effect assessment	167
Table 7-1	Viewpoints and methods for normalisation	183
Table 7-2	Tentative evaluation of weighting methods	197
Table 8-1	Correspondence between terms from decision analysis and terminology used in LCIA	213
Table 8-2	Areas of protection and underlying societal values	216
Table 8-3	Classification of AoPs according to societal values	218

About the Editors

Helias A. Udo de Haes

From 1960 to 1968, Helias A. Udo de Haes studied biology in Leiden, The Netherlands. He performed his PhD research in the Max Planck Institut für Verhaltensphysiologie, Seewiesen, Germany. In 1970, he became a staff member of the Environmental Biology Department at Leiden University, and in 1978, he founded the Centre of Environmental Science (CML) of Leiden University, of which he is still scientific director. In 1986, he earned a chair in environmental science.

The main topics of Udo de Haes' work are in the field of what is now called 'Industrial Ecology', and they concern methodology development of life-cycle assessment (LCA) and substance flow analysis (SFA). From the beginning, he joined the Society of Environmental Toxicology and Chemistry (SETAC) Europe Steering Committee for LCA and was its chairman for several years. He was also chairman of two consecutive SETAC Europe working groups on life-cycle impact assessment (LCIA). From 1996 onwards, he contributed to the International Organization for Standardization (ISO) process in the development of the 14040 series of standards on LCA. In 1998, he led a global initiative on the identification of best practice in LCA. Together with colleagues in SETAC, and in co-operation with the United Nations Environment Programme (UNEP) office in Paris, this has led to the recent start of the UNEP–SETAC Life-Cycle Initiative, aimed at enhancement of the application of LCA and life-cycle thinking in business and policy practice.

Göran Finnveden

Göran Finnveden is a research leader at the Environmental Strategies Research Group (fms) and the Swedish Defence Research Agency (FOI). He received an MSc in chemical engineering at the Royal Institute of Technology in 1989 and a PhD in natural resources management at Stockholm University in 1998. He was a co-author of the Nordic Guidelines of LCA (1995). He has worked for IVL, the Swedish Environmental Research Institute, and has been active in several SETAC Europe working groups. He is a member of the International Expert Group on LCA and Integrated Solid Waste Management. Current research interests include LCA and other environmental systems analysis tools such as strategic environmental assessment, environmentally extended input–output analysis, and the connections between different tools. Application areas include energy, waste, and defence sectors.

Mark Goedkoop

Mark Goedkoop holds an MSc in Industrial Design Engineering from Delft University of Technology (1993). He worked as independent design consultant until 1990, when he moved his focus to the field of ecodesign. In the process, he established PRé Consultants and pioneered the field of LCA. His focus is and was the development of practical yet scientifically sound tools to improve the environmental performance of products and services, including the Eco-indicator projects (1995 and 1999), which resulted in a methodology that is used to translate the many different environmental impacts of a product into a single score, the product services project that resulted in tools that can be used to combine the environmental assessment both on life-cycle and societal levels, the dematerialisation assessment tool, and the LCA software SimaPro.

Michael Hauschild

Michael Hauschild is associate professor in environmental assessment of products and systems at the Technical University of Denmark (DTU), at the Innovation and Sustainability Group of the Department of Manufacturing Engineering and Management. With a background as a chemical engineer, he received his PhD in terrestrial ecotoxicology from DTU in 1992 and then embarked on his work on LCIA as one of the main individuals responsible for the development of the EDIP LCA methodology. For this work, he and his colleagues received the Nordic Council of Ministers' Great Nature and Environment Award in 1997. He has been active in several consecutive SETAC Europe working groups on LCIA, the latest as chairman of the task group on ecotoxicity indicators, and he is an appointed member of the Danish delegation to the ISO committee ISO 207/SC 5 on LCA. His latest work is on spatial differentiation in LCIA and on the development of methodology for assessment of ecotoxicity and human toxicity in LCA.

Edgar G. Hertwich

Edgar G. Hertwich is Adjunct Professor for Energy and Environmental Systems Analysis with the Institute for Thermal Energy and Hydropower, Norwegian University of Science and Technology (NTNU) in Trondheim. He is also a research scholar at the International Institute for Applied Systems Analysis (IIASA) in Laxenburg, Austria. He has a PhD in Energy and Resources from the University of California, Berkeley, USA. Hertwich has worked on risk analysis, multimedia fate modelling, and LCA. The toxic equivalency potentials he developed for his dissertation are now used by www.scorecard.org, the web page of Environmental Defense, and by the U.S. Environmental Protection Agency (USEPA). Hertwich is currently working on projects that deal with the LCA of energy facilities in Norway and the environmental impacts of consumption.

Patrick Hofstetter

Patrick Hofstetter holds a master degree in mechanical engineering from Eidgenössichen Technischen Hochschule (ETH), Zürich, Switzerland, has operated his own consultancy 'Büro für Analyse & Ökologie', and conducted research for the report and database 'Ökoinventare für Energiesysteme' (Life Cycle Inventories for Energy Systems) at the Energy Systems Laboratory, ETH Zürich. He has done research on LCIA since 1994 and has been a member of the LCA Steering Committee of SETAC Europe. He completed his PhD thesis in 1998 on 'Perspectives in Life Cycle Impact Assessment; A Structured Approach to Combine Models of the Technosphere, Ecosphere, and Valuesphere' at the Chair of Environmental Sciences: Natural and Social Science Interface, ETH Zürich. From 1999 to 2002, he was an ORISE Research Fellow in the Systems Analysis Branch, NRMRL, USEPA Cincinnati, Ohio, USA, and since 1999 he has been a visiting scientist at the Harvard School of Public Health, Boston, Massachusetts, USA.

Olivier Jolliet

Olivier Jolliet is a physicist and Professor at the Swiss Federal Institute of Technology (EPFL), Lausanne, Switzerland. He heads the group for Life Cycle Systems within the Institute of Environmental Science and Technology. He obtained a master, then a PhD in physics in 1988 and did postdoctoral work at the Silsoe Research Institute, Great Britain, and as a visiting scholar at Massachusetts Institute of Technology, USA. He is presently program manager for impact assessment within the UNEP–SETAC Life-Cycle Initiative.

Walter Klöpffer

After his studies of chemistry and his PhD in physical chemistry at the Karl Franzens University in Graz, Walter Klöpffer joined the Battelle Institute, Frankfurt/Main, Germany, in 1964. Since 1975, he has been a professor for Physical Chemistry at the Johannes Gutenberg University in Mainz. In 1995, three years after joining C.A.U. GmbH, he was appointed editor-in-chief of *The International Journal of Life Cycle Assessment*. The renewed international interest in persistent organic pollutants (POPs) induced a revival of his earlier research work on exposure modelling and on the application of the precautionary principle to the assessment of organic chemicals.

Wolfram Krewitt

Wolfram Krewitt currently is project manager, German Aerospace Research Center (DLR), Institute of Technical Thermodynamics, Department of System Analysis and Technology Assessment. His work covers ecological and economic aspects of integrating new decentralised energy technologies (renewable energies, fuel cells) into the existing energy system and LCA of new decentralised energy systems. His education includes energy engineering at the University of Stuttgart and mechanical engineering at the Technical University of Aachen.

Erwin Lindeijer

Erwin Lindeijer is a senior consultant with broad experience in environmental issues, which he applies in performing environmental LCAs within TNO Industry, Department of Sustainable Product Innovation (DPI), Eindhoven, The Netherlands. He has more than 10 years' experience in performing LCA and developing and reviewing LCA methodology. Recently he has led some large interdisciplinary projects, in which he bridged the experts' with the generic decision-makers' positions. He has chaired the SETAC Europe first LCIA subgroup on normalisation and weighting, and has chaired the LCIA subgroup on resources and land use in the SETAC Europe second LCIA working group. He has also chaired the COST E9 subgroup on land use in forestry and LCA.

Ruedi Müller-Wenk

A native of Switzerland, Ruedi Müller-Wenk received a degree in electrical engineering at the Swiss Federal Institute of Technology, Zürich. His professional activities involve the computer and automobile industries. He has been head of finance and control in the food industry and is Professor for Environmental Technology at the University of St. Gallen, Switzerland.

Stig I. Olsen

Stig I. Olsen holds an MSc in biology from the University of Copenhagen, Denmark, and an Industrial PhD in LCA from the Technical University of Denmark (DTU). He is currently a scientific associate and consultant in LCA and methodology development with the Institute for Product Development, DTU, Lyngby. He is chairman of the SETAC Europe Working Group on Impact Assessment subgroup on human toxicity and past member of a support group at the Danish Standardisation Association for LCA standards.

David W. Pennington

David W. Pennington's research focuses on the development and analysis of multimedia chemical fate, exposure, and (eco)toxicological models to calculate indicators that support environmental decision-making. This research particularly addresses the concept of estimating the time and space integrated likelihood (risk) and potential consequences of toxicological impacts associated with low level or disperse chemical emissions. Such indicators are essential when comparing many chemicals and emissions in application domains such as LCA and may even have a role of growing importance in regulatory risk assessments.

José Potting

José Potting received her MSc degree in Environmental Sciences at the Agricultural University in Wageningen in 1988, and in 2000, she successfully defended her PhD thesis on method development for spatial differentiation by introducing techniques from integrated assessment modelling in LCA. First at Utrecht University and now within the Dutch National Institute for Public Health and the Environment (RIVM), she shifted her focus from LCA to integrated assessment modelling. During all her appointments, Potting actively participated in international working groups, amongst others as chair of the Scientific Task Group on Global And RegionaL Impact Categories (STG-GARLIC) that was established under the European branch of SETAC.

Bengt Steen

Bengt Steen is an adjunct professor in the Department of Environmental System Analysis and the research manager at the Centre for Environmental Assessment of Products and Material Systems (CPM) at Chalmers University of Technology, Göteborg, Sweden. He holds a PhD in chemical engineering and has been active in environmental research since 1968 and in LCA research since 1989. He has focused his research on environmental evaluation methods in product development and is one of the creators of the Environmental Priority Strategies (EPS) system.

Contributors*

Jane C. Bare
Systems Analysis Branch
Sustainable Technology Division
National Risk Management Research Laboratory
U.S. Environmental Protection Agency (USEPA)
Cincinnati, Ohio, USA

Lauren Basson
Department of Chemical Engineering
University of Sydney
Sydney, Australia

Andreas Ciroth
TU Berlin
Institute for Environmental Engineering
Department of Waste Minimization and Recycling
Berlin, Germany

Pierre Crettaz
Office Fédéral de la Santé Publique
Bundesamt für Gesundheit
Section Chimie et Toxicologie
CH-3098 Köniz, Switzerland

Göran Finnveden
Environmental Strategies Research Group (fms)
Swedish Defence Research Agency (FOI)
Stockholm, Sweden

Mark Goedkoop
PRé Consultants B.V.
Plotterweg 12
3821 BB Amersfoort, The Netherlands

Michael Hauschild
Department of Manufacturing Engineering and Management
Technical University of Denmark (DTU)
DK-2800 Lyngby, Denmark

Edgar G. Hertwich
Industrial Ecology Programme and Department of Thermal Energy and Hydropower
Norwegian University of Science and Technology
NO-7491 Trondheim, Norway

Patrick Hofstetter
ORISE Research Fellow
U.S. Environmental Protection Agency (USEPA)
Cincinnati, Ohio, USA

Mark Huijbregts
Department of Environmental Studies
University of Nijmegen
NL-6525 ED Nijmegen, The Netherlands

Jessica Johansson
Department of Systems Ecology
Stockholm University
Stockholm, Sweden

Olivier Jolliet
Life Cycle Group for Sustainable Development (GECOS-DGR)
Ecole Polytechnique Fédérale de Lausanne (EPFL)
CH-1015 Lausanne, Switzerland

Walter Klöpffer
C.A.U. GmbH
Dreieich, Germany

Wolfram Krewitt
System Analysis and Technology Assessment
German Aerospace Research Centre (DLR)
Pfaffenwaldring 38-40
D-70569 Stuttgart, Germany

Erwin Lindeijer
TNO Industry
Department of Sustainable Product Innovation (DPI)
Eindhoven, The Netherlands

Thomas Mettier
Natural and Social Science Interface (UNS)
Eidgenössischen Technischen Hochschule (ETH)
Zürich, Switzerland

Ruedi Müller-Wenk
Institut für Wirtschaft und Ökologie (IWÖ)
University of St. Gallen
CH-9000 St.Gallen, Switzerland

Greg Norris
    Sylvatica
    North Berwick, Maine, USA

Stig I. Olsen
    Institute for Product Development (IPU)
    DK-2800 Lyngby, Denmark

David W. Pennington
    Life Cycle Group for Sustainable Development (GECOS-DGR)
    Ecole Polytechnique Fédérale de Lausanne (EPFL)
    CH-1015 Lausanne, Switzerland

José Potting
    National Institute for Public Health and the Environment (RIVM)
    Bilthoven, The Netherlands

Carsten Schulze
    Institut für Umweltsystemforschung
    Universität Osnabrück
    D-49069 Osnabrück, Germany

Jyri Seppälä
    Finnish Environment Institute
    Helsinki, Finland

Bengt Steen
    Chalmers University
    Göteborg, Sweden

Stephan Volkwein
    C.A.U. GmbH
    Dreieich, Germany

Helias A. Udo de Haes
    Centre of Environmental Science (CML)
    Leiden University
    Leiden, The Netherlands

Additional contributors**

Martin Baitz
Almut Beck
Magnus Bengtsson
Joris Broers
Christel Cederberg
Suzanne Efting
Beate Escher
Maarten ten Houten
Gjalt Huppes
Norihiro Itsubo
Thomas Köllner
Helena Mälkki
Yasunari Matsuno
Berit Mattsson
John May
Llorenc Mila i Canals
Willie Owens
Jerome Payet
Isa Renner
Arnold Tukker
Kristin Becker van Slooten
Stephan Volkwein
Bo Weidema

* Affiliations were current at the time the text was written.
** See chapter acknowledgements.

Foreword

This book presents the results of the Society of Environmental Toxicology and Chemistry (SETAC) Europe Second Working Group on Life-Cycle Impact Assessment (LCIA). In the first phase of the working period, from 1994 to 1996, there were 2 groups active in this field: one in Europe, aiming to define a scientific basis for LCIA, and one in North America, aiming to identify critical issues in this area. The present working group was founded by SETAC Europe and was active in the period 1998 to 2000. The focus has been on input from members coming from European countries. However, the working group also involved members from other countries, for instance, the U.S. and Japan, assuring that literature from countries outside Europe also has been considered. The total number of active participants amounted to about 50, coming from 15 countries.

This book aims to make the first steps towards the identification of best available practice in the field of LCIA and to build on the work of the International Organization for Standardization (ISO). The ISO 14042 standard on LCIA defines the relevant terminology, establishes a general technical framework, sets a number of important requirements on LCIA application, and specifies procedural requirements such as those for a critical review of the results. The ISO Technical Report TR 14047 contains examples, clarifying the different elements of the LCIA process. The identification of best available practice, the aim of the present working group, clearly goes beyond these 2 ISO reports. What types of impact categories should be defined? What are the best methods and what are the best data? To which extent is there a general best practice, or to which extent will it be dependent on the type or location or application? How should we deal with the choice between average and marginal modelling? And where should we choose category indicators (CIs) in the environmental mechanism of an impact category: at midpoint or at endpoint level?

During the work period, it became clear that the tasks of the working group should focus on the further development of the technical framework, an overview of the existing data and methods for different impact categories, together with a discussion on the most relevant critical issues. The results provide a broad overview in several new areas, and this can guide present-day applications of LCIA. Evaluation criteria also have been established and applied to the method discussion, including the scientific validity of the results, the transparency of the methods involved, the environmental relevance of the results, the feasibility for practical use, and the link with other phases in LCA, in particular the link with life-cycle inventory (LCI) methods and data. Also, a final chapter has been included about the conceptual structure of LCIA. However, the real establishment of best practice in LCIA largely remained outside the scope of the working group.

The working group thus became a first basis for a proposal for a structural co-operation between the United Nations Environmental Programme, represented by its Division of Technology, Industry and Economics (UNEP-DTIE) in Paris, and SETAC, represented by its International Council (ICS). This envisaged co-operation is called the 'Life-Cycle Initiative' and deals with the identification of best available practice in LCA, including both LCI, LCIA, and possibly also life-cycle management. The present results from the working group on LCIA constitute a key contribution to the start of this UNEP–SETAC initiative. A first activity in the Life-Cycle Initiative will be to perform definition studies in the area of LCI and LCIA; the results presented here will be an important input to the latter study. It is also an explicit aim of the Life-Cycle Initiative to incorporate working groups into the programme, in order to continue to stimulate bottom–up scientific input. In this way, a new phase of the process can start, be it in another formal setting.

We would like to thank warmly all those involved in the process. Contributions to this process were fully on a voluntary basis, without any financial support. The results will therefore all the more stand as a clear beacon for the further development of knowledge in this interesting and important area.

—— Helias A. Udo de Haes, chairman, SETAC Europe Second Working Group on LCIA
—— Olivier Jolliet, vice-chairman, SETAC Europe Second Working Group on LCIA

Introduction

Edgar G. Hertwich, David W. Pennington, Jane C. Bare

Life-Cycle Assessment

Developed from chemical engineering principles and energy analysis, life-cycle assessment (LCA) is a method to account for the environmental impacts associated with a product or service. The term *life cycle* indicates that all stages in a product's life, from resource extraction to ultimate disposal, are taken into account. One precursor of this cradle-to-grave evaluation was the comparative environmental assessment of energy technologies. A photovoltaic (PV) cell, for example, requires no fossil fuel and produces no emissions during its operation, but its production requires more energy per unit installed power than does the construction of a fossil fuel power plant. Viewed over the entire life cycle, the delivery of electric power from a PV solar cell causes emissions of CO_2, emissions connected to the production of the materials as well as to the manufacturing and installation of the cell. When we compare energy technologies, it is therefore important to account for upstream and downstream inputs and emissions. LCA developed out of an effort to systematise and interpret the results of such investigations.

This book summarises the contributions of the Society of Environmental Toxicology and Chemistry (SETAC) Europe Second Working Group on Life-Cycle Impact Assessment (WIA-2) to the development and recognition of best available practice in impact assessment (Udo de Haes et al. 1999). This effort builds upon the efforts of the SETAC Europe first working group (WIA-1) and its document, *Towards a Methodology for Life-Cycle Impact Assessment* (Udo de Haes 1996), and the SETAC North America working group, which composed *Life-Cycle Impact Assessment: The State-of-the-Art* (Barnthouse et al. 1997). The efforts of these groups were likewise extensions of *A Technical Framework for Life-Cycle Assessment* (Fava et al. 1991) and *A Conceptual Framework for Life-Cycle Impact Assessment* (Fava et al. 1993). The work reported in this book also draws on the standards for LCA, the International Organization for Standardization 14040 series (ISO 1997, 2000).

Today, LCA consists of 4 phases:
1) goal and scope definition,
2) inventory analysis,
3) impact assessment, and
4) interpretation.

The 4 phases of LCA are displayed in Figure 1-1, in relation to LCA applications (ISO 1997). The first phase, goal and scope definition, serves to define the purpose and extent of the study, and it contains a description of the system studied. An important issue is the basis for comparison of different systems: the functional unit of a product or the service delivered. In the case of energy technologies, should the comparison be based on the

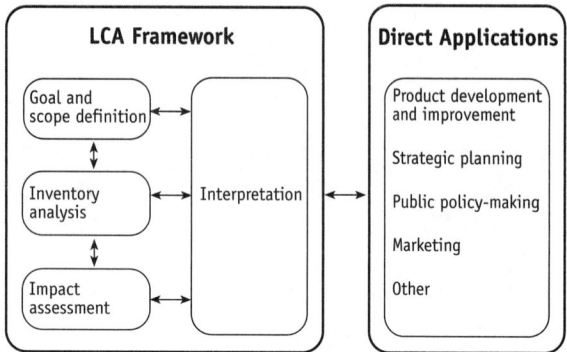

Figure 1-1 Phases of LCA
(© International Organization for Standardization [ISO]. This material is reproduced from ISO 14040:1997 with permission of the American National Standards Institute on behalf of ISO. No part of this material may be copied or reproduced in any form, electronic retrieval system or otherwise or made available on the Internet, a public network, by satellite or otherwise without the prior written consent of the American National Standards Institute, 25 West 43rd Street, New York, NY 10036.)

amount of energy produced per kWh or on a certain time profile of power demand that needs to be satisfied? In comparing stand-alone electricity generation, for example, we must recognise that a PV solar system would need to include batteries in order to obtain the same functional unit as a diesel generator.

The second phase of LCA, inventory analysis, consists of data collection and analysis. Data on the environmental interventions or stressors (e.g., emissions, land use, resource use, noise) connected to each process in the life cycle are collected, often guided by a process flow diagram (United Nations Environment Programme [UNEP] 1996). Data are obtained from the main producers and their suppliers (process owners) and from LCA databases on standard inputs such as energy and materials. Data are then processed to produce an inventory of environmental interventions per functional unit, containing, for example, the number of square metres of land occupied for a year per kWh of electricity produced and the kilograms of CO_2 emitted per kWh produced. This data processing is not always straightforward. For processes that produce more than one output, for example, combined heat and power production, decisions must be made about how to allocate environmental burdens to each output. For materials that are recycled, environmental burdens associated with mining must be allocated among primary and recycled materials.

The third phase of LCA, impact assessment, serves to evaluate the significance of the environmental interventions contained in a life-cycle inventory (LCI). In practice, an inventory will contain a long list of emissions and resource uses. Its purpose is to determine the relative importance of each inventory item and to aggregate interventions to a small set of indicators, or even to a single indicator. This is done in order to identify those processes that contribute most to the overall impact, or to compare products.

According to the ISO, life-cycle impact assessment (LCIA) consists of 2 mandatory elements, classification and characterisation, and 3 optional elements, normalisation, grouping, and weighting. These elements are described in Figure 1-2, which indicates the sequence of the elements as imagined by the ISO.

On the basis of ISO 14042 (ISO 2000), we define the mandatory elements as follows:
- Classification — Assignment of inventory data to different impact categories, such as global warming, ozone depletion, toxicological human and ecosystem impacts, …
- Characterisation — Calculation of category indicator (CI) results for each impact category using characterisation factors, which are estimated using characterisation models.

We define the optional elements for LCIA as follows, on the basis of ISO 14042:
- Normalisation — Calculating the magnitude of CI results relative to reference information, such as the emissions in a reference area characterised by the same characterisation method.
- Grouping — Assigning of impact categories to groups of similar impacts or ranking categories in a given hierarchy, for example, high, medium, and low priority.

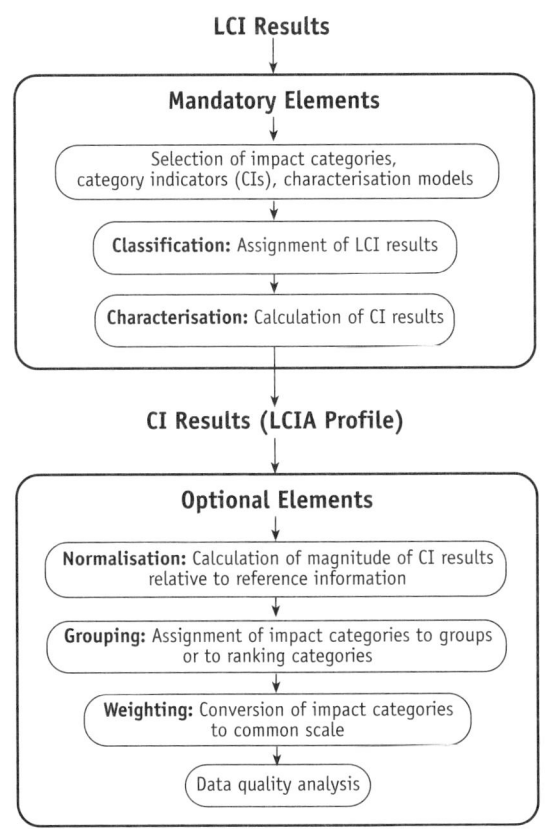

Figure 1-2 Elements of the LCIA phase
(© International Organization for Standardization [ISO]. This material is adapted with modifications from ISO 14042:2000 with permission of the American National Standards Institute on behalf of ISO. No part of this material may be copied or reproduced in any form, electronic retrieval system or otherwise or made available on the Internet, a public network, by satellite or otherwise without the prior written consent of the American National Standards Institute, 25 West 43rd Street, New York, NY 10036.)

- Weighting — Converting indicator results of different impact categories to a common scale, based on value choices. This can potentially include a final aggregation to a single indicator.

In practice, impact assessment often consists of 2 or 3 sequential multiplication steps (performed by LCA software tools), as illustrated in Figure 1-3. In the first step, the inventory items are multiplied by characterisation factors. Characterisation factors express the contribution of each inventory item to a specific environmental problem (assessment endpoint). Emissions of methane and laughing gas, for example, are multiplied by their global warming potentials (GWPs). GWPs express the effectiveness of each gas in capturing infrared radiation, compared to CO_2. They can then be added to the CO_2 emissions to obtain the CI value for climate change in terms of CO_2 equivalents. In the second step, different CIs are normalised, for example, by dividing each by a CI value for the total amount of emissions in a country or region. In a third step, the normalised CIs are multiplied by a set of weighting factors that express the relative importance of each environmental problem. According to the international standard on LCIA, ISO 14042, normalisation and weighting are optional.

Characterisation

$$\text{CI values} = \begin{pmatrix} & CO_2 & CCl_4 & CBrH_3 \\ GW & 1 & 1400 & \\ OD & & 1.1 & 0.6 \\ HT & & 14000 & 12000 \end{pmatrix} \times \begin{pmatrix} CO_2 & 2100 \text{ kg} \\ CCl_4 & 0.5 \text{ kg} \\ CBrH_3 & 0.3 \text{ kg} \end{pmatrix} = \begin{pmatrix} GW & 2800 & \text{kg } CO_2\text{-equ} \\ OD & 0.73 & \text{kg CFC-11-equ} \\ HT & 10600 & \text{kg toluene-equ} \end{pmatrix}$$

Characterisation factors   LCI results   CI values

Weighting

$$\text{Weighted results} = \begin{pmatrix} GW & OD & HT \\ 0.2 & 800 & 0.4 \end{pmatrix} \times \begin{pmatrix} GW & 2800 & \text{kg } CO_2\text{-equ} \\ OD & 0.73 & \text{kg CFC-11-equ} \\ HT & 10600 & \text{kg toluene-equ} \end{pmatrix} = 560 + 584 + 4240 = 5384$$

Category weights   CI values   Weighted indicator results

Figure 1-3 LCIA as a number of matrix multiplications, based on a hypothetical LCI (To demonstrate the elements, row and column names are included in the matrices; they are omitted in the actual calculation. In this example, normalisation has been folded into the category weights and is not conducted as a separate element. GW = global warming, OD = ozone depletion, HT = human toxicity)

Alternatively, the inventory items can be multiplied by a single impact factor that expresses the environmental importance of each stressor, for example, as ecopoints per kilogram of nitrate emitted or in terms of external costs ($ per kg CO_2).

The fourth phase of LCA, interpretation, is to evaluate the study in order to derive recommendations and conclusions. LCA is seen as an iterative process (see Figure 1-1), and interpretation may lead to an adjustment of the goal and scope or to further investigations of the inventory and associated impacts. In interpreting the LCA result, we take into account not only numerical results produced in the previous 2 phases but also qualitative issues such as the sensitivity of the results to assumptions and value choices, the quality of the underlying data, and the strength of evidence for various claims.

Approaches to Impact Assessment

Life-cycle impact assessment aims to quantify the importance of environmental stressors tabulated in an LCI and to aggregate the stressors to a small number of CIs and, in some cases, to one final indicator. It draws connections between stressors and valued items that are potentially affected, such as human health or ecosystem functioning. Such a connection is expressed in a stressor–impact chain (cause–effect chain), describing environmental mechanisms. Figure 1-4 displays the stressor–impact chain for acidifying substances and the terms used in ISO 14042. At the endpoint of a cause–effect chain is an impact on a valued item, for example, material damage to historical structures from acid rain or lung cancer from asbestos. These impacts or endpoints can be organised into areas of protection (AoPs, also called *safeguard subjects*).

Chapter 8 proposes a conceptual structure for LCIA that addresses the connection between stressors, endpoints, and AoPs. Because a stressor can have multiple consequences, that is, can affect multiple endpoints, LCIA actually addresses a web of sometimes parallel, sometimes overlapping or interacting stressor–impact chains. The importance of an individual stressor depends on both the magnitude of the impacts it causes and the importance of these impacts.

Top–down and bottom–up approaches

In the traditional approach to LCIA, the assessment of the magnitude of impacts is called *characterisation*, and the assessment of the importance of impacts *weighting* or *valuation*. This approach of organising the work starting from the stressors has been called *bottom–up* (Hofstetter 1999). It was developed in previous SETAC work (Heijungs et al. 1992; Fava et al. 1993) and codified by the ISO 14042 standard on LCIA (see Figure 1-2). An alternative, top–down approach organises impacts according to which AoPs are affected by recognisable stressors, starting from the valued items (Steen and Ryding 1991; Hofstetter 1998). This approach recognises that it is concerns about valued items that drive LCA, and hence takes them as the starting point of LCA method development. It complements the demand that impact assessment should drive inventory analysis, and not vice versa (Barnthouse et al. 1997).

Our working group debated which approach to take. The bottom–up approach prevailed in this initial debate because it was more familiar and reflected the work of many group members. However, the top–down approach insights influenced the chapter discussions, and a call is made for the development of compatible top–down and bottom–up approaches in Chapter 8. The initial preference for bottom–up approaches is reflected by the organisation of the chapter task forces according to impact categories, not according to

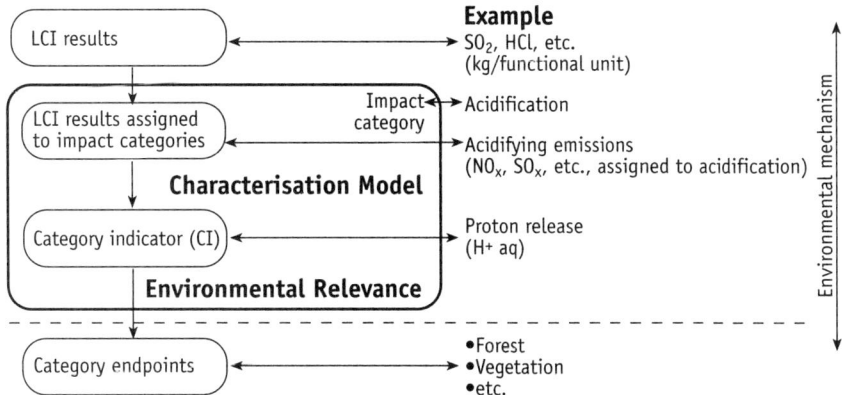

Figure 1-4 Concept of CIs according to ISO 14042
(© International Organization for Standardization [ISO]. This material is adapted with modifications from ISO 14042:2000 with permission of the American National Standards Institute on behalf of ISO. No part of this material may be copied or reproduced in any form, electronic retrieval system or otherwise or made available on the Internet, a public network, by satellite or otherwise without the prior written consent of the American National Standards Institute, 25 West 43rd Street, New York, NY 10036.)

AoPs such as human health or ecosystems. Figure 1-5 displays the organisation of this book, which also reflects this structure. The work of our group, however, shows a certain degree of convergence of the 2 approaches, as the chapter content shows.

The distinction between bottom–up and top–down approaches is related to another distinction, that between approaches following environmental themes and those using damage function methods. *Environmental themes* (Heijungs et al. 1992) are problem areas: acid rain, ozone depletion, global warming, for example. *Damages* (or consequences) are impacts on valued items, such as the death of fish in an acidified lake or skin cancer from increased exposure to ultraviolet radiation. *Damage functions* are the mathematical relationships between stressors and damages. The top–down approach leads to the development of damage function methods, while the bottom–up approach organises the assessment according to indicators for each environmental theme. These indicators may, however, also be based on damage functions.

Midpoint and endpoint indicators

In the development of the environmental themes approach, the main goal of a characterisation method is to describe the importance of a stressor

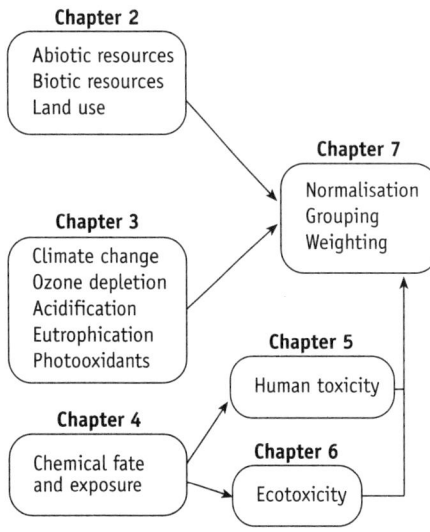

Figure 1-5 Structure of the book according to assessment steps and impact categories

within a category, that is, relative to other stressors in the same category, and to aggregate stressors within a category to a CI. The point of comparison can be at the end of the stressor–impact chain, or at a point within the impact chain where the mechanism of stressors is similar. In evaluating the contribution of different greenhouse gases to climate change, for example, the GWP compares the time-integrated radiative forcing of the gases, not the damages caused by climate change. The connection between the midpoint, radiative forcing, and the endpoints, increased storm frequency or sensitive habitat destruction, is presumed to be the same because all gases act according to the same mechanism of action.

While the environmental themes approach can contain assessments at both the midpoint and the endpoint levels, the damage function approach requires an assessment of endpoints. For ISO, it is important that CIs be environmentally relevant, that is, that they relate in a defendable manner to a valued item or AoP. Debates about the relative merits and relevance of CIs at the midpoint versus at the endpoint influenced the discussions of the working groups in this book. In particular, the Brighton 2000 workshop (Bare et al. 2000) reached a consensus that both midpoint- and endpoint-level indicators have complementary merits and limitations. Both approaches should be developed further, in a compatible framework. Hence, both are addressed in this book.

The choice of a specific impact assessment approach influences both application and method development. In Environmental Priority Strategies (EPS), for example, inventory items are simply multiplied with their damage factor (e.g., environmental load units per kg CO_2; Steen and Ryding 1991). This is simpler than the 2-step procedure of the environmental themes approach (see Figure 1-3). For method development, it is convenient because it offers a way to directly evaluate nuisances, such as noise or a haze that prevents a clear view, using the methods of environmental economics. The intermediary results of the environmental themes approach, however, may be of significant interest to decision-makers, representing, for example, the importance of different inventory items in an impact category.

In method development, damage function methods begin from the perspective of the valued items, and the natural scientific analysis is integrated into the developing method by a weighting system that indicates how the valued items are affected. The environmental themes approach, on the other hand, develops characterisation methods according to environmental problem areas and hence follows the organisation of scientific inquiry into environmental issues. Characterisation methods that follow the environmental themes approach can also include elements of weighting. Chapter 5, for example, discusses options for including a weighting of the severity of toxic effects, that is, a value-based judgement of the desirability of each human health effect, based on a specification of the consequences and their duration.

While it may seem desirable to forecast all endpoint effects of ozone depletion, Figure 1-6 illustrates that, currently, related damage models are unable to comprehensively quantify all impacts with high levels of certainty. Existing data may allow a prediction of potential impacts, such as skin cancer or cataracts, but the data do not currently support the inclusion of effects such as crop damage, immune system suppression, materials damage, and marine life damage. However, particularly in LCA studies that require the analysis of tradeoffs between and/or aggregation across impact categories, damage approaches have recently been gaining popularity. Damage modelling may facilitate more explicit and informed weighting, in particular defendable aggregation across categories in terms of common parameters. For example, human health impacts associated with climate change can be compared with those of ozone depletion using a common basis such as disability-adjusted life years (DALYs; Hofstetter 1998).

Temporal and spatial differentiation

Within the past 10 years, there has been increasing discussion centred on various issues associated with sophistication, accuracy, and complexity in LCIA, including the consideration of temporal and spatial dimensions of potential impacts (Fava et al. 1993; Udo de Haes 1996; Barnthouse et al. 1997; Bare et al. 1999). Potting (2000) stated that one of the

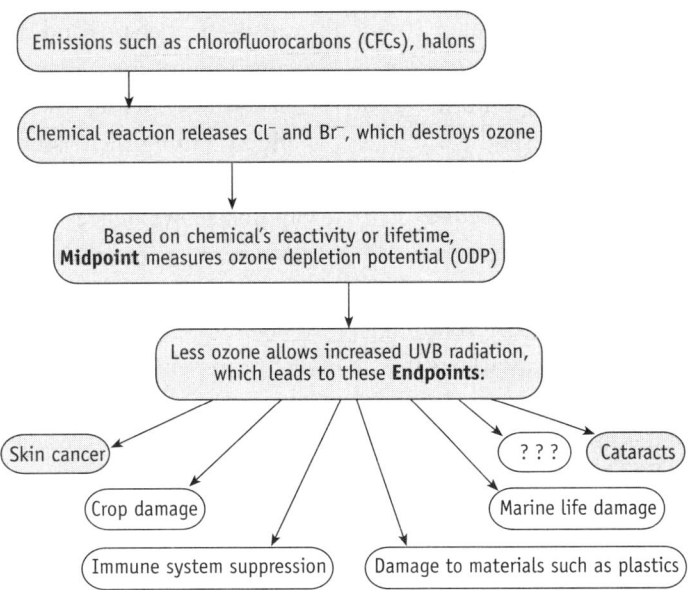

Figure 1-6 Ozone depletion midpoint or endpoint (damage) modelling
(Shaded boxes indicate items that currently can be quantified.)

major problems facing LCA was its inability to adequately model expected impacts, particularly because of a common disregard for the spatial differences in the effects of environmental stressors. This concerns all those impact categories that are not global in nature: biodiversity, land use, biotic resources, toxicity, photochemical oxidation, eutrophication, and acidification. In these categories, we now distinguish between site-dependent and site-generic impact assessment methods.

Accounting for spatial differentiation in LCA remains complicated by the lack of spatial distinction in most emissions and resource consumption inventory databases. Generic characterisation factors are often proposed, but with a growing recognition of the need to quantify the uncertainty of not accounting for spatial differentiation.

Discussions of temporal distinction have received less attention. The WIA-2 background document (Udo de Haes et al. 1999) recognised the value choices involved in choosing to focus attention on a finite time basis, suggesting that different time horizons for impact effects could be used for sensitivity analysis. Some researchers suggest considering all time-integrated impacts and discourage discounting that can result in short-term impacts being penalised relative to more persistent impacts.

The appropriate level of sophistication and comprehensiveness for each impact assessment depends on the type of application and is key to effective environmental decision-making. If the selected methodology is too complex and has very broad boundaries, particularly for screening-level assessments, limited resources may prohibit completion of the study. On the other hand, if a narrow and/or too simplistic framework is defined, the assessment can lead to incorrect decisions, either by relying on gross, inaccurate modelling and data assumptions (lack of sophistication), or by failing to address important factors relating impact to stressors (lack of comprehensiveness). The chapters in this book, and the UNEP–SETAC Life-Cycle Initiative, focus primarily on generic application dependency. This recognises both the dependency of the method on the type of application of LCA and the fact that there are only a limited number of different application types. For each type of application, the aim is to develop a generic method, hence the term *generic application dependency*. This is midway between a uniform method on the one hand and one method per application on the other. One of the key choices, for example, is whether and to what degree the application allows for a spatial differentiation of the stressors and hence a site-specific assessment. In practice, this means we aim for 2 to 4 variations of the same method. It would be advantageous if these different variations could still be related to each other. Chapters 3 and 4 present a discussion of the options for addressing this choice.

Towards Best Available Practice in LCIA

The purpose of LCA is to foster an understanding of the causes and types of environmental impacts related to a product or service and to provide a factual basis for decision-making. LCA will be used in companies — and will be successful as an environmental decision support tool — only if it is straightforward to use, provides unambiguous answers, is scientifically defendable, and uses environmentally relevant indicators. Companies and practitioners hence require off-the-shelf characterisation and weighting factors. They often express the desire for standardisation because it would help them to communicate the results internally and to external stakeholders, and because it would make it easier to compare different LCAs or exchange aggregated LCA results. It is frequently stated that standardisation would also remove some of the ambiguity created by the existence of 'competing' LCA methods.

Because of the invariable presence of value judgements in any LCIA, there is no unique, best method for conducting an impact assessment (Hertwich et al. 2000). In our view, some elements in the competing methods represent different judgements about both what goes on in nature (the cause–effect chain) and the importance of different valued items. We accept, however, that decision-makers have neither the need nor the capacity for understanding these issues in their full complexity, and that they would be better off with a single approach. We hence strive to develop a best available practice for impact assessment. The work presented in this book reviews and advances impact assessment methods, but it also presents an effort to help achieve consensus or compromise among method developers. These are necessary steps towards a best available practice. We hope our work will provide an important contribution to the UNEP–SETAC Life-Cycle Initiative.

In many cases, methods that more accurately model damage pathways, and thereby better reflect our concerns, require more data, both from the life-cycle practitioner (e.g., on place and timing of the environmental interventions) and from method developers (e.g., site- and time-specific meteorological conditions for all possible release sites). There is hence a difficult tradeoff between certainty and feasibility. As method developers, we tend to offer improved assessments by reducing uncertainty, even if this leads to methods that are impractical for most situations. We recognise that the more demanding LCIA is, the less it will be used, leading to missed opportunities for environmental improvements and sometimes to misleading results (Hertwich et al. 2000). Nonetheless, method development should not stop. We will achieve better assessments, which are not necessarily always more complex, if we try to make them better. Attempting better assessments implies that some proposals may turn out to be impractical for many, if not all situations, but history shows that this changes with time. The aim of this book is to advance method development so that it will, within a foreseeable time, lead to a best available practice, even if some of the methods are not yet complete or practical.

Outline of *Life-Cycle Impact Assessment: Striving towards Best Practice*

The contents of the chapters in this book are aligned with the focuses of individual task forces, which have been in operation since 1997. Many of the issues mentioned above were discussed at length within the individual task forces, as outlined below:

Chapter 2 — Principles and methods of 3 types of environmental interventions: abiotic extractions, biotic extractions, and land use. Several newer methods are presented for each of the 3 issues. An elaboration is given of important definitions and principles, which are necessary for working out practicable methods.

Chapter 3 — Best available practice regarding CIs and sets of concomitant characterisation factors for climate change, stratospheric ozone depletion, ground level ozone formation or photo-oxidant formation, acidification, and aquatic and terrestrial eutrophication. Particular attention is given to a number of issues relevant for the non-global impact categories (such as models and model domains, substances covered, and spatial and temporal aspects).

Chapter 4 — Recognition that there is currently no one best available model for use in all applications and situations for chemical fate and human exposure modelling in LCIA. The features and capabilities of multimedia and single-medium models for characterising toxicological impacts are discussed. The importance of issues such as spatial variability is highlighted.

Chapter 5 — Suitability of human toxicological impact measures for use in characterisation in LCIA. Promising approaches are addressed in detail and evaluated on the basis of a set of selected criteria. Human health indicators are classified and presented as toxicological potency-based indicators (reflecting the risk, probability, or likelihood of an effect, while not implying actual effects, by definition) and severity-based indicators (reflecting both the likelihood and the consequences or resultant damage).

Chapter 6 — Ecotoxicity, the use of multiple-species measures, and the implications of the working point on the dose–response curve. Discussions centre on how mixtures are taken into account, as well as on the choice of assessment endpoint: the affected versus the disappeared fraction of species. To help evaluate the different options, the suitability of criteria for LCIA is presented and a preliminary evaluation of the approaches is performed.

Chapter 7 — Methods and practice for normalisation, grouping, and weighting within LCIA. Differences between conducting these steps after an endpoint characterisation versus a midpoint characterisation are noted, with some discussion on how to handle aggregation when some impacts are unknown or unquantified. The authors provide guidelines to what types of normalisation and weighting are appropriate in combination.

Chapter 8 — Issues associated with the conceptual structure and scope of LCIA: the role of LCA as a systems analysis tool versus a decision support tool, a discussion of potential AoPs, a discussion of life support functions (LSFs) as an impact category, and the selection of the boundaries of impact categories included within LCIA. This chapter developed not as a specific task force discussion paper, but from the authors' proposals and a series of email debates.

References

Bare JC, Hofstetter P, Pennington DW, Udo de Haes HA. 2000. Life cycle impact assessment workshop summary; Midpoints versus endpoints: The sacrifices and benefits. *Int J LCA* 5(6):319–326.

Bare JC, Pennington DW, Udo de Haes HA. 1999. Life cycle impact assessment sophistication international workshop. *Int J LCA* 4(5):299–306.

Barnthouse L, Fava J, Humphreys K, Hunt B, Laibson L, Noesen S, Norris G, Owens J, Todd J, Vigon B, Weitz K, Young J. 1997. Life-cycle impact assessment: The state-of-the art. Pensacola FL, USA: Society of Environmental Toxicology and Chemistry (SETAC).

Fava J, Consoli F, Denison R, Dickson K, Mohin T, Vigon B, editors. 1993. A conceptual framework for life-cycle impact assessment. Pensacola FL, USA: Society of Environmental Toxicology and Chemistry (SETAC).

Fava J, Denison R, Jones B, Curran M, Vigon B, Selke S, Barnum J, editors. 1991. A technical framework for life-cycle assessment. Pensacola FL, USA: Society of Environmental Toxicology and Chemistry (SETAC).

Heijungs R, Guinée JB, Huppes G, Lankreijer RM, Udo de Haes HA, Wegener Sleeswijk A, Ansems AMM, Eggels PG, van Duin R, de Goede HP. 1992. Environmental life-cycle assessment of products. Guide. Leiden, NL: Centre of Environmental Science (CML). 96 p.

Hertwich EG, Hammitt JK, Pease WS. 2000. A theoretical foundation for life-cycle assessment: Recognizing the role of values in environmental decision making. *J Ind Ecol* 4(1):13–28.

Hofstetter P. 1998. Perspectives in life cycle impact assessment: A structured approach to combine models of the technosphere, ecosphere and valuesphere. Boston MA, USA: Kluwer.

Hofstetter P. 1999. Top–down: Arguments for a goal-oriented assessment structure. Landsberg, D: LCA Global Village, Ecomed. http://www.ecomed.de/journals/lca/village. Accessed 31 Mar 2002.

[ISO] International Organization for Standardization. 1997. ISO 14040: Environmental management—Life cycle assessment—Principles and framework. Geneva, CH: ISO.

[ISO] International Organization for Standardization. 2000. ISO 14042: Environmental management—Life cycle assessment—Life cycle impact assessment. Geneva, CH: ISO.

Potting J. 2000. Spatial differentiation in life cycle impact assessment. Utrecht, D: Utrecht Univ, Dept of Science, Technology and Society. ISBN 90-393-2326-7.

Steen B, Ryding S-O. 1991. The EPS Environmental Accounting Method: An application of environmental accounting principles for evaluation and valuation of environmental impact in product design. Göteborg, S: Swedish Environmental Research Institute (IVL).

Udo de Haes HA, editor. 1996. Towards a methodology for life-cycle impact assessment. Brussels, B: Society of Environmental Toxicology and Chemistry (SETAC) Europe. ISBN 90-5607-005-3. Report of the SETAC Europe First Working Group on Life-Cycle Impact Assessment.

Udo de Haes HA, Jolliet O, Finnveden G, Hauschild M, Krewitt W, Müller-Wenk R. 1999. Best available practice regarding impact categories and category indicators in life cycle impact assessment. *Int J LCA* 4(2):67–74. Background document for the Society of Environmental Toxicology and Chemistry (SETAC) Europe Second Working Group on Life-Cycle Impact Assessment.

[UNEP] United Nations Environment Programme. 1996. Life-cycle assessment: What it is and how to do it. Paris, F: UNEP, Industry and Environment. ISBN 92-807-1546-1.

Impact Assessment of Resources and Land Use

Erwin Lindeijer, Ruedi Müller-Wenk, Bengt Steen

With written contributions from Martin Baitz, Joris Broers, Christel Cederberg, Göran Finnveden, Maarten ten Houten, Thomas Köllner, Berit Mattsson, John May, Llorenc Mila i Canals, Isa Renner, and Bo Weidema

Abstract — This chapter is concerned with the elements of the input side of a life-cycle assessment (LCA) inventory: abiotic and biotic resources that are transferred from nature to human-controlled processes and land areas that are occupied by such processes, possibly after a preliminary adaptation of their properties to the intended purpose. Why is the use of natural resources and land areas an impact on nature that merits being treated within LCA? A short answer: Because use of abiotic resources may lead to an impoverishment of nature to the detriment of future human generations, because use of biotic resources may lead to a decline or extinction of species being components of biodiversity as well as of resource dotation, and because human land use threatens biodiversity and the proper functioning of life support systems essential for nonhuman and human life. Above all, land use has been identified as the main cause of biodiversity degradation, so that its adequate inclusion in LCA methodology appears indispensable.

In this chapter, the reader will not find best available methods for assessing the impacts of resource and land use. Such a choice would be unsuitable in view of the current state of the art. Instead, this chapter lays down important principles, definitions, distinctions, and objectives. These contribute to a better understanding of the impact problem of resource and land use, and to a common platform for the further development of appropriate impact assessment methods.

Earth is endowed with almost illimited abiotic resources (oil, copper ore, etc.). But the progressive extraction and subsequent use of them can lead to situations in which their availability for future generations risks being substantially lessened because of the low quality of remaining stocks. Methods have been proposed for assessing this type of degradation of nature, but an open question is how to express loss of quality. Some abiotic resources, such as fresh water, are not locked below ground, as ores are, but are important components of living ecosystems. Consequently, the impact assessment of their extraction, in addition to representing future scarcity of the resource, also should represent the current degradation of biodiversity and of certain life support functions (LSFs).

Biotic resources are wild plants and animals; they must be distinguished from plants and animals grown in man-controlled cultures because the environmental impacts of the extraction of wild animals or plants are entirely different from those of the culture processes. Marine fish and precious woods are cases of biotic resources for which impact assessment methods for LCA are immediately needed because of the increasing population declines. In addition to resource depletion, impact assessment methods must account for degradation of the natural environment because biodiversity and LSFs are always involved.

With respect to land use, we first introduce the fundamental distinction between land occupation and land transformation, as well as how the decrease in environmental quality is allocated to these 2 cases and ultimately to the functional unit as the object of an LCA study. We address the multitude of elementary land use activities, like ploughing or irrigating, which has led to the synthesis of combined land use types. Land is not used up by land use; we describe that the impacts of land use rather consist of the deterioration of habitats for nonhuman life, as well as the degradation of soil qualities, with a damaging influence on important life support systems such as the freshwater or carbon circuits. Land use also leads to increased competition for land amongst contemporary humans; we propose to treat this important issue outside of LCA.

We briefly discuss, but do not qualify or disqualify, a rather comprehensive choice of published methods. It is our opinion that the LCA community should discuss the proposed principles, definitions, distinctions, and objectives. On the basis of an agreed framework, valuable elements of the methods described here could be combined to improve future impact assessment procedures.

Life-Cycle Impact Assessment Framework for Resources and Land Use

Aim of this chapter

Man-controlled processes may cause not only output flows to nature but also input flows from nature towards the process. This chapter deals with the latter. The input side of life-cycle inventories (LCIs) essentially contains 3 types of environmental interventions:

1) Extraction of abiotic resources — the transfer of nonliving material (e.g., metal ores, fossil fuels, lime) from nature into man-controlled processes
2) Extraction of biotic resources — the capture of organisms living freely in nature (e.g., fish, deer), in order to introduce them to man-controlled processes
3) Allocation of land areas to man-controlled processes — activities necessary to change or maintain the properties of such areas in view of optimising the intended man-controlled process.

It is a fact that the impact assessment of these input-oriented environmental interventions is much less developed than the impact assessment of emissions. In the course of the last few years, assessment methods have been published, but these start from different positions and focus on different aspects of the general problem, so that comparison becomes difficult and of limited use. In consequence, the aim of this chapter is not to propose best available methods for the impact assessment of resource extraction and land use. It appeared to be more constructive to work out a best available set of principles, definitions, distinctions, and objectives, in order to clarify the problems of impact assessment of resources and land use, and to create a common framework for further development of existing and new assessment methods, as well as for a selection of best available methods in the future.

Extraction of biotic or abiotic resources and land use are discussed together in the same chapter because they have common elements and because it is useful to fix the framework for the corresponding impact assessment methods as a whole. In short, the following common elements merit mention at the beginning of this chapter:

- Use of biotic or abiotic resources and land means that objects of nature with a limited supply are allocated exclusively to a human-controlled process, at least for a certain time. Consequently, these objects may be destroyed (fossil fuel resources) or degraded (metal resources, land). This represents a damage to nature.
- The available quantities of certain objects of nature may be increased by natural regrowth: This is typical for biotic resources such as fish or trees. In contrast, there is practically no regrowth in the case of most abiotic resources (metal ore deposits, fossil energies) and in the case of land area.
- A degraded quality of certain objects of nature may be recovered by spontaneous action of nature (spontaneous renaturalisation of fallow land) or by man-controlled recovery processes (recycling of waste materials, active land improvement).

Areas of protection

It is important to link the different impacts of land and resource use to areas of protection (AoPs; International Organization for Standardization [ISO] 1999). The most relevant AoPs for land and resource use are discussed below (see Chapter 8 for further discussion of AoPs).

AoP natural resources

'Natural resources' are extractable entities with implications for their present, but mainly future, availability. The AoP natural resources relates to all natural resources: abiotic resources such as oil, ore deposits, fossil and fresh surface water, as well as biotic resources such as wild populations of fish and trees. An item cannot be a natural resource unless, after extraction, it is useful for human purposes.

AoP natural environment

In Chapter 8, the authors suggest splitting the AoP 'natural environment', as described by Udo de Haes (1999), in two. This is suggested in order to distinguish the functional and nonfunctional (intrinsic) values of nonhuman life and its natural prerequisites. The functional values of the natural environment are represented by so-called 'life support functions' (LSFs). Examples of LSFs are production of biomass, regulation of climate, generation or regeneration of soil, pollination of crops, mitigation of floods and droughts, and natural purification of air and water (de Groot 1992). Such functions support human and nonhuman life.

The nonfunctional values of the natural environment are represented by 'biodiversity' and 'natural landscapes'. Diversity of life and uniqueness of natural landscapes provide intrinsic values; in addition, they may be important elements of systems providing LSFs.

AoP manmade environment

The AoP 'manmade environment' consists of those elements that are produced and/or maintained by humans (e.g., crops, plantations, domestic animals, buildings, manmade landscapes). The AoP manmade environment is distinguished from natural resources and the natural environment because the valuation of manmade objects is different from that of natural ones.

AoP human health

'Human health', in contrast to the others above, is an AoP whose direct impact from land and resource use is very limited: Humans do not get sick or die because of land use or because of the use of natural resources. However, human life may be seriously threatened indirectly by land and resource use because of their dramatic degradation of the natural environment.

Interventions and their impacts on AoPs

Extractions of abiotic or biotic resources result mainly in direct impacts on the corresponding natural resources: Extraction and subsequent use of a material or organism may reduce the quantity (e.g., reserves of mineral oil or salmon) or the quality (e.g., concentration of copper ore) of the stock. If the item to be extracted is not buried in the underground but forms a part of nonhuman life or its natural prerequisites, the extraction can also result in an impact on the natural environment, by damaging LSFs and biodiversity: Extracting freshwater from a river, or precious trees from a tropical forest, causes not only a degradation of the corresponding resource but also important changes in ecosystems.

In addition, the whole process of extraction (to be clearly distinguished from simple extraction as an elementary intervention) may cause impacts on the natural environment from emissions, waste deposits, energy use, etc.

Land use can have direct impacts on various endpoints or AoPs, mainly biodiversity, LSFs, natural resources, and manmade environment. For example, using land as a motorway surface may exert impacts on biodiversity through loss of habitats, on LSFs through loss of buffer capacity for rain water, and on natural resources through disappearance of deer because of habitat segmentation.

Environmental mechanisms of resources and land use

The environmental mechanism between the interventions and impacts on midpoints or AoPs (endpoints) can be visualised by a cause–impact network (see Figure 2-1). As with most other

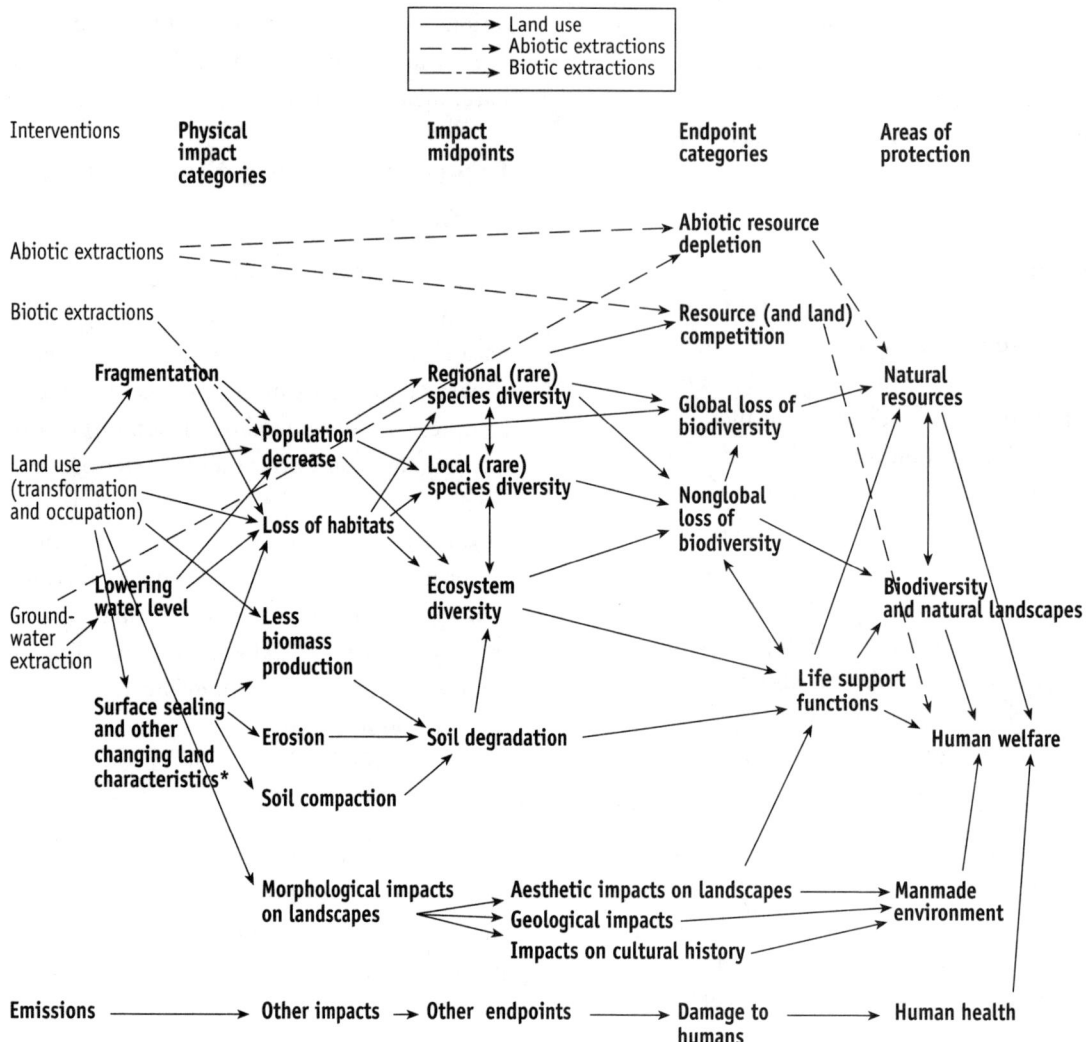

Figure 2-1 The cause–impact network, focusing on resources and land use. (The * indicates characteristics such as erosion resistance, filter and buffer capacity, humus quality, and water regulation [see the proposal from Baitz et al. 1998].)

impact categories, much is still uncertain about the cause–impact network related to resources and land use. The cause–impact network is meant only to illustrate the complexity and levels of assessing impacts in life-cycle assessment (LCA). Links from biotic or abiotic extractions and land use to the AoP human health are not shown for simplicity reasons.

The causal relationships between the interventions and the various levels of impact categories in Figure 2-1 are depicted by arrows. These relationships cannot be quantified fully, especially the steps from midpoints onwards to the AoPs. On the other hand, it may be expected that a reasonably comprehensive impact assessment is possible without the inclusion of all the cause–impact paths shown in Figure 2-1.

Characteristics of impact assessment methods for resources and land use

In order to describe all relevant approaches to land use and resource impact assessment, we use a general descriptive scheme. The basis for this scheme is given below.

There are various degrees of sophistication to impact assessment. The lowest degree would exclude regionalisation completely and would use only one indicator or only classification sets. The highest degree of sophistication would require full quantification, full regionalisation, and a comprehensive, relevant, and consistent set of indicators that refer to the various AoPs. These levels of sophistication have been mentioned as core criteria for distinguishing between life-cycle impact assessment (LCIA) approaches in Udo de Haes et al. (1996).

In Table 2-1, we list the ISO 14042 requirements (compulsory for those LCA users who desire to comply with ISO norms) and recommendations for selecting impact categories and indicators. Based on that list, a limited set of criteria (Table 2-2) is used to describe the various methods.

Impact Assessment of Abiotic Resources

Starting point

Impact assessment methods for abiotic resources have been suggested by several authors and are reviewed in an earlier Society of Environmental Toxicology and Chemistry (SETAC) document (Udo de Haes 1996). In this document, several types of abiotic resources were identified. The 3 main types are these:
1) Deposits or stocks that are irreversibly depletable, such as ores, fossil energies, and many other mineral deposits
2) Funds, temporarily or locally depletable, such as peat and nutrients from soil minerals
3) Flows, nondepletable but with a limited availability at a certain time, such as solar radiation and freshwater precipitation.

Within the AoP natural resources, the emphasis is on the impacts on future availability as a functional value for humans, whereas the impact of an extraction on present availability for other humans (leading to competition amongst persons of the present generation) is considered part of economic or social welfare.

The resource concept is tightly linked to human valuation and technology: Iron was not a resource until the Iron Age. An object of nature is a natural resource only when humans show an interest in extracting and subsequently using this object in human-controlled processes, mainly processes of the economic system.

The depletion of abiotic resources may be dealt with in different ways in the inventory, depending on how the resource is defined. Depletion could be seen as being caused by the mining company if the resources are understood to be the ore in the ground. If the ore is already extracted, we can no longer speak of resource extraction, only of resource use. If on the other hand, the global availability of concentrated substances were considered to be the resource, the mining company could be seen as an actor who increases the available quantity of the resource, and the user who causes the dissipation of the material into the environment would be the one who depletes the resource. In this chapter, we consider the availability of minerals in ores (that is, in nature) as the resource aspect to deal with. For the sake of simplicity, we assume that stocks of concentrated abiotic resources in the economic system are more or less constant in time and small in comparison to the deposits in nature. This means that the decrease of the stock of a natural resource is considered equal to the decrease of the stock in nature, and that extracted quantities of the resource are considered to be dissipated sooner or later. In the future, more sophisticated concepts will consider stock changes of concentrated materials in the economic system, as well as a differentiation between dissipative and nondissipative uses.

These and some other elements of the definition of abiotic resources are dealt with in the SETAC Europe First Working Group on Life-

Table 2-1 ISO requirements and recommendations for selecting impact categories and indicators[a]

ISO Nr.	Requirements and recommendations
14042	**Requirements**
	a) The selection of impact categories, category indicators and characterisation models shall be consistent with the goal and scope of the LCA study
	b) The sources for impact categories, category indicators and characterisation models shall be referenced
	c) The selection of impact categories, category indicators and characterisation models shall be justified
	d) Accurate and descriptive names shall be provided for the impact categories and category indicators
	e) The selection of impact categories shall reflect a comprehensive set of environmental issues related to the product system being studied, taking the goal and scope into consideration
	f) The environmental mechanism and characterisation model which relate the LCI results and category indicator and provide a basis for characterisation factors shall be described
	g) The appropriateness of the characterisation model used for deriving the category indicators in the context of the goal and scope of the study shall be described
14042	**Recommendations**
	a) The impact categories, category indicators, and characterisation models should be internationally accepted, i.e., based on an international agreement or approved by a competent international body
	b) The impact categories should represent the aggregated emissions or resource use of the product system on the category endpoint(s) through the category indicators
	c) Value choices and assumptions made during the selection of impact categories, category indicators and characterisation models should be minimised
	d) The impact categories, category indicators and characterisation models should avoid double counting unless required by the goal and scope definition
	e) The characterisation model for each category indicator should be scientifically and technically valid, and based upon a distinct identifiable environmental mechanism and/or reproducible empirical observation
	f) The category indicators should be environmentally relevant
	g) It should be identified to what extent the characterisation model and the characterisation factors are scientifically and technically valid

[a] ISO 1999. (© International Organization for Standardization [ISO]. This material is reproduced from ISO 14042 with permission of the American National Standards Institute on behalf of ISO. No part of this material may be copied or reproduced in any form, electronic retrieval system or otherwise or made available on the Internet, a public network, by satellite or otherwise without the prior written consent of the American National Standards Institute, 25 West 43rd Street, New York, NY 10036.)

Cycle Impact Assessment (WIA-1) report (Finnveden 1996); some additional aspects are discussed in the following sections.

Currently, public discussions of abiotic resource depletion as an environmental problem are quite controversial. Many say that the total quantities of accessible abiotic resources can be expected to be practically illimited, so that their reduction by human extraction and dissipation is negligible as an environmental problem. Others believe that the current rate of destroying fossil fuel resources and dissipating other abiotic resources is a significant impoverishment of nature, to the detriment of future generations. The authors of this chapter

Table 2-2 General items to address when discussing category indicators (CIs) in subsequent tables

Item nr.	Items to address regarding CIs
1)	Essentials (extent of quantification, regionalisation, and comprehensiveness, including environmental relevance) of the indicator set
2)	Sensitivity of the indicator to the intervention changes
3)	Description of environmental mechanism and characterisation model, including extent of reliance on reproducible empirical observations and scientific or technical validity
3)	Extent of representation of category endpoints through aggregated interventions
5)	Extent of and description of value choices and assumptions (on the levels of model and data)
6)	Extent of double counting and other consistency issues
7)	Applicability of available data

share the opinion that there is a large uncertainty with respect to current stocks of abiotic resources, as well as with respect to future human needs. We propose that lack of knowledge should not be filled by instinctive optimism, and we consider it justified to monitor the change in nature's quality that is due to current extraction rates of abiotic resources. Working out and further improving impact assessment methods for abiotic resources is therefore desirable, and the results of LCA studies can provide a reasonable basis for judging whether environmental damage from abiotic resource use is important, for example, compared to environmental damages from emissions.

Definitions and distinctions

When we discuss abiotic resources, we must define certain terms and make certain distinctions, which may also be important in the context of biotic resources (and shall not be repeated in the section on biotic resources).

Distinction between abiotic and biotic resources

Biotic resources are alive, at least up to the moment of extraction, whilst abiotic resources are dead, although they may originate from past life. Wood is a biotic resource, but coal and peat are abiotic resources, although they consisted of biomass a long time ago. Because biotic resources are living objects before extraction, they are always a part of the biosphere, and their removal not only reduces the natural stock of the resource but also exerts a change on ecosystems. In contrast, abiotic resources are often buried outside the biosphere, so that their stock reduction has no direct influence on ecosystems.

A second difference is that most abiotic resources have fixed stocks, whilst biotic stocks may grow: They are fund or flow resources. It is normal for biotic stocks to show replenishment rates that are nonnegligible compared to the usual extraction rates because of the birth of new individuals and the biomass growth of already-living individuals. This means that future availability of biotic resources is conditioned mainly by the relation between the extraction rate and the replenishment rate. Of course, the replenishment rate of biotic resource stocks may be influenced negatively by excessive extraction rates or unfavourable changes in environmental conditions. Examples from the few abiotic resources that show nonnegligible replenishment rates are (in addition to the already mentioned solar radiation and freshwater precipitation) sand, gravel, and volcanic outputs.

In some cases, natural resources are a mix of biotic and abiotic components. An important example of this is topsoil, consisting of minerals, dead organic matter, and an enormous variety of life. In consequence, impact assessment of soil extractions is a particularly difficult problem.

Defining the environmental intervention extraction of biotic or abiotic resources

The term *extraction* here means that an object is removed from nature and brought into someone's possession for a certain use. Possession is not the same as legal ownership; it is possible that areas of wilderness, with their ore deposits, wild plants, and wild animals, are legally owned by someone, so that the act of extraction does not change this legal ownership. But the act of extraction means that the extracted objects are transferred from nature into the extractor's control and cannot maintain their previous behaviour in their ecosystems.

In principle, extraction is an intentional action. The extractor wants to transfer the object from nature into his possession in order to introduce it into a subsequent process. But extraction may also happen unintentionally: A car driver kills a deer crossing the road, or a fishery team intending to harvest herrings causes an unintentional bycatch of unwanted fish species, which subsequently die. The impact on nature is the same, whether the action of the extractor is intended or not. In consequence, an LCI of road transport or of herring fishing should contain unintentional removals of objects of nature.

Further, it is necessary to distinguish between extraction and the extraction process. The environmental intervention *extraction* represents the removal of a certain quantity of a resource that was part of nature before its extraction. Theoretically, this removal could be made without any additional environmental interventions. But in normal practice, the removal is done by an extraction process, which includes the use of machinery, transportation equipment, buildings, etc. In consequence, the extraction process gives origin to a multitude of environmental interventions such as energy use, air pollution, noise, etc., in addition to the 'naked' extraction. In this text, the word *extraction* always means only the environmental intervention, and not the entire extraction process.

The term *extraction* generally expresses the fact that humans physically transport an object from nature into the manmade environment. For example, a fish is transported from the open ocean into the body of a fishing vessel, for further processing towards a marketable product. But instead of transporting the object from nature into buildings or vehicles, it is also possible that the extraction process consists of expanding the borderlines of the manmade environment so that an object formerly in nature now becomes a part of the manmade environment. In this enlarged sense, we could consider it a special case of extraction if a piece of land, with its mineral deposits, its vegetation cover, and its soil-bound life, is reallocated from a natural area into a mining site or an area of intensive agriculture. Instead of transporting objects of nature towards the manmade environment, equipment and procedures of the manmade environment are then transported into a former area of nature.

Competition and depletion

Depletion and competition of natural resources are stated to be separate issues of concern in Heijungs et al. (1997).

Depletion means that the availability of a natural resource is decreasing because the quantities and/or quality of the item in nature are reduced. The focus here is on the development of the resource availability over lengthy time periods: Depletion of a natural resource means that for future generations of humans, the access to this resource is expected to be more onerous, or even that future generations will have to substitute other types of resources. If a natural resource is on the path to depletion, this does not necessarily mean that the present generation of humans notices anything like higher prices or reduced supply.

Depletion can be caused by any rate of extraction, if the natural resource is of the stock type in which the reproduction rate is negligible. Natural resources of the funds type, such as ground water or marine fish, may be reduced temporarily but will not be depleted in the long run, as long as the natural replenishment processes are not damaged, which normally sets upper limits to the extraction rates. No depletion is possible in the case of flow-type resources because extraction cannot be higher than replenishment.

Competition over a natural resource represents a different problem. *Competition* means that the current supply of a certain resource is limited, so that an exclusive use of that resource implies less availability of the same resource for other contemporary users. The issue of competition was introduced in Finnveden (1996) as 'decreased availability for a limited period'. It is further defined by Heijungs et al. (1997), who state that competition measures how much of a certain resource or land is occupied by the product system for a certain time period, reflecting the reduced availability for use by others at the same time. Resource competition differs from resource depletion, but the one does not exclude the other. In the case of solar radiation, several individuals of the same generation might compete for using the energy on a given surface, but this will not influence the availability of sunshine for future generations. But if competition for fossil oil increases because oil companies lower production, the resulting increase of competition between contemporary humans parallels a lowering of the depletion speed in favour of future generations.

It is a matter of debate whether competition should be considered an environmental impact at all. Finnveden (1996) has stated that 'a choice has to be made…if competition of resources should be considered'. It may be seen as an economic issue, because present competition for resources between individuals or communities is related to economic and political choices, such as supply and demand for industrial goods. The severity of the competition impact is, in the end, determined by the amount of welfare related to the possession (or lack of possession) of that resource. This impact is thus solely defined in the present social arena. Competition can therefore be seen as an economic issue but also as a social partitioning issue (an aspect of sustainability in the broad sense), with very important implications, at least locally. Despite its importance, it is our opinion that the competition aspect should not be included in present LCA, as it would open LCA to a wide spectrum of issues, which may better be addressed by a separate tool for decision-making. Competition will therefore receive marginal attention in this chapter.

Regional and global resources

Natural resources are situated over the whole surface of the earth, including continents and oceans. Should the impacts of using biotic and abiotic resources be studied on a regional level, or is it better to consider the global situation? It appears necessary to distinguish the 2 cases.

1) If a natural resource is hidden in the ground and therefore not connected to the biosphere, its location is unimportant, as long as the impact assessment focuses on the 'naked' extraction and its consequences for the magnitude of the resource stock instead of on the entire extraction process and the subsequent transportation process, with their various environmental interventions. Consequently, the depletion problem of such 'locked-in' resources is normally considered at the global level, and the stock quantities as well as the extraction quantities are summed over the whole globe. If regional subdivisions are made, this is due to political problems and not to environmental problems. The global approach is typical for natural resources such as fossil energies and metal ores, even if they are produced partially by open pit operations. In practice, these products are transported around the globe, and the availability of the resource is

not a function of the precise distribution of stocks over the globe.

2) If the stock of a natural resource is in close contact with the biosphere, any reduction of stock by extractions may also influence biodiversity and LSFs. It is generally agreed that biodiversity and several LSFs must be maintained at the regional level: It is not sufficient to ensure a high diversity of plants and animals, as well as an intact freshwater circuit, in South America only, for example. Rather, degradation should be avoided in every region. This means that stock reductions from extractions of biotic resources such as fish or wood must be studied at the regional level, as far as effects on biodiversity and life support systems are concerned, although the depletion aspect is still studied at the global level.

Natural resources now and in the future

Natural resources are objects of nature which are extracted by man from nature and taken as a useful input to man-controlled processes, mostly economic processes. If such extractions increase the chances for depletion of a natural resource within the next centuries or millennia, this is seen as an impoverishment of nature to the detriment of future generations.

This gives rise to a problem: Will the needs of future generations for natural resources be the same as the needs of the present generation? It appears highly probable that future generations will maintain the need for energy and fresh water. But will there be a continuation of the current requirements for mercury, phosphates, and wild salmon? Or will future generations instead require large quantities of hafnium and ferns? We cannot exclude the possibility that any object of nature might become an important input to man-controlled processes of the future. Every object of nature, then, could be called a natural resource and treated accordingly in LCA. But we believe that this goes too far.

For practical reasons, we define natural resources on the basis of current requirements. If we avoid depletion of these natural resources, future generations will have the option of maintaining our generation's pattern of needs. If they want to change their pattern of needs in comparison to ours, they will have the opportunity to dispose of all those abiotic materials that currently are not extracted in substantial quantities because we show no interest in them. If, in addition, the current generation avoids a decrease of biodiversity, future generations will have the option of changing their needs for biotic resources.

Operationalisation of impact assessment of abiotic resources

Abiotic deposit resources

The impact assessment of depleting abiotic resources was reviewed by Finnveden (1996) as falling broadly into 4 options.

Option 1: Aggregation of energy and materials on the basis of energy and mass. If used, it was suggested that the full weight of the ores and minerals be taken as the basis for aggregation, rather than just the amount of refined products, that is, the metal (Finnveden 1996:42).

Option 2: Aggregation based on measures of deposits, D, and current consumption, U. The following characterisation factors Q have been suggested (Müller-Wenk 1978; Heijungs et. al 1992; Fava et. al 1993; Guinée 1995; Guinée and Heijungs 1995):

2a) $Q = 1/D$
2b) $Q = U/D$
2c) $Q = 1/D \times U/D$

Option 3: Aggregation based on environmental interventions caused by future, hypothetical processes. Pedersen (1991) suggests a characterisation for reversibly used abiotic resources, based on the environmental burdens associated with a recovery process that would bring the resource back to its original state. Müller-Wenk (1999) uses the surplus energy, required for extracting and processing lower-grade deposits in future, as an indicator of the severity of the depletion impact. In the Eco-indicator 99 (Goedkoop and Spriensma 1999), this approach has been used, aggregating the major resources based on surplus energy requirements.

Option 4: Aggregation based on exergy consumption or entropy production. Finnveden and Ostlund (1997) suggested a method based on exergy.

The older proposals were discussed in Finnveden (1996) and Heijungs et al. (1997). Below, we comment on more recent proposals.

Comments referring to Option 1: This option is not seriously considered a meaningful impact category indicator because the aggregation of interventions is performed without any impact model or rationale. It may, however, be considered a coarse indicator for resource use, although the aggregated figure should be interpreted carefully.

Comments referring to Option 2: In methods that use the total quantity of resource deposits, D, uncertainty is large. The earth's crust contains huge amounts of resources in low grades, and it is almost impossible to fix meaningful figures for a quantity D that goes beyond the 'proven and currently extractable reserves'. Apart from this general problem, the formula that includes both stock and yearly extraction (Option 2c) was considered the best approach by Guinée and Heijungs (1995) and Heijungs et al. (1997) because both the abundance factor $1/D$ and the inverse of the stock lifetime at the present extraction rate (U/D) are included. In the new LCA guide from the Centre of Environmental Science (CML; Guinée et al. 2001), recent data for Option 2c is given, based on ultimate resource stocks (presence in the earth's crust). The main uncertainty lies in the relation between this type of stock and its availability for human needs (a model uncertainty). In absolute figures, nearly all these stocks have very long lifetimes, given the present extraction rates. Focusing on their relative abundance neglects differences in technical availability, which implies a large range in relative depletion value.

Comments referring to Option 3: The approach of Weidema (Pedersen 1991; Weidema 2000) actually uses scenarios for future resource extraction to eliminate altogether the impact category of abiotic resource depletion. This approach can in principle be used for several impact categories, perhaps leaving only one: land use (Weidema 2000). This approach therefore leans very much on scenario-building in which resources are used to maintain availability in the future, and therefore will imply large data uncertainties whenever operationalised. The Ecological Footprint approach is an operationalisation which goes in this direction, although it may not be completely consistent with LCA. Alternative approaches based on future scenarios (similar to Option 3) have been used by Müller-Wenk (1999) and Steen (1999). Steen defines *indicator* as the mass of an element in the form of ore or mined concentrates. His approach means that the characterisation factor becomes equal to the concentration. The rationale for defining one CI for each element is that each element has some unique qualities. Different elements are combined in the weighting phase. His default weights are derived from the estimated costs to extract and produce the same resource from sustainable sources. This includes use of technical scenarios that introduce uncertainty. Müller-Wenk (1999) starts from the fact that the stocks of practically all abiotic resources in the earth's upper crust are abundant, but the easily usable high-grade portion of the overall stock is comparatively small. The consequence of extracting a certain amount of stock is therefore that future generations will have to use lower-grade stocks, whereby the decrease of concentration per 100 years will be steeper or flatter, depending on the type of resource. Current extraction of high-quality ores can therefore be charged with the extra environmental interventions caused by future extractions of low-grade stocks of this resource. In order to simplify things, the increased energy requirement for future use of low grades may be taken as an indicator for all environmental interventions caused by the future mining and refining processes. His approach has been applied with slight modifications by Goedkoop and Spriensma (1999) in the co-indicator Eco-indicator 99. However, in both cases, the estimates of future energy consumption for extraction are rather uncertain.

Comments referring to Option 4: As the main proposer of this option, Finnveden suggested an energy-based indicator (Finnveden and Ostlund 1997). But it is questionable whether loss of exergy is really a suitable indicator for the environmental problem of abiotic resource depletion. Finnveden uses the following arguments in favour of the Finnveden–Ostlund method:

1) It can be argued that the usable energy (*exergy*) is the ultimate limiting resource.[1] This can be claimed because each material resource has an associated energy cost, so that every potentially limited resource is limited in part because its energy costs are too high. Given sufficient amounts, a society can divert exergy to the acquisition or mobilisation of whatever material is in short supply.
2) The problem about resource consumption is the reduction of usable material. For a material to be usable, it must normally be concentrated, structured, and ordered, compared to the surroundings. A well-known scientific quantity that is often interpreted as a measure of the disorder of a system is *entropy*. The entropy production may therefore be a useful indicator of resource consumption. Exergy and entropy are related to each other, so instead of entropy, exergy can be used.
3) A possible indicator of the resources used can be a measure of the 'costs' to bring the resource back to the state from which we took it. The exergy loss is the minimum amount of energy that is necessary to bring a material back to the starting state. Usable energy (exergy) can thus be used as an indicator of the environmental costs of getting the resource back.

Because exergy is expressed in terms of the availability of energy to 'perform work for humans', this indicator seems indeed an interesting option. However, in the examples given, the chemical entropy dominates the exergy calculation. This type of exergy does not seem to represent the type of availability we are concerned with when we discuss resource depletion because it is availability on a microscopic level. Another factor to consider is the situation in which a chemical composition is more dispersed, including the energy for transporting and processing ore as part of the exergy content. Practical examples up to now do not show this *dispersion sensitivity*, which is the basis for our doubts about the sensitivity of this approach.

Conclusions and survey of methods. Uncertainty in the models is thus an important characteristic, which we could not assess in detail.

Next to model and data uncertainties, another characteristic of these proposals is the choice of midpoints or endpoints. There are at least 3 aspects that may be addressed with impact categories and CIs for abiotic resource depletion. These are
1) competition or present availability,
2) future availability, and
3) LSFs and biodiversity.

It can be stated that the methods of Option 2 are closer to present competition, although Options 2b and 2c do give some information in the direction of future depletion with the *U/D* factor. The methods according to Options 3 and 4 more explicitly express future resource availability and therefore could be considered as better focused towards future availability as the main problem of resource extraction.

---

[1] Because neither matter nor energy can be produced or consumed, it may be questioned what is lost when a resource is used. A reasonable answer may be that usable matter and energy are consumed and may become depleted.

However, a judgement on which approach is the best available practice was not possible. To make this judgement, the intended use of resource indicators should be well formulated and discussed. The same questions as are raised at the end of Finnveden (1996) still stand: What is to be defined as the core problem (depletion, competition) and how to deal with substitutability and dependence on (future) technology. All methods are summarised in Table 2-3.

Abiotic fund resources

Ground water, aquifers (except fossil water), and lakes, but also gravel, sand, and river clay, may be considered to belong to the class of abiotic fund resources.

The difference between fund resources and stock resources is that natural replenishment is negligible in the case of a stock-type resource, while natural replenishment may avoid or slow down depletion in the case of a fund-type resource. It is therefore possible to treat the depletion problem of fund resources in the same way as for stock-type resources, but with replacing the current consumption U by the net current consumption $U - R$, where R is replenishment. In consequence, there is no depletion as long as R at least equals U, and depletion is slowed down if R supplies a substantial fraction of U. If the consumption U exceeds the replenishment R for a long time, the natural stocks of the resource may come to an end, and the resource then changes from the fund type to the flow type because consumption can no longer exceed replenishment. This would be, of course, a diminution of natural richness. For example, the annual consumption of fresh water may exceed the annual inflow, as long as a basin of fossil water can be tapped. If this volume of fossil water is used up, fresh water becomes a flow-type resource and the annual consumption is limited to the annual inflow.

But the main environmental problem of high consumption rates in the case of fund-type abiotic resources is probably the fact that replenishment rates R do not remain constant and that the natural environment is unfavourably influenced with respect to biodiversity and LSFs. This cannot be taken into account by a simple adaptation of the methods given in the preceding section.

Practical experience shows that continuous increase of freshwater use by man-controlled processes affects the biodiversity of rivers and lakes and disturbs the LSF of the water circuit. This means that depletion of fund types of natural resources can be much more than merely a problem of dwindling resource availability. Assessment methods to deal with this complexity are not yet in a developed stage.

Abiotic flow resources

Inflow of solar radiation, flowing fresh water supplied by precipitation, and nutrient flow from weathering processes belong to the group of abiotic flow resources.

Natural resources of the flow type cannot be depleted. In contrast, competition for these resources among contemporary humans may be a serious social problem, but we suggest dealing with this problem outside of LCA.

But abiotic flow resources are typically in close contact with the biosphere. This means that, even more than in the case of fund-type resources, extractions may have a significant influence on the AoP natural environment. If a river loses a part of its natural water flow because of a hydraulic power plant or an agricultural irrigation system, this may have serious consequences for nonhuman life in the river's floodplain. If water is extracted from a given river area, the environmental effect is comparable to the effects of land use. In LCIA, it appears practicable to treat human activities in a floodplain area (such as water extractions, landfills, damming, and erection of buildings) in a similar way as land uses on 'solid' land surfaces. Thus, the flows become midpoints in another environmental mechanism.

Table 2-3 Present options for operationalisation of deposit abiotic resources

Items according to Table 2-2	Properties of model
Fava et al. 1993	
Essentials[a,b]	Full quantification through characterisation factor ($Q = 1/D$), no regionalisation, no specification of resource types
Sensitivity	High, on any use of scarce resources
Mechanism and model	Simple, based on inventory flows and amounts of deposits
Extent of representation	Limited because of unclear endpoint concept
Choices or assumptions	Resources of different substances are exchangeable with regards to 'scarcity aspect'
Consistency	Unclear which substances are included (if Cu ore is treated as 1 type of deposit, e.g., 1 result is obtained; if dealt with as 2 types [sulphides, oxides], same resource flow gives different result)
Applicability	Reasonably good
Guinée and Heijungs 1995	
Essentials[a,b,c]	Full quantification ($Q = U/D$), no regionalisation, no specification of resource types, environmentally relevant only on a short-term basis
Sensitivity	Low if endpoint is total abiotic resource depletion, as deposits tend to be prospected and determined in relation to yearly consumption; high if endpoint is deposit depletion
Mechanism and model	Simple, based on inventory flows and amounts of total use and deposits
Extent of representation	Limited because of unclear endpoint concept
Choices or assumptions	Resources of different substances exchangeable with regards to scarcity aspect; present deposit amounts correlated to total resource amounts
Consistency	Less sensitive to dividing resources into subgroups than method by Fava (1993)
Applicability	Reasonably good
Heijungs et. al. 1992; Guinée et al. 2001	
Essentials[a,b,c]	Full quantification ($Q = 1/D \times U/D$), no regionalisation, no specification of resource types
Sensitivity	High, on any use of scarce resources
Mechanism and model	Simple, based on inventory flows and amounts of total use and deposits
Extent of representation	Limited because of unclear endpoint concept
Choices or assumptions	Resources of different substances are exchangeable with regards to scarcity aspect
Consistency	Unclear which substances are included (if Cu ore is treated as 1 type of deposit, e.g., 1 result is obtained; if dealt with as 2 types [sulphides, oxides], same resource flow gives different result)
Applicability	Reasonably good
Pedersen 1991; Weidema 2000	
Essentials	Resources are not an endpoint; can be dealt with through system expansion making an inventory with scenarios on future resource extraction
Sensitivity	High, on any use of scarce resources
Mechanism and model	Include interventions from future resource extraction in inventory
Extent of representation	Land use explains only part of impacts on category endpoints
Choices or assumptions	Future scenario of resource extraction is chosen
Consistency	Good
Applicability	Some scenario data available from the author; some lacking

continued

Table 2-3 *continued*

Items according to Table 2-2	Properties of model
Müller-Wenk 1999	
Essentials	Quantitative, characterisation factor based on increased energy demand caused to future generations, no regionalisation
Sensitivity	High, on any use of scarce resources
Mechanism and model	Based on impacts from anticipated future resource extraction processes; resources per se not a safeguard subject; low extent of empirical observations as base
Extent of representation	Energy explains only part of impacts on category endpoints
Choices or assumptions	Future scenario of resource extraction is chosen
Consistency	Good
Applicability	Some data available from author; some lacking
Finnveden and Ostlund 1997	
Essentials	Quantitative, no regionalisation; exergy by itself is of low environmental concern
Sensitivity	High for ore exergy, but perhaps low for scarcity and depletion
Mechanism and model	Based on exergy; clear scientific concept
Extent of representation	If ore exergy is endpoint, then representation is good
Choices or assumptions	Exergy is good indicator of resource availability
Consistency	Fair; use of lower-grade ore may result in higher exergy consumption
Applicability	Some data available from author; some lacking

[a] Q = Characterisation factor.
[b] D = Measure of deposits.
[c] U = Yearly consumption.

The direct use of solar energy by solar cells may be treated in a similar way in LCIA. Because humans cannot use more than the current inflow rate of solar radiation, there is no depletion. If human competition is treated outside of LCA, there is no scarcity problem in connection with the use of solar energy. But like water, the inflow of solar energy is closely connected to the biosphere. As long as humans use only an insignificant part of solar radiation, there will be practically no influence on life and on life-support systems. But if an area is substantially covered by solar cells, the corresponding surface is under permanent shadow, and the consequences on the natural environment (biodiversity and life-support systems) are comparable to the erection of a 'normal' building on that surface. In LCIA, then, human use of solar radiation as an energy source is therefore to be treated as a kind of land use.

Conclusions

The aggregation of extractions of different resources is desirable if we seek to reduce the amount of information about these extractions. This aggregation requires a preliminary agreement on the environmental issue at stake. No best available practice could be identified. A lack of consensus in the LCA community exists. To some degree, this results from the character of the impact category, which differs from that of, for example, global warming potential (GWP) in 2 ways:
1) It is dependent on human technology, and
2) it is not a clear physical entity.

For abiotic fund resources, it is possible in principle to adapt the assessment methods of stock-type resources by replacing the consumption rate with the net consumption rate (consumption minus replenishment). But because of the natural process of replenishment, fund resources are closely connected to the biosphere. This means that impacts on biodiversity and life-support systems may become more important than resource availability. Methods to assess such impacts are in a very early stage of development, particularly in the case of abiotic flow resources.

Impact Assessment of Biotic Resources

Starting point

Until now, LCA practice was essentially oriented towards abiotic objects whenever the resource problem was taken into consideration: Inventories typically give quantities of fossil energies or ores used but not quantities of wild animals or plants. In this chapter, we express the view that mankind exerts an influence on nature, mostly judged as adverse, by removing living (also called *biotic*) objects from nature. Humans extract such objects from nature because they qualify them as a resource, that is, as a useful input to economical or other man-controlled processes. But as a consequence of the extraction, the population of the extracted object is reduced, at least temporarily. This population decrease may be seen as a problem from 2 different perspectives.
1) From a resource perspective, this decrease worsens the chances for future human generations to cover their needs for this biotic resource.
2) From the overall perspective of the natural environment, the population decrease amplifies the risk of ecosystems degradation and of species extinction, so that life-support systems and biodiversity are at stake.

The need for impact assessment of biotic resource extractions

For practical reasons, LCA cannot express the full complexity of a functional unit's impacts on nature, and reasonable simplification is mandatory. Is it necessary to work out LCA methods for the use of biotic resources, in order to open the way for reasonably complete LCAs in the case of products with biotic ingredients?

The number of species living now is estimated to be more than 10 million (United Nations Environment Programme [UNEP] 2000:16). Only a tiny fraction thereof is extracted and used as a resource by our developed societies, mainly forest trees for wood, fish for food, and various plants for medical use (UNEP 2000:68). But for this small fraction of species, the extractions by humans may be a dominant cause of their endangerment. This is certain in the case of many species of precious woods and marine fish; in these cases, the trade demand influences the population dynamics of the species more than do other stressors such as habitat degradation from human land use or emissions (World Conservation Monitoring Centre [WCMC] 1992:365; Food and Agriculture Organization of the United Nations [FAO] 1997, 2000; UNEP 2000:112).

In order to satisfy the completeness criterion of LCA, it is therefore necessary to include extractions of biotic resources in LCIs, and concepts are required for the impact assessment of such inventory items. The practical importance of this requirement is growing, because in the future, more and more biotic materials will be supplied by man-controlled cultures, supplementing the extractions of 'wild' animals and plants. Comparing the environmental damages from a biotic extraction in nature to the environmental damages from the creation of a similar item by a man-controlled culture will become a relevant application in LCA practice. For example, is it environmentally preferable to eat wild salmon or salmon from aquacultures? Further, is it environmentally preferable to buy a piece of furniture made from wild teak wood or from plastic materials?

Definitions and distinctions

Before proposals for the impact assessment of biotic extractions can be discussed, a number of preliminary questions and statements must be clarified. The definitions and distinctions given in 'Distinction between abiotic and biotic resources' (p 17) also apply to biotic resources. Further definitions and distinctions are presented below.

Distinction between products from biotic resource extraction and products from man-controlled production

Many biotic items (e.g., wood, fish, meat, or mushrooms) may be extracted from nature but can also be products of a manmade culture. In certain cases, the user does not notice any difference in quality between the 2 origins. But it is obvious that the impact on nature is totally different, whether a product is extracted as a biotic natural resource or produced by a manmade culture such as aquaculture, agriculture, or silviculture. It is therefore necessary to indicate the origin of a biotic material in an LCA inventory, in order to ensure the appropriate environmental assessment.

Biotic materials from man-controlled cultures. If human society wants to use animal or plant species as food or as inputs to technical processes, the supply from nature may be considered insufficient or noncompetitive, so that man-controlled cultures of such species are developed. With the use of appropriate production methods, many species can be produced in quantities that meet requirements of the markets. If produced in man-controlled cultures, species are not depleted by increasing demand from the human economy; instead, their populations may even increase. If consumers eat more pork, for example, the total number of pigs in all farming operations will increase. However, market demand may develop in such way that certain varieties of domestic animals and cultivated plants become neglected and eventually disappear. This may be considered a loss of biodiversity, but it is not caused by biotic extraction. It is, rather, a side effect of the disappearance of consumer demand.

Here, the use of biotic materials from man-controlled cultures is not considered a case of biotic resource extraction: The desired product is not extracted from nature but is the economical output of manmade production facilities such as cattle, deer, and fish farms; crop productions; greenhouse operations; or forest plantations. The environmental problem of such man-controlled cultures is not the depletion risk of the cultivated species. But such production processes typically involve energy and materials inputs, emissions, and land uses that must be listed in an LCI of the respective production process.

Biotic materials from nature. Here, *wild animals or plants* are those extracted from nature by human activities, such as gathering, catching, and hunting, because humans want to use the corresponding species or parts of it. The growth of wild species and their populations is essentially controlled by natural processes, and human interference with the life of the species is essentially restricted to the extraction process. However, from a practical point of view, it may be useful to include cases of low-level human support that improves a species' ability to survive, such as the occasional feeding or protecting of young individuals. In spite of such small support, the regeneration rate of the species remains essentially controlled by nature and not mainly by man, as in man-controlled cultures. The natural stock of the species may increase, decrease, or come to a final extinction, conditioned by the magnitude of the human extraction rate relative to the natural regeneration rate. Species population decline and extinction is the environmental problem to be considered here.

What does low-level human support mean? For LCA practice, it is desirable to fix the limits of low-level human support, in order to minimise the grey area between biotic materials from nature and biotic materials from man-controlled cultures, and to clarify whether the impact assessment must be done on the basis of the man-controlled production process or on the basis of biotic extraction from nature.

The FAO distinguishes between aquaculture and catches of wild organisms in the marine and freshwater environment. According to the FAO, *aquaculture* is 'the farming of aquatic organisms, including fish, molluscs, crustaceans, and aquatic plants. Farming implies some form of intervention in the rearing process to enhance production, such as regular stocking, feeding, and protection from predators, etc. It also implies ownership of the stock being cultivated...' (World Resources Institute [WRI] 1999:317). In other words, if the replenishment side of the population dynamics is substantially influenced by human intervention, be it releasing young individuals from a nurturing operation or increasing the life expectancy of individuals through feeding or protection, this is aquaculture, even if the animals live in a natural river.

Applying the FAO definition of aquaculture to all kinds of man-controlled cultures, we propose that low-level human support to a species is exceeded if one of the following cases of intentional human action happens:
1) Increase of reproduction rate by favouring insemination and by nursing of young individuals
2) Increase of the mean life expectancy of the population by feeding or fertilising or by defending against predators or competitors.

According to this principle, we suggest that the following borderline cases are cases of biotic materials from man-controlled cultures and not cases of biotic resource extraction:
- Wood from a forest where the origin of trees consists mainly of planted seedlings
- Fish from water bodies where the high population density requires feeding or medical care
- Fish living under natural conditions, except young fishes are regularly supplied by hatcheries
- Plants from an area with mechanical or chemical weed control.

At first glance, it appears somewhat strange that the removal of trees from a lovely forest and the catching of fish out of a natural river should not be cases of extractions of biotic resources, if the trees are planted and if the young fish are regularly released from a hatchery. But it is obvious that, from a purely natural resources point of view, these removals are generally not an impact: The essential argument is that the removal of trees and fish from these man-controlled cultures does not lead to lower populations because they are in general balanced or even overcompensated by man-controlled input of young individuals. This does not exclude that the removal may show undesirable impacts outside of the AoP natural resources.

The extinction of species

If a species is overexploited as a biotic resource, or if it is threatened by any other environmental intervention, its population may decrease, either by a reduction of the population density or by a reduction of the geographic extension of the species. The decrease of a species population is a process that lasts decades or centuries or even longer, until it eventually leads to a regional or global extinction of the species. From a resource availability perspective as well as from an ecosystem functioning perspective, not only final extinction but also the previous phases of population decrease are considered damages, and the impact assessment should take into account this stepwise development towards extinction in one way or an other.

To monitor the steps of population decrease towards final extinction, the International Union for Conservation of Nature (IUCN) established a system in which a species' population dynamics are classified into the following levels:
- extinct,

- extinct in the wild,
- critically endangered,
- endangered,
- vulnerable, and
- lower risk.

These levels are defined by data elements such as the percentage of population reduction in a reference period, low counts of individuals, or low geographical extension (IUCN 2001).

As an example, the northern bluefin tuna fish (*Thunnus thynnus*) is classified by the IUCN as critically endangered, CR A1 (at extremely high risk of extinction in the wild in the near future, due to population decrease of more than 80%), with respect to its Western Atlantic stock. The Eastern Atlantic stock is classified as endangered, EN A1 (at very high risk of extinction in the wild in the near future, due to population decrease between 50% and 80%) (IUCN 2001).

Although the IUCN classification does not yet cover all known species and contains substantial uncertainties, it can be a useful element in impact assessment methods for biotic resources extraction.

The twofold impact assessment of biotic resources extractions

Because biotic resources are always a part of the biosphere, an extraction of a corresponding species always causes impacts on 2 AoPs:
1) natural resources, and
2) natural environment.

An example may illustrate the need for a 2-fold impact assessment of a biotic extraction. If 1000 tons of tuna fish are extracted from the North Atlantic, causing a temporary or permanent population reduction, the 2 types of impact to be modelled are as follows:
1) The impact of the extraction on natural resources should express the decreased global availability of tuna fish to cover future human needs, resulting from the current reduction of the tuna population.
2) The impact of the extraction on the natural environment should express the change in biodiversity and ecosystem functioning in the North Atlantic, resulting from the current reduction of the tuna population.

It is important to note that this 2-fold assessment is not a double counting: If the tuna fish were not edible, the reduction in the tuna fish population could still be considered to cause the same damage to the AoP natural environment, determined by the second assessment above. But because the tuna are edible and therefore a natural resource, the reduction of the tuna fish population is, in addition, a damage to the AoP natural resources, determined by the first assessment above.

The justification of the 2-fold assessment becomes even clearer if the extraction of rainbow trout (*Oncorhynchus mykiss*) from European rivers is considered: Such an extraction can be assessed as a damage to natural resources, whilst the same extraction is seen by many biologists as a benefit to the natural environment.

Objects of nature and manmade artefacts in the natural environment

A *biotic resource* is an object of nature extracted by man and used subsequently in man-controlled processes. This leads to the question of whether living objects released into the environment by man are also objects of nature. A similar question is related to biodiversity: Does biodiversity increase if humans release new organisms into the natural environment?

There is an old human practice of transporting species from one part of the world to another, where these species did not exist before. Such transfers were intentional if the plant or animal was considered to be useful, that is, a biotic natural resource, or if it was considered to be interesting or beautiful. In other cases, worldwide transportation has caused unintentional transfers of organisms. Such transported plants and animals were concentrated mainly in the manmade environment, but sooner or later, the natural environment was invaded by these organisms. It is common practice in biology to consider organisms as *natural* if their man-assisted migration happened in the remote past. In contrast, newcomers in the natural environment of a region are considered to be *artefacts*.

In addition to invasion by transfer from other regions, the natural environment of a region may be invaded by plants and animals that are the product of human selection and cultivation. Here again, biologists distinguish between events in the remote past and releases in recent times, the latter being considered as artefacts. In the future, the technique of genetic recombination will produce more new organisms, the release of which will probably not be seen as an increase in biodiversity.

We assume here that plants and animals are objects of nature and could be biotic resources, if they are not considered artefacts by biologists.

Guidelines for impact assessment of biotic extractions

Category indicators for the impact category extraction of biotic resources

A CI is the sum of the product of environmental intervention A_i and the characterisation factor Q_i, added over all members i of an impact category. If the members of an impact category are considered too heterogeneous, so that no characterisation factors can be found that permit a summation over all members i, it may be helpful to split the impact category into subcategories. Each subcategory then includes a cluster of members with increased homogeneity, and the CI is calculated per subcategory only, leaving the weighting between subcategories to a later phase of LCA. As an extreme, the i members could be so heterogeneous that each member is a subcategory of its own, which means that LCIA then yields no aggregation of LCI analysis positions at all.

Is there a need for aggregation of biotic extractions in LCIA? The answer is yes, if we want to develop LCA into a decision-support system that is really helpful in cases like these:
- Comparison between fishing methods with broad bycatch and low bycatch
- Comparison of buildings made with various materials extracted as biotic resources
- Comparison of food products with various ingredients extracted as biotic resources.

From WCMC (1992), we can conclude that humans currently extract a few thousands of species in order to use them as biotic resources, mainly food plants, wood, rattans, medical plants, ornamental plants, and terrestrial and aquatic animals for food. An elevated number of these species may appear as positions in inventories for LCA cases mentioned in the preceding paragraph. Therefore, aggregating biotic materials in LCIA is desirable, in order to reduce the complexity of LCA interpretation to a practicable level. However, the heterogeneity of biotic resources will probably enforce the formation of subcategories, each of these having its own CI.

Apart from the need for subcategories, the findings of the preceding section imply that extraction of biotic resources will lead to 2 types of CI, each of them having its own set of characterisation factors:
1) A first set focuses on the impact of extractions on the future availability of a species as a resource (AoP natural resources).
2) A second set focuses on the impact of extractions on a species as an element of the current ecosystems (AoP natural environment).

A third set of characterisation factors would be necessary if we chose to include competition for currently available biotic resources in LCIA. Reasons not to include competition have already been given in the last paragraph in 'Competition and depletion' (p 18).

Minimum list of species to be included in extraction of biotic resources

In view of the practical difficulties of providing the necessary data for the impact assessment of all possible biotic resource extractions, it is useful to ask which types of biotic resources merit a first priority to be treated in LCA. We propose that a species should be contained in a corresponding minimum list of biotic resources if both of the following 2 conditions are fulfilled:

1) The species has substantial importance as a global resource. This type of information is collected and distributed by the FAO, for instance, by the FAO forestry database (FAO 2000) and by the FAO review of the state of world fishery resources (FAO 1997).
2) The species is classified as vulnerable, endangered, or critically endangered. This type of information is collected and distributed by the IUCN in its Red List 2000 database (IUCN 2001).

If a species is listed in either database, it may be considered a scarce biological resource, and data are then available on the current annual extraction quantities (FAO) and on the current probability of extinction (IUCN).

As a provisional minimum list, we have selected species that are mentioned in WCMC (1992:331–406) as wild resources and that are qualified as threatened because of human extractions. This list contains roughly 100 species, with precious woods and marine fish as the largest groups (see Table 2-4).

At a later stage, it is of course desirable to include in LCIA all species extracted by humans as biotic resources. But this will involve large additional data collection efforts.

The relation between biotic extraction and its impacts

Discussion is ongoing about the best position for CIs along the cause–effect chains that link environmental interventions with AoPs. For the case of the environmental intervention 'extraction of biotic resources', the question at stake could be phrased as follows: Is it preferable to express the CI as an entity situated near the environmental intervention, for instance, kilograms or pieces per extracted object, or as an entity representing the damage to the AoPs 'biotic resources' and 'natural environment', for instance, reduction of worldwide resource availability until year Y or reduction of species diversity until year Y?

We do not intend to continue this midpoint and endpoint discussion or to anticipate its final findings. But it seems clear that, irrespective of the precise positioning of CIs anywhere between the environmental intervention and the AoPs, these CIs should at least be oriented towards damage at the AoP level. If CIs are developed without a background concept of damage to AoPs, there is a risk that CIs of different impact categories expressing damages exerted on the same AoPs will be incoherent and will become a complication instead of a support for the interpretation phase of LCA practice.

Consequently, CIs on different impact pathways pointing to the same AoP should not be determined independently; they need a common orientation. For the case of biotic resource extractions, this means the following:

- Because damage to natural resources is caused not only by biotic extractions but also by abiotic extractions, it is desirable to operate with a common definition for the term *damage to natural resources*. The CIs for biotic extractions and for abiotic extractions should therefore be the same or at least different proxies for the same background entity.
- Because damage to the natural environment is caused not only by biotic resource extractions but also by land use and various chemical emissions, it is desirable to operate with a common definition for the term *damage to the natural environment*. If the 3 CIs for land use, chemical emissions, and biotic extractions are

Table 2-4 Minimum list of important 'wild' biotic resources threatened by high extraction rates

Resource type	Species name	
Food plants, including fodder plants for domestic animals[a]	Brazil nut (*Bertholletia excelsa*) Cardamom (*Elettaria cardamomum*)	Chili pepper (*Capsicum annuum*) Olive (*Olea laperrinei*)
Timber[b]	*Astronium urundeuva*	*Gossweilerodendron balsamiferum*
	Aspidosperma polyneuron	*Machaerium villosum*
	Ilex paraguaiensis	*Mimosa verrucosa*
	Didymopanax morototoni	*Pericopsis elata*
	Araucaria angustifolia	*Plathymenia foliosa*
	Araucaria cunninghamii	*Pterogyne nitens*
	Araucaria hunsteinii	*Cedrela fissilis*
	Zeyhera tuberculosa	*Cedrela odorata*
	Bombacopsis quinata	*Entandrophragma angolense*
	Cordia milleni	*Khaya senegalensis*
	Atriplex repanda	*Lovoa swynnertonii*
	Brachylaena huillensis	*Milicia excelsa*
	Cupressus atlantica	*Eucalyptus deglupta*
	Cupressus dupreziana	*Abies nebrodensis*
	Juniperus procera	*Abies numidica*
	Joannesia principes	*Cedrus libani*
	Irvingia gabonensis	*Pinus armandii*
	Aniba duckei	*Pinus koraiensis*
	Ocothea porosa	*Pinus pentaphylla*
	Bertholetia excelsa	*Pinus pseudostrobus*
	Acacia albida	*Balfourodendron riedelianum*
	Acacia caven	*Esenbeckia leiocarpa*
	Acacia tortilis	*Nesogordonia papaverifera*
	Anadenanthera macrocarpa	*Ulmus wallichiana*
	Dalbergia nigra	*Tectona hamiltoniana*
	Dipterix elata	
Rattans[c]	*Calamus* species	
Medicinal plants[d]	*Berberis vulgaris*	*Pilocarpus jaborandi*
	Cephaelis ipecacuanha	*Rauvolfia* spp.
	Ephedra sinica	*Silybum marianum*
	Pausinystalia yohimbe	*Urginea maritima*
	Physiostigma venenosum	
Ornamental plants[e]	*Dendrobium* genus	*Paphiopedium* genus
	Calanthe genus	*Phragmipedium* genus
Terrestrial animals for food[f]	Large number of species hunted or gathered, but no comprehensive information available.	
Aquatic animals for food[g]	Alaska pollock (*Theragra chalcogramma*) Anchoveta (*Engraulis ringens*) Japanese pilchard (*Sardinops melanostictus*) South American pilchard (*Sardinops sagax*) Chilean jack mackerel (*Trachurus murphyi*) Atlantic cod (*Gadus morhua*) Chub mackerel (*Scomber japonicus*) Atlantic herring (*Clupea harengus*) European pilchard (*Sardina pilchardus*) Silver carp (*Hypophthalmichthys molitrix*) Skipjack tuna (*Katsuwonus pelamis*) Common carp (*Cyprinus carpio*)	
Non-food uses of animals[h]	Large number of species hunted and gathered, but no comprehensive information available.	

[a] WCMC 1992:332–337.
[b] WCMC 1992:345–348.
[c] WCMC 1992:351.
[d] WCMC 1992:352.
[e] WCMC 1992:353–358.
[f] WCMC 1992:359–365.
[g] WCMC 1992:369. The 12 most important fish species, representing 35% of annual world catch.
[h] WCMC 1992:374–389.

positioned at endpoint level, they should be of the same format. If the CIs are positioned at midpoint level, they may be different but should at least be fixed in a coordinated manner, because they represent proxies for damage to the same nonhuman life. For example, trout populations in rivers are damaged by toxics in the effluents, by land-use activities that denaturalise the riverbed, and by fishing. If an LCIA method results in figures that represent the effect on the fish population from each of these 3 sources in a totally different way, this results in very poor decision support, compared to an LCIA method in which the effect of each damaging factor is expressed, for example, as a percentage of population reduction.

Proposals for CIs that have already been developed in the context of abiotic extractions, land use, and chemical emissions should therefore be examined in view of their application to biotic extractions, with the aim of coordinating the respective CI development.

Operationalisation

Category indicators for extraction of biotic resources associated with AoP natural resources

Here, the focus is on the extracted object as a natural resource: To what extent does the current extraction worsen the possibilities for human society to cover future needs for the extracted object? A biotic stock reduction that results from an extraction may have consequences for nature and humans other than merely the reduction of future supply, but these additional consequences must be clearly separated from the resource issue and will be treated in the next section.

In analogy to proposals for abiotic resources, a characterisation factor, Q, for a species extracted as a biotic resource could have the following form (Heijungs et al. 1992):

$$Q = (U - P)/D^2 \qquad (2-1),$$

where U is the worldwide or regional current annual use of a species, P is the worldwide or regional current annual replenishment of this species, and D is the worldwide or regional current stock of this species.

The dimension may be number of individuals or kilograms of total body mass of the species. If the quantities of A extracted, as tabled in an LCI, are multiplied by the factor $(U - P)/D^2$, the resulting term represents the impact on the corresponding natural resource. If $U < P$ for a species, Q can be taken as 0; this means that no depletion is expected because the current production capacity of nature is not fully used by man. But if $U > P$ for a species, the population decreases by an annual amount of $U - P$, which may lead to a future scarcity or even disappearance of the resource. An extraction quantity A is considered more serious, if A is large in relation to the stock D of the species, and if the ratio of the annual decrease of the population, $U - P$, divided by the stock D, is large (Heijungs et al. 1992). It is immediately plausible that the extracted quantity A represents a high impact on the natural resource if the resource stock D is only a small multiple of A. When we compare resource cases with equal A/D ratios, it becomes also plausible that cases with higher $(U - P)/D$ ratios represent a higher impact on the resource than cases with lower $(U - P)/D$ ratios. This leads to a characterisation factor $Q = (1/D)\,(U - P)/D$ or $(U - P)/D^2$.

In contrast to the above proposal, there are also voices (Klöpffer and Renner 1995) in favour of a characterisation factor Q that contains only the ratio of the annual stock reduction $(U - P)$ to the total amount of stock D:

$$Q = (U - P)/D \qquad (2-2).$$

According to this alternative proposal, 2 different resources having the same $(U - P)/D$ ratio are considered equally damaged if the extraction quantity A of the first resource is 10% of D, whilst the extraction quantity A of the second resource is only 0.0001% of D. Klöpffer's argument in favour of this is as follows: The fraction A/D is arbitrary because the size of the functional unit is arbitrary. It could be argued that the size of the functional unit is arbitrary, but the arbitrariness is the same for all environmental interventions of a given LCI; hence, the A/D expresses the relative ratio for all resources considered, and this may give additional information. However, in Klöpffer's view, this is a part of normalisation, not of characterisation, which must be done separately for each type of resource.

At this stage, it is interesting to remember the current data availability, based on IUCN and FAO databases. From FAO databases, it is generally possible to obtain U, the yearly use of the species. From the IUCN Red List database, it is possible to obtain the code for the level of endangerment of the species. The following definitions connect the IUCN endangerment codes to a coarse information on $(U-P)/D$ per species (http://redlist.org/info/categories_criteria.html):

- Critically endangered CR A1 means $(U-P)/D$ is > 80%, with U and P estimated for a period of 10 years (or 3 generations of the species if this is longer than 10 years).
- Endangered EN A1 means $(U-P)/D$ is > 50% but < 80%, with U and P estimated for a period of 10 years (or 3 generations of the species if this is longer than 10 years).
- Vulnerable VU A1 means $(U-P)/D$ is > 20% but < 50%, with U and P estimated for a period of 10 years (or 3 generations of the species if this is longer than 10 years).
- Lower Risk means $(U-P)/D$ is < 20%, with U and P estimated for a period of 10 years (or 3 generations of the species if this is longer than 10 years).

The population reduction range over 10 years (or 3 generations of the species if this is longer than 10 years) can be calculated back to an annual $(U-P)/D$. This means that the Klöpffer and Renner approach can be implemented with IUCN and FAO databases if normalisation of the extraction quantity of the LCI is made with the annual use, U. In contrast, the term $(U-P)/D^2$ is not directly available from the IUCN database as presented on the Internet; but we assume that U, P, and D exist as separate figures in IUCN internal documentation.

Regionalisation. Irrespective of the equation used, the question arises whether U, P, and D should be taken as global or regional figures. Is biotic resource depletion a global problem, or should it be split into a number of regional problems? Geographical distribution is definitely important for ecosystem aspects, but resources are transportable in principle, so that LCIA of biotic extractions with respect to the AoP natural resources can be done at the global level, analogous to LCIA for abiotic resources such as metals and fuels. As a matter of fact, the dominant biotic resources like fish or wood are not only transportable in principle, but they are actually transported worldwide: The driving forces of the global market economy ensure that regional demands, at least from developed economies, are covered by supplies from any provenance. This is an argument for a treatment of biotic resource depletion at the global level. However, replenishment aspects of species population may also justify a treatment on the regional level, for instance, the FAO fishery regions.

How many subcategories? The next question to be answered refers to aggregation: Should we treat as a separate subcategory each of the species being extracted as a biotic resource? This would mean that the characterised extraction quantities $A \times Q$ are not summed up over groups of species, and the $A \times Q$ of each species extracted stays on as a separate CI. We now know that in this case a number on the order of 100 or several hundreds of subcategories for biotic resources extraction would result as a minimum. Such an inflation of CIs is hardly acceptable in LCA practice. In contrast, a summation of the $A \times Q$ over groups of species or all species is defendable, for the following reasons:

- A term $A \times Q$ summed up over all extracted species, or summed up over relatively homogeneous species groups such as fish or trees, can give a reasonable overall picture of the damage to biotic resources caused by the functional unit of an LCA study.
- Chapter 6 shows that the methods proposed for the assessment of ecotoxics do not distinguish between more important and less important species, but treat all species alike. It is not unjustifiable to proceed in the same way with LCIA of extractions.

Category indicators proposed. At least as a first approach, from the resource point of view, 2 biotic resource subcategories, fish and precious woods, may be formed with a CI of the subcategory determined by the equation

$$CI = \sum A_i \times b_i \times (U_i - P_i)/D_i^2 \qquad (2\text{-}3)$$

(according to Heijungs et al. 1992) or

$$CI = \sum A_{in} \times b_i \times (U_i - P_i)/D_i \qquad (2\text{-}4)$$

(according to Klöpffer and Renner 1995), where the summation is over all biotic resources i of the respective subcategory, and b_i is the relative scarcity of biotic resource i. The value of b_i is equal to 1, as long as we assume that all biotic resources i can be considered approximately equally scarce if their term $(U_i - P_i)/D_i^2$ or $(U_i - P_i)/D_i$ is the same. A_i is the extraction quantity of the LCI, A_{in} is the extraction quantity of the LCI, normalised by the annual use, U_i. Annual replenishment and current stock of the resource are represented by P_i and D_i. The dimension of A, U, P, and D is kilograms.

However, in a later phase, the assessment expressed by the above equations could be refined by determining factors b different from 1. There may be cases where the damage to the AoP natural resources could be considered unequal, if 1 kg is extracted from 2 different resource species with the same magnitude of $(U - P)$ and D. The reasons for such inequality could originate from biological or from economical properties of a given resource: The future recovery of a depleted stock of a given species might require more or less time than the recovery of another species (biological properties), and the unavailability of a given resource species might cost a future human society more or less than the unavailability of another resource species. The question of biological property differentiation merits at least some additional comments here.

Because of biological properties, different biotic resources with the same magnitude of $(U_i - P_i)/D_i^2$ or $(U_i - P_i)/D_i$ should be considered as different, if a fast or slow recovery has to be expected upon a temporary relief from the extraction stress: A low stock D is a less severe degradation of the resource if the corresponding population is biologically capable of recovering rapidly during a ban on extractions. In WCMC (1992:370), good examples are given for marine fish stocks with fast or slow recovery potential. Because of overfishing, for example, the North Sea herring reached a minimum stock of 75 000 t in 1975, but after a 5-year ban on fishing, the population recovered to a stock weight of 1.4 million t, which was as high as in the 1950s. In contrast to this, overfishing of Western Atlantic bluefin tuna led to a reduction of the adult population by nearly 95% from 1970 to 1990, although catch quotas were fixed after 1982. One reason for the nonrecovery of the bluefin tuna could be that fishing was not stopped totally, and catch quotas were too high. But the main reason is biological: Species with a short reproduction cycle like herring's are recovering rapidly, whilst species with a long reproduction cycle like the bluefin tuna's need longer recovery periods (WCMC 1992). Consequently, the average time between 2 generations of a species is scientific information that should be used to determine the factor b_i. Of course, recovery of overexploited populations is also conditioned by other factors of the complex ecosystem dynamics, but as a first approach, the time between 2 generations of a species is a reasonable indicator of recovery time.

In conclusion, the impact of biotic extractions on the AoP natural resources can be, as a first step, treated by working out, on the basis of Equations 2-3 or 2-4, 2 CIs for the 2 subcategories of 'wild' fish and 'wild' forest trees, each group containing some dozens of members that are aggregated to the CI on the basis of global figures for annual use U, annual replenishment P, and current stock D. The factor b can be set to 1, or equal to the time between 2 generations of the species as a further refinement. Other objects of biotic resource extraction, such as food plants, medicinal plants, ornamental plants, and terrestrial animals for food, could be treated in a later step because extraction in this case seems to a lesser extent the reason for population decreases. The necessary data for fish and precious woods as wild resources are largely available from the FAO and the IUCN.

In Sas (1997), another approach for the development of impact indicators is proposed, which also makes use of the species-specific time between 2 generations. At the inventory level, 2 types of biotic extraction are distinguished:
1) harvesting individuals of a species, which is expressed in kilograms of fresh weight, and
2) degradation that is due to complete extraction, which is expressed in square metres of degraded area (Sas 1997:13).

In our view, the latter type of environmental intervention is oriented to land use. In contrast, a biotic extraction that is expressed as body weight extracted per species fits well into the context of this section.

At the assessment level, 2 types of impact are distinguished:
1) impact on biodiversity and
2) impact on LSFs (Sas 1997:24).

Although Sas does not distinguish between the AoPs natural resources and natural environment in the same way as we do in this chapter, it seems defendable here to interpret Equation 1a of Sas (1997:25) as the Sas proposal of a CI for extraction of biotic resources associated with the AoP natural environment:

Risk of extinction
of biotic resources = $\Sigma_i A_i \times (t_{i,\text{reprod}}/D_{i,\text{current}})$ (2-5),

where A_i is kilograms of biomass of species i extracted, according to LCI, t_i is reproduction time in years of species i, and D_i is the current worldwide biomass of species i.

It is remarkable that the current worldwide overshoot of annual extractions over annual replenishment $(U - P)$ does not play any role in the Sas proposal.

Finnveden (1996:43) mentions that characterisation methods for extractions of biotic resources could also be based on adding the mass content, the energy content, or the exergy content of the extracted species. So far, we have not encountered elaborated LCIA methods that use these principles; neither have we seen arguments that support expressing the impact of living species' extraction on biotic resources by simply adding their mass or their energy content.

A short characterisation of proposals in shown in Table 2-5.

Category indicators for extraction of biotic resources associated with AoP natural environment

If a species is overexploited by man as a biotic resource, there is not only the risk that the future availability of the species diminishes but also the risk that the natural environment is influenced immediately because the corresponding species cannot maintain its previous function in the ecosystems connected to it, and biodiversity may be reduced. We have already mentioned the fundamental difference between viewing a given species from the resource depletion point of view and viewing it from an ecosystems point of view. It is therefore clear that a CI associated with the AoP

Table 2-5 Proposals for operationalisation of biotic resources

Items according to Table 2-2	Properties of model
Heijungs et al. 1992	
Essentials[a,b,c,d]	Full quantification, characterisation factor $Q = (U - P)/D^2$, where $U > P$. CI = unweighted sum over all species with $U > P$.
Sensitivity	High, on any use of scarce resources
Mechanism and model	Simple; consumption weight per functional unit related to global amounts of living stock and net yearly stock decrease
Extent of representation	Acceptable
Choices or assumptions	All species treated alike, based on kg body mass. Possibilities for substituting resources not included.
Consistency	Adequate
Applicability	Probably applicable with currently available databases
Klöpffer and Renner 1995	
Essentials[a,b,c,d]	Full quantification, characterisation factor $Q = (U - P)/D$, where $U > P$. CI = unweighted sum over all species with $U > P$.
Sensitivity	High, on any use of scarce resources
Mechanism and model	If extracted quantity from LCI is normalised with U, model is rather similar to Finnveden 1996 and Heijungs et al. 1992, but contribution of a species i to indicator value according to Kloepffer and Renner (1995) becomes bigger than according to Finnveden and Heijungs et al., if $U \ll D$.
Extent of representation	No judgement
Choices or assumptions	All species treated alike, based on kg body mass. Possibilities for substituting resources not included.
Consistency	Adequate
Applicability	Applicable with currently available databases
Sas 1997	
Essentials[a,d,e]	Full quantification, characterisation factor ($Q = $ (reproduction)$/D$). It is understood that $Q = 0$ if species population is stable in spite of extractions. CI is unweighted sum over all species, but summation may be restricted to groups of similar species.
Sensitivity[b,c]	High, on any use of resources with long reproduction time; no sensitivity to speed of decrease of stock (D) caused by relative magnitude of $U - P$.
Mechanism and model	Based on simple idea that impact of a given extraction quantity can be expected to be more serious if species concerned has relatively low level of living stock and if it takes a long time until young individuals are able to reproduce.
Extent of representation	No judgement
Choices or assumptions	All species treated alike, based on kg body mass. Possibilities for substituting resources not included. Reproduction time assumed to be good indicator of species' capacity to rebuild decreased living stock.
Consistency	Adequate
Applicability	Applicable with available databases

[a] Q = Characterisation factor.
[b] U = Yearly global consumption.
[c] P = Yearly global replenishment.
[d] D = Measure of living stock.
[e] Reproduction = Reproduction time of a given species.

natural environment will differ from a CI associated with the AoP natural resources.

Wild species being used by man as biotic resources are an extremely small part of the totality of species, perhaps some 1000 out of millions. Impacts on the large majority of species originate mainly from land use and from ecotoxicological emissions. A CI referring to the AoP natural environment should therefore not be developed in an isolated way for biotic resources. CIs for population reduction and extinction risk of species, caused by different interventions such as ecotoxicological emissions, land use, and resource extraction, should be similar or at least should be determined in a coordinated way.

Now, Chapter 6 and this chapter do not conclude with a clear-cut proposal for CIs. But we can notice that concepts like the potentially affected fraction of species (PAF) or the potentially disappeared fraction of species (PDF) are mentioned in the chapter on ecotoxics. These express the impact on the AoP natural environment by counting the percentage of species threatened in a certain geographical region, whereby all species are considered of equal importance for the quality of the natural environment. In the context of these approaches, characterisation factors (Q) and CIs could be built up for biotic extractions, as demonstrated by the following case.

Entandrophragma trees are now threatened in Ghana because the current annual extraction of N cubic metres is considered too high. A decrease of the annual extraction to the level of M m³ would eliminate the threat to this species. In consequence, a reduction of the annual extraction by a quantity of $(N - M)$ m³ would decrease the number of threatened species in Ghana by 1. An annual extraction of A_a m³ of Entandrophragma therefore causes the fraction of $A_a/(N - M)$ of the continuous threatening of one species in Ghana, according to the concept of marginal damage. If an LCI contains a one-time extraction of A m³ Entandrophragma wood, this extraction is responsible for a fraction of $A/(N - M)$ of the continuation by 1 year of the threatening of one species in Ghana.

The characterisation factor (Q) for any threatened species i in Ghana could then be fixed to $1/(N_i - M_i)$, and the CI summing up all biotic extractions referring to all tree species i in Ghana is then

$$CI = \sum A_i \times 1/(N_i - M_i) \qquad (2\text{-}6).$$

This CI represents the total number of threatened species in Ghana during one additional year, as caused by all extraction quantities A_i of different species, as used for a certain functional unit (or entity of functional units). A_i cannot exceed $(N_i - M_i)$; otherwise, the concept of marginal damage would not be applicable. In consequence, CI is a sum of fractions. For example, a functional unit might be the cause for 1/100 of the continuation by 1 year of the threat to species S_1 in Ghana, and for 1/500 of the continuation by 1 year of the threat to species S_2 in Ghana. If all species are weighted the same way (as in the PAF concept of ecotoxicology), the amounts can be summed up to a CI of 0.012.

In case the terms $(N_i - M_i)$ refer to different levels of threat in the sense of the IUCN classification, this would have to be taken into account (in Equation 2-6) by an additional factor, because it is more serious if an extraction A is responsible for 1% of the threat to a critically endangered species than if it contributes 1% of the threat to a vulnerable species. Even if all species are considered of equal importance, the probability of future extinction is higher in the case of the critically endangered species, so that the corresponding damage to biodiversity caused by the extraction quantity A is higher.

The above CI may be divided by the total number of tree species in Ghana, in order to build up a term of the PAF type, as currently used in the context of ecotoxicological emissions.

A CI according to Equation 2-6 represents the biodiversity part of damage to the AoP natural environment. We leave open the question of how the damage to life support systems, the other part of the same AoP, could be built up.

Conclusions

Out of many millions of living species of the earth, only a few thousand species are used as resources by humans, and serious problems of resource depletion seem to be limited primarily to some 100 fish species and precious woods.

Resource-oriented CIs for these 2 subcategories can be worked out on the basis of FAO and UNEP–WCMC databases.

In addition to this resource aspect, biotic extractions have an impact on biodiversity and LSFs, which is comparable to but probably less dramatic than the same impact originating from land use and ecotoxic emissions. Modelling of this second type of impact should be well coordinated and the solution worked out in the context of land use or ecotoxic emissions. A possible approach has been proposed for this type of biotic extraction impact.

Impact Assessment of Land Use

Starting point

In the remote past, humans used certain areas of land for such purposes as harvesting food and providing shelter, in such a way that there was only a very low interference with other organisms living in the same area, and the landscape did not show traces of human activities. In the meantime, this has changed: Humans have drastically modified the shape and properties of large land areas according to their requirements, and they exclude wild animals and plants from coexisting on such land areas and in their neighbourhoods. This has led to a substantial reduction of regional biodiversity, to a far-reaching change in the natural landscapes, and to noticeable disturbances in systems like the freshwater circuit. *Land use*, in the sense of this chapter, refers to such intensive human activities, aiming at exclusive use of land for certain purposes and adapting the properties of land areas in view of these purposes. In short, land use means allocating land to the erection of buildings or the planting of cornfields, levelling out and draining such areas, and preventing weeds and wild boars from living on them.

Although land use has been acknowledged for decades to cause degradation of the natural environment, this has not resulted in equally rapid integration of land-use impacts in LCA, so that the latter has been concerned mainly with emissions until now. The main reasons for this are as follows:

- Land-use impacts seem very dependent on the regional or even local situation, which is generally not known in LCA.
- Land use as an environmental intervention is much more complex than, for example, the emission of CO_2 or fine particles with an aerodynamic diameter less than 10 μm (PM10), because it may consist of many different elementary activities such as excavating, ploughing, draining, or fertilising. Describing types of land use and studying the corresponding impacts on nature is therefore a very difficult task for developers of LCA methodology.
- Land use data are not included in traditional environmental data collection schemes because such data are not easily available.

However, in policy documents such as the Dobris Assessment (EUROSTAT 1995), land use is taken into consideration as a separate and important environmental issue. It is generally recognised that the land use practices of the last few decades have been a main cause for the observed serious reduction and threatening of biodiversity all over the globe, amply documented at the level of population size and geographical occurrence of plant and animal species. In fact, in the 21st century, land-use changes are expected to have the largest impact on biodiversity of all possible interventions (Sala et al. 2000).

In LCA, land-use methodologies have developed from the first rough classification of a few land-use types in 1992 to a widespread research activity and a sizeable number of proposals in 2000 (Lindeijer 2000). Methodological discussions have been held around this environmental aspect in the SETAC WIA taskgroup for the past few years, contributing considerably to the necessary methodological framework.

In this section, we discuss the major methodological issues for land use (both inventory and impact assessment), and we mention recent proposals. In contrast, we do not yet propose a best available practice because, in our opinion, further developments are needed in this new field of LCA.

Definitions and distinctions

First, we define the general term *land use*, after which we identify a number of distinctions in the land-use model. (See also the corresponding sections under the headings 'Impact Assessment of Abiotic Resources', p 15, and 'Impact Assessment of Biotic Resources', p 26; parts of them also apply to land use.)

*Land use i*s an environmental intervention identified as an entry in an LCI. It has the dimension of area × duration of use if the occupation of a certain area for a certain purpose is to be expressed. If the entry in the inventory refers to change or transformation of the properties of the land area in view of an intended use (or non-use), we propose the dimension area. We reserve the term *land use* for these 2 types of intervention:
1) land occupation and
2) land transformation.

In order to distinguish clearly between environmental intervention and impact, we call the former *occupation process* and *transformation process*, and the latter *occupation impact* and *transformation impact*. The dimension of the impacts is determined by the choice of the CI.

Distinction between occupation and transformation of land areas

The *occupation process* is the use of a land area for a certain man-controlled purpose, such as growing corn, piling up solid wastes, or erecting buildings. It is assumed here that the state of the area is already suitable to the intended use; in consequence, the occupation process essentially includes a set of environmental interventions that maintain the flora, fauna, soil, and soil surface in this state. Examples of environmental interventions that maintain an area's properties for its intended use are irrigation, weed control, and repairing a road surface. It is obvious that environmental interventions influence not only the suitability of the land area for the intended use but also the environmental quality of the land, for example, its value as a habitat for wild plants and animals or its capacity as a buffer storage and filter for fresh water. This environmental quality (hereafter called, in short, *quality*) usually stays at more or less the same level during the period of the occupation process, while a given number of functional units are produced on this area. Figure 2-2 expresses the occupation process by the continuous horizontal line between times t_1 and t_2. On the vertical axis, the environmental quality of the land is shown. This quality can be expressed in various indicators for flora, fauna, soil surface, or structure and should indicate the magnitude of the impact on nature; details will be discussed later.

The figure shows that the environmental quality between times t_1 and t_2 is equal to B. If the use of the land area ends at time t_2 so that the area becomes fallow land, the quality will generally increase after t_2 along the dotted line and arrive finally at a higher level D after some time. The

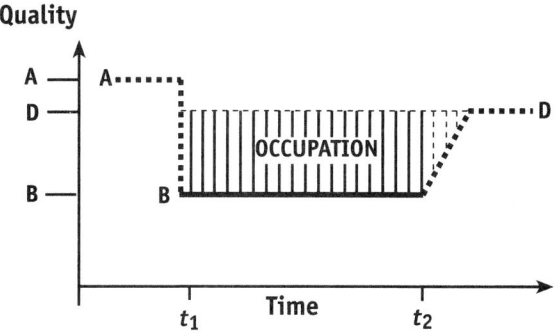

Figure 2-2 Land occupation with renaturalisation

time needed for such a *renaturalisation* may be long in the case of spontaneous renaturalisation by forces of nature, and may be abbreviated in the case of a man-assisted renaturalisation process. The possible magnitude of quality D is mainly influenced by the biogeographical conditions at the location of the land area, but to some extent also by the use history of the area and its surroundings. If the biogeographical region of the area is not known from the inventory, general assumptions are necessary for the determination of quality D.

Occupation impacts are environmental impacts, originating from the occupation process, and potentially damaging AoPs. Typically, the occupation impacts grow in parallel to the duration of the occupation process. The quality levels B and D represent the magnitude of environmental impacts for current use and for abandonment of the land plot, respectively. In consequence, the occupation impact can be represented by the vertically hatched area in Figure 2-2. This means that the prevention of potential renaturalisation is seen as the occupation impact. The occupation impact takes place primarily inside the occupied land area, but there are also impacts outside this area. Examples include the change of vegetation due to interrupted migration of animals and seeds, or the change of river dynamics due to reduced buffer capacity and increased erosion of a plot inside the river's catchment area.

The *transformation process* implies the change of a land area according to the requirements of a given type of occupation process. Examples are draining and filling of a marshy area for subsequent use as cropland, or deforesting an area for subsequent use as an airport runway. Renaturalisation may be seen as a transformation process in reverse. Typically, the duration of human activities carrying out a land transformation is short, compared to the duration of subsequent chains of land occupation processes.

We propose to describe the transformation process, in the inventory, only by its type (defining the situation before and after the transformation) and the size of the area, but not by the duration of the corresponding physical activities.

Transformation impacts are environmental impacts originating from the transformation process. Such damages occur primarily inside the transformed land area, but they can also exceed the perimeter of the transformed land area. Typically, the transformation impacts are independent of the duration of the transformation process. They can be considered as net land-quality changes due to the transformation processes involved. The magnitude of impacts depends on the extent of modification of flora, fauna, soil, or soil surface caused by the transformation process, and on the size of the transformed area.

Figure 2-3 shows a vertical line at t_1, indicating a transformation process preliminary to the occupation process. The environmental quality of the land is hereby reduced from A to B. Quality levels A and B are determined by the type of transformation process and are influenced also by the biographical region of the location. If this biogeographical region is not known from the inventory, it is necessary to go back to general assumptions. If the land is no longer used after t_2, the quality rises to D, this level again being influenced by biogeographical conditions.

A special problem in LCA is the link between transformation impacts and the number of functional units produced. The numbers of functional

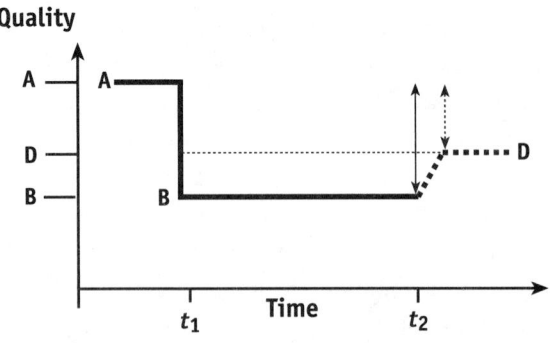

Figure 2-3 Transformation

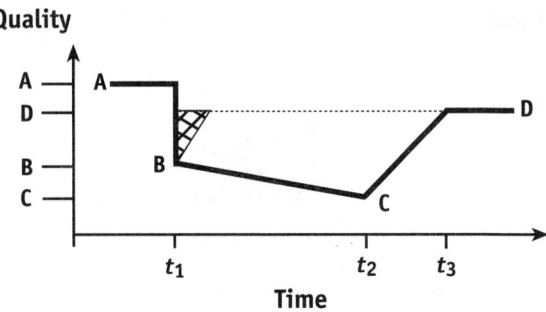

Figure 2-4 State of flora, fauna, soil, or soil surface before, during, and after a human activity

units depend on the chain of occupations after the initial transformation, but not on the transformation itself. For example, the quantity of corn produced or coal mined is related to the type and duration of the occupation process, whilst the preliminary clearing of forest and the environmental impacts of this clearing have no immediate connection to the quantities produced thereafter. Additional assumptions are necessary to create this connection, for instance, on the expected lifetime of a mining operation.

In Figure 2-4, the more complex situation of a combined transformation and occupation is shown. A practical case could be the open-pit mining of minerals. The initial transformation at t_1 (this could be the removal of overburden from the field to be mined) reduces quality from A to B. If there were no subsequent occupation processes, spontaneous renaturalisation would start immediately at t_1, to bring back quality to level D (straight line rising from point B). During the time period t_1 to t_2 of the occupation process (this could be the period of continuous mineral extraction), it is assumed here that the quality level falls gradually from B to C, for instance, caused by a progressing erosion of the occupied area. After t_2, renaturalisation increases quality to level D. If renaturalisation were prevented, the final quality D would be equal to C, unless erosion continues.

The difference between quality levels A and D is the net transformation. The transformation impact is represented by area × net quality difference ($A - D$). It is a remarkable property of this concept that the net quality difference $A - D$ lasts forever, in contrast to the quality difference given below for the occupation impact, which is of limited duration. This means that the damage from transformation can be seen as infinite, because quality loss lasts forever, whilst the damage from occupation remains temporary and can therefore be considered finite.

The occupation impact is represented by area × quality difference × time, whereby time is $t_2 - t_1$, and the quality difference is approximated by $D - (B + C)/2$. This means that the occupation impact depends on the momentary quality state of the area during the period of the occupation process.

It is important to remember at this point that impacts often exceed the perimeter of the transformed or occupied area. If the area in the terms above is the transformed or occupied area, the enlarged perimeter of the impact must be included in the amount of the quality difference.

In practice, knowledge about the various quality levels before and after transformation is limited. A simplified approach could then be based on the following simplifying assumptions: If nature has enough time to act, the final quality state D could become roughly comparable to the quality A just before transformation, and further, the quality decrease during the occupation interval t_1 to t_2 could be assumed to be negligible in comparison to the quality drop $(A - B)$. Under these conditions, the term $(A - B) \times (t_2 - t_1)$ area could be taken as the occupation impact. And in contrast to the proposal above, the transformation impact could be declared to be, in Figure 2-4, the difference between the dotted and the straight line, from t_2 to t_3; this triangle changes to the same size as the hatched triangle in Figure 2-4 under the simplifying assumptions mentioned above. This means that land transformation would be held responsible for the time it takes to re-create a land quality roughly comparable to the quality before transformation, whilst land occupation would be held responsible for delaying, during the time of occupation, the start of this process to re-create a better environmental quality of the land (see Köllner 2001).

The choice of the reference state for the measurement of occupation impacts[2]

The occupation impact caused by land use is measured in relation to a reference state, that is, it is expressed as the difference between the actual state and the reference state of environmental quality. The choice of reference state is not arbitrary but relates to the distinction between transformation and occupation impacts as defined in the previous section. The occupation impact should be defined in such a way as to avoid any overlap with the transformation impact and to give full expression to the impacts not captured in the transformation impact. This can be done by using the final steady state (the renaturalisation potential, level D in Figure 2-4) as the reference state, since the transformation impact is defined as the difference between the original and the final steady state (levels A and D in Figure 2-4).

Note that the final steady state is not necessarily fixed (as shown in Figure 2-4) but may itself change as a result of the human activity in question, for instance, as a gradual degradation of the renaturalisation potential (as shown in Figure 2-5).[3] Figure 2-5 could represent the cutting of primary forest for agriculture, where erosion occurs during the subsequent agriculture practice.

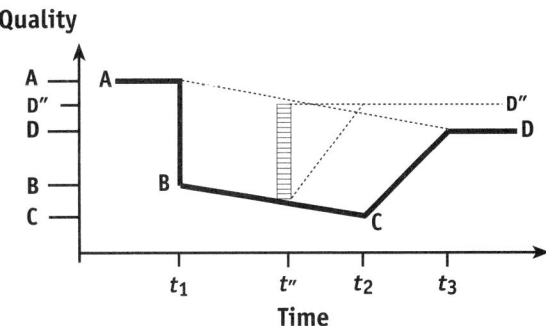

Figure 2-5 State of flora, fauna, soil, or soil surface before, during, and after human activity, indicating gradual and continuous degradation of renaturalisation potential (reference state) and reference state (D'') if human activity were terminated at t''.

Thus, the occupation impact is measured as the difference between the actual level (the fully drawn curve in the figures) and the level of the current renaturalisation potential, that is, the final steady state if the occupation were to end immediately and the land were to recover (with or without human aid). This is level D'' for the activity shown by the rectangle in Figure 2-5. Other options for choosing a reference state are discussed below.

It has been suggested (Blonk and Lindeijer 1995) that the original state before any human intervention (level A in the figures) could be used as reference state. This would imply that all transformation impacts since the initial human use would be allocated over the subsequent activities in relation to their duration, while disregarding the

---

[2] For transformation, the reference state may also be important, to identify the biogeographical location of the transformation (because in more vulnerable areas, the impact from a transformation with the same initial and final state may be considered different). This choice of reference follows the same reasoning as for occupation.

[3] Note that this reasoning implies that no distinction is made between transformations that are not allowed to renaturalise and transformations that cause a physically irreversible or permanently changed situation. Considering the difficulty of distinguishing between both situations in practice, this seems unavoidable. See also the preceding section.

actual renaturalisation potential resulting from these activities. While the original state is relevant as a reference level for an initial transformation impact, it does not have any relation to the occupation impacts of subsequent activities[4], and its use would therefore lead to suboptimisation of the current land use.

Likewise, it has been suggested (Baitz et al. 1998) that the state immediately before the studied activity could be used as the reference state (e.g., state *B* in Figure 2-6). This would imply that the occupation impacts of the area between the 2 dotted lines and between t' and t_2 in Figure 2-6 would not be ascribed to any activity, while any subsequent activity would be ascribed occupation impacts relative to the preceding activity only (the area below the lowest dotted line in Figure 2-6). This would also eliminate the distinction between transformation impacts and occupation impacts. Occupation impacts would be attributed only to land use that includes changes. Continuation of land use as it is would not be ascribed any impact. This implies ignoring the impacts from this occupation and prevention of renaturalisation. Alternatively, the first human activity in the area (the one with level *B* in Figure 2-6) would—to be consistent—continue to be ascribed occupation impacts until all human activity in the area ends, as if it were not terminated at all. It may be questioned whether this expresses a satisfactory causal relationship between the human activities and their impacts.

For the same reasons as those for selecting the state just before an activity, the state immediately after the studied activity (i.e., before renaturalisation) would not be a relevant reference state.

Based on the above reasoning, we can draw these conclusions:
- If the reference state is chosen at a level above the current renaturalisation potential, the

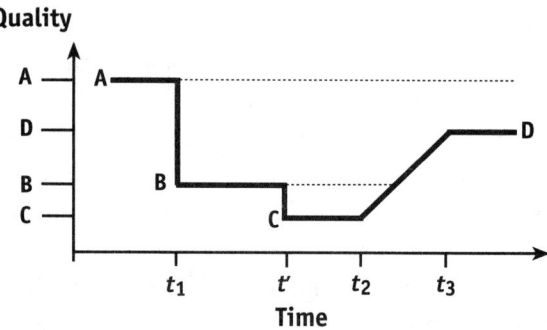

Figure 2-6 State of the flora, fauna, soil, or soil surface before, during, and after 2 human activities (in the periods t_1 to t' and t' to t_2, and its (mis)attribution over the 2 activities when the state immediately before the studied activity is used as the reference state. Note that in this figure, the reference state is chosen as *A* for the activity $t_1 - t'$.

impact measured would include a part of the permanent transformation impact, that is, it would not be representative of the temporary occupation impact alone.
- If the reference state is chosen at a level below the current renaturalisation potential, the impact measured would exclude part of the occupation impact of the current land use.

Thus, according to our reasoning, using the current renaturalisation potential as a reference seems to be the most consistent choice. Other reasonings may be possible; however, we did not assess whether these alternative reasonings lead to consistent results.

How to determine this reference state depends on the indicators selected to express the land quality and impacts on it. For biodiversity, Lindeijer et al. (1998) have chosen to define it as 'the maximum biodiversity of the region', whereas Köllner and Jungbluth (2000) have chosen the maximum biodiversity in recent history. The discussion on the choice for a detailed reference level under the above framework is not finished and needs further development.

---

4 Whether the preceding level is high or low depends solely on the preceding process, whereas the defined prevention of renaturalisation depends on the potential quality level obtained after the preceding process.

Practical attribution of transformation impacts to functional units

Occupation impacts are easily allocatable to the functional units of an LCA because the impact is proportionate to the occupation time, and the output in functional units is also proportionate to this occupation time. The relationship to produced functional units is less evident with respect to the impact of a preceding transformation.

In principle, a transformation impact should be attributed to subsequent land users who take an economical profit from the transformation process. There is a problem related to attributing land transformations to functional units, which can be seen at 2 levels:

1) If transformation is followed by a succession of various types of occupation, how can we allocate the environmental impact of the preliminary transformation to the members of the following chain of occupations? For example, a swamp is transformed first into a dry area. This area is then used as a cropfield, at a later stage it is used for sheep grazing, and finally it is used as a campground. How can we split the environmental impact of transformation on the 3 subsequent cases of occupation?

2) At the level of a single occupation in this succession chain, how should we allocate the single occupation's share of the environmental impact of the preliminary transformation to one functional unit produced during this case of occupation? For example, if sheep grazing is executed during a 5-year period, how large a share of the environmental impact from transformation is allocatable to 1000 kg of produced meat?

At the level of the chain of occupations, we propose that the impact of the preceding transformation be fully allocated to the first case of occupation, because the transformation was made exactly for this first case in the chain of subsequent occupations and not for later cases of other occupation types.

At the level of the single occupation type in a chain of succession, the problem is to determine the total number of functional units produced during the whole duration of this occupation, so that the transformation impact allocated on this occupation can further be split up between these functional units. This is comparatively easy in the case of a mining operation, where geologists can give an estimate of the total metal output until the mine is exhausted. However, the amount of functional outputs is often unknown: How many tons of grains will be produced on the area of a cleared forest? How many vehicles will drive on a new road? Two possibilities are given for the attribution of transformation impacts to functional units in such difficult cases.

1) Attribution by economic depreciation — In analogy to the practice in economics, we could envision working with 'standard depreciation times' (Vogtländer 2002). In certain cases, such depreciation times are rather arbitrary, whilst in other cases, they can be well justified. For example, desiccation of a swamp must be repeated after 20 to 25 years, if the soil level is lowered by oxidation of organic material; in consequence, the transformation impact can be attributed to the expected production output of 20 to 25 years of occupation.

2) Attribution of trends — An activity performed for a certain functional unit may itself not show any significant relationship to land transformations (e.g., the road transport of 1000 refrigerators over a distance of 600 km, as registered in an LCA inventory, does not require the construction of any new road). However, the total area of land made available for transportation may increase over time, due to a long-term trend of the corresponding total of transportation activities. The transformation associated with this land transformation can then be attributed to the trend. As an example, in 10 years' time, the Netherlands transformed 9500 km of agriculture area into roads; during the same time period, the inland traffic

increased by 16×10^9 car-km and 5.2×10^9 to-km of trucks. If such trend data are available, the land transformation impact can be attributed to functional units of traffic or transport, in addition to the normal allocation of land occupation impact. It is conceivable to say that the vehicle-kilometres required according to any LCA inventory are a part of this road transport trend, leading to a certain additional area transformed annually per 10^9 additional vehicle-kilometres annually (marginal approach).

A completely different way to tackle this land transformation impact is to disregard the life-cycle perspective. This implies 'forgetting' that the service-delivering activity (involving transformation) is not only produced but also used for an uncertain amount of time, and may be adequate when performing an environmental impact assessment of the activity instead of an LCA.

Elementary activities of land use and land-use types

At the LCI level, it is possible to register as separate entries each of the elementary activities of land use, such as ploughing, asphalting, or draining. It is obvious that the number of such elementary activities may be very high: Producing corn or mining clay on a certain area includes a multitude of individual steps, each of them having an influence on the state of the land area. As an alternative, the inventory may consist of predefined sets of such elementary activities, *land-use types*. The latter is present practice, especially for agricultural and silvicultural land use, where land-use types are used like 'high intensity monocrop agricultural production'. Land-use types, as used in Ecoinvent 2000, the Swiss Eidgenössischen Technischen Hochschule (ETH) inventory database on systems for energy supply and similar basic materials, are described in Jungbluth and Frischknecht (2001).

If such predefined land-use types are used at inventory level, it happens that not only physical activities such as ploughing but also applications of chemicals such as fertilisers or pesticides are included in the sets of elementary activities. If a certain land-use type includes the distribution of 20 g nitrate fertiliser per square metre of the cultivated area, this is essentially the same chemical as the nitrate emitted by the farmer's combustion motors. The only difference is that the fate and exposure model of the 2 sources is different: The emission of the fertiliser is focused on the cropfield, whilst the nitrates from the exhaust fumes go into the atmosphere and are widely distributed in low concentrations.

For the sake of clarity, it would be preferable to show chemical emissions separately, like fertiliser and biocide applications, so that land-use types contain physical activities only. The possibility would then be open to assess fertiliser and biocide application under the impact categories of nutrification and ecotoxicity, if desired. However, splitting land-use types into their physical and chemical fractions is a problem because it may be hard to determine which impacts are caused by the physical part of a certain land management method and which by chemical outputs. Impacts on biodiversity or soil fertility may be due in part to airborne nutrifying agents, but to what extent? Non-intended nutrification by NO_X deposition adds to this problem.[5] In practice, separation between impacts by physical and chemical interventions may be impossible.

If land-use types also contain distribution of chemicals, special attention should be paid to the problem of double-counting impacts. One solution is to disregard nutrification and ecotoxicity from unintended use of chemical inputs for land use, with the argument that land-use impacts are further in the cause–impact network and thus include

---

[5] A distinction should be made here between intended and non-intended impacts. Intended impacts relate to the economic system (fertiliser impact on crop yield) and may be assessed as such, and unintended impacts are to be assessed as environmental impacts (nutrification by excess use of fertiliser).

both items.[6] This may not be correct for impacts caused outside the specific land-use area. Another solution is to allow for double-counting when the impact assessment method has the possibility to show these double-counts, albeit in different terms. This is especially acceptable when the CIs for nutrification and ecotoxicity are much more sensitive to these chemical interventions than is the land-use impact indicator.

Distinction between impacts on ecosystem functions and co-products of land-use

Land surface, including the natural environment on and below the land surface, has a number of different functions (de Groot 1992; Marks et al. 1999)[7]:
- Habitat for humans
- Habitat for nonhuman life
- Base for food and other biotic production
- Element in the freshwater circuit
- Element in the global energy circuit
- Place of abiotic resources deposits.

The intended or unintended effect of a land-use activity can be such that one or several of the abovementioned functions are influenced in an environmentally undesirable way. All changes to these functions may be considered impacts. However, if certain land functions are positively influenced by land use, the corresponding improvement is sometimes claimed to be a co-product of the land-using activity, comparable to food or wood as main output products of agriculture and forestry. For example, if the steep meadows in alpine regions are used for cattle grazing or hay collection, agriculture not only produces meat but also prevents avalanches from falling on settlements and roads down in the valleys. In consequence, preventing avalanches could be called a *co-product* of alpine cattle farming.

We propose here that the dividing line between impact and co-product is drawn by this question: Is the output part of the environment or part of the economic system? If the output is a part of the economic system and therefore a co-product, a price can generally be associated with it, whereby this price is possibly paid by the community. In alpine regions, the farmers 'sell' their service of protecting the region against avalanches and landslides to the state. In such a situation, it is clear that the inputs and outputs of the farmer's land-use activity are allocated amongst meat production and protective services as co-products.

As a special case, a change in land or ecosystem functionality might be seen as a co-product and not a (damage-reducing) impact, if a prospective study investigates the environmental effects of different ways to manage land use. For instance, CO_2 reduction may be the object of a study in which global warming from CO_2 is not treated as an impact but prevention of global warming is the intended product, while all other impact categories are still impacts. This is a reasonable approach only when the output in terms of CO_2 avoidance is explicitly part of the functional unit. Such studies can be seen as preparative work for subsequent price negotiations with those who take a profit from the land-use activity.

In all other cases, a change in land or ecosystem functionality should be considered as an impact of the land-using activity. It is inadequate to pretend that a co-product is produced whenever a more damaging alternative of land use on the same land area is conceivable.

---

[6] For instance, in the Eco-indicator 99, the choice was made to exclude the nutrification and ecotoxicity impacts of traditional agriculture instead of the land-use impacts related to it.

[7] The distinction between land-use functions and ecosystem functions is worth mentioning here. Ecosystem functions are a subset of the total set of land-use functions. The possibility of providing resources is generally considered a land-use function, not an ecosystem function. Use for human habitation is also related to the land, not to ecosystems.

Distinction between 'solid' land areas and intracontinental water bodies

So far, the term *land use* is used in the sense of the use of 'solid' land. But in addition to this solid land, continents contain a few percent of intracontinental surfacewater areas (rivers and lakes, including adjoining areas flooded from time to time). Although this is only a small part of the total continental surface, rivers, lakes, and adjoining wetlands are of comparatively high importance as habitats of nonhuman life and as parts of certain life-support systems.

It is desirable that LCIA methods account for the specific aspects of water area use. For the water as a substance, this may be done by impact assessment of the abiotic resource surface water. The same can be done for the biotic resources living in the water. For seas, the water body changes due to physical changes may be considered irrelevant, except for coastal areas where the water flow may be altered.[8]

For land use in rivers, lakes, and their floodplains, a specific type of land-use assessment should be made. This assessment may be derived from the one used for solid land use. Since sea benthos is sometimes fluid and the regeneration or renaturalisation time may be high, impacts may be more temporal than on land (see Lindeijer et al. 1998, Appendix 2 to Annex 2). At present, no impact assessment methods have been proposed for this situation.

The relation between land use and its impacts on AoPs

According to the framework and definitions stated in the preceding sections, the different types of land-use impacts are discussed here.

Land use mainly has impacts on the endpoints or the AoPs natural resources, natural environment, and manmade environment. Until now, these types of impacts have been divided into 3 or 4 subcategories in the LCA community (Finnveden 1996; Udo de Haes et al. 1999):
 1) increase of land competition,
 2) degradation of biodiversity,
 3) degradation of LSFs, and
 4) degradation of cultural values (including landscape impacts).

The relationship between the environmental interventions of land use and these impact types is discussed in some detail below. For clarity, the impacts that result from occupation are separated from those that result from transformation of land.

Increase of land competition

Land occupation by one user leads to the impact that the possibility for other humans to use the same land area during the same time period is reduced, or is even nonexistent in case of fully exclusive land occupation. This reduced availability leads to increased competition for land areas of the corresponding type. Competition over land in this strict sense relates to the availability of land to contemporary potential users. In this view, land areas are like abiotic resources, or could even be considered as a kind of abiotic resource (land area as a flow resource that cannot be depleted by the current generation because the total area will still be available in the future, if the focus is restricted on square metres only and not on quality).

Land transformation is the process of changing the quality of a given land area, whereby this quality change involves the suitability of the area for the user's purposes, as well as the accompanying change of the environmental quality. Focusing here on the former aspect, land transformation decreases the supply of land of the pretransformation use quality, whilst increasing by the same number of square metres the land supply with the post-transformation use quality. In practice, land trans-

---

8 Changes due to input of substances should be considered under toxicological or acidification and nutrification impacts.

formations tend to increase the share of lands that enjoy high market demand, to the detriment of lands with qualities less preferred by potential users. This is illustrated by the global trend from agricultural areas towards building areas. In consequence, land transformation reduces the intensity of competition amongst contemporary land users because the supply is better adapted to the demand.

As already mentioned in the sections on abiotic resources, we propose to consider competition between contemporary humans as an issue of the economic system and not of LCA. Nevertheless, we shall present below some published proposals for the implementation of the land competition impact in LCA.

Degradation of biodiversity and life-support functions

Land occupation may cause an impact on biodiversity and LSFs because the purpose of the land use generally necessitates maintaining the area in a nonnatural state, so that habitats are degraded or even destroyed. The impact of the corresponding human activities can be interpreted as postponing a spontaneous renaturalisation to the potential environmental quality of the land. The impact of land occupation on biodiversity is not limited to the occupied area: At least in highly stressed regions like Central Europe, occupying an area in a nonnatural mode has a degrading influence on the biodiversity of the whole region, whereby time lags of many decades may be involved (Müller-Wenk 1978). The notion of a time lag in these impacts is well known among biodiversity experts (e.g., de Groot 1992; Angelstam 2000; Niemelä 2000). Whilst impacts on the occupied area are called *local impacts*, impacts exceeding the perimeter of the occupied area are called *regional impacts* (Köllner 2001). Impacts of land occupation on LSFs are somewhat less evident than impacts of land occupation on biodiversity. However, impacts on climate regulation through changed albedo effect or evaporation rates are significant (Intergovernmental Panel on Climate Change [IPCC] 2001). Further, it is well known that today's pattern of land occupation, with large sealed surfaces and limited wetlands, has affected the freshwater circuits and caused new types of river floods. Obviously, these impacts are not limited to the perimeter of the area occupied.

Land transformation leads to impacts that are more visible because the situation before and the situation after the transformation often show dramatic differences. If land with a high quality of nature (for instance, in terms of biodiversity and life-support potential) is transformed into land with a lower quality, such events are therefore frequently the object of public debates. Although visible destruction of habitats by current land transformation is a serious impact, it is important to note that the currently stated biodiversity loss in Europe and other intensively used regions is caused mainly by past transformations (National Institute of Public Health and the Environment [RIVM] 1992). But because land transformations made 100 years ago cannot be the object of a current LCA, it appears reasonable to charge in LCA the current occupation of the corresponding land, because this occupation prevents, or at least postpones, the reversal of the historical land transformation, as deemed necessary for reducing the environmental damage.

Degradation of cultural values and landscapes

Land transformations may exert an impact on the manmade environment through degradation of recreational and historical values related to manmade landscapes and buildings. Changing the geography of a manmade landscape can also be considered here. Land transformations may also reduce the aesthetic value of natural and manmade landscapes.

Land occupation, in contrast, is often seen as an instrument for maintaining cultural values and manmade landscapes. This explains certain public payments to traditional agriculture, which are justified by cultural and aesthetic concerns.

The obvious difficulty in assessing these impacts comes from the fact that definition and valuation of cultural and aesthetic goods varies widely amongst the members of human society. In consequence, authors of land-use assessment methods have not given much attention to this aspect.

Operationalisation of land-use impact assessment for AoPs

It would be desirable to have 'comprehensive' impact assessment methods for land use that cover all important links from land use to the various AoPs. But for the time being, available assessment methods focus on one type of impact that is considered to be of particular importance in the eyes of a method's author. We therefore group published methods with respect to the AoP or the endpoint that is their primary concern.

There is a certain degree of correlation between the various impacts of land-use types. If a method tries to model the impact of land use on biodiversity, it is conceivable that the proposed indicators of biodiversity degradation turn out to be reasonable indicators of the degradation of certain LSFs as well, and vice versa. As a matter of fact, if a given type of land use is good for a high biodiversity on the used area as well as on the surrounding region, there is a fair probability that this land-use type is also good for life-support systems like the freshwater circuit or the regeneration of soil. But it may not be the best land use alternative with respect to the system of climate regulation. On the other hand, a method that models the influence of land use on climate regulation may produce useful indicators for biodiversity, but not for supporting the system of solar energy fixation.

It is therefore a reasonable first step that authors of land-use assessment methods start by giving priority to the most important damage link from land use to the AoPs. But as a second step, they should check whether the proposed link also yields a reasonable indicator for the impacts of land-use types to other important AoPs or endpoints, and the method should then be complemented if necessary. As long as this is not done, we do not want to recommend a best available method, but at the same time, we are of the opinion that most of the following methods merit attention as possible contributors to a future comprehensive method that merits being called *best available method*.

Competition for land area by humans of the same generation

Land use influences land competition amongst contemporary human users; as mentioned before, this important problem should be treated by instruments other than LCA.

However, if an LCIA of land use with respect to land competition were considered desirable, methods are available that shall be discussed here in short. There are different levels of sophistication. We could consider land (actually land under human use[9]) a homogeneous resource. This allows for a very simple impact assessment: Add up all land occupations (Heijungs et al. 1997) and optionally divide this by the total area of land, excluding natural land (van Ewijk et al. 2000). This option allows a comparison with competition for other flow resources such as solar energy. If the aim is to compare land competition with competition for abiotic deposits such as metals or fossil fuels,

---

[9] Generally speaking, natural land is not under competition amongst humans, at least as long as recreational use without changing its quality is not considered competition through land use. However, when the absence of other humans is considered part of the recreational function, this could be captured under competition (whereas changing its physical quality is transformation, leading to other damages). Also, one can argue that as soon as it is institutionally possible to confer property rights on a land surface to an individual owner, giving him the power to exclude others from using the area, there is competition for such land. An example is natural lands that have a potential to become future building areas.

we should be aware of the difference between deposits and flows. Fava et al. (1993) have proposed a similar comparison using the formula $1/R \times P/R$, where P is the amount of land used yearly and R is the total amount of land available.

One further degree of sophistication is to assess land competition at the regional level, by taking the amount of land in a region as a divisor, or possibly the number of its inhabitants. This allows comparison of regions with different land scarcity. A next step in sophistication would be to divide the total land surface of a region into the different land-cover types (e.g., Coordination of Information on the Environment [CORINE]), with a given quantity of surface per land-cover type per region. The number of such different land-cover types varies depending on the chosen level of differentiation, but will certainly not be below some dozens. But this further level of sophistication leads to the complication mentioned before: At the level of one land-cover type, the total supply within a region is no longer fixed but is influenced by land transformation trends. The authors seem to give no guidance for valuing and offsetting increased competition for land of pretransformation quality with decreased competition for land of posttransformation quality.

In Table 2-6, published proposals for the operationalisation of land competition are summarised.

Until now, only the lowest level of sophistication has been operationalised, in square metres by Baumann et al. (1992), but more recently, in the correct units of square metres per year (m²·y) according to the update of the CML guide (Guinée et al. 2001) and in a recent update of the IVAM ER database (van Ewijk et al. 2000). In van Ewijk et al. (2000), however, the aggregation is divided by the total land area under human use.

Impacts on the natural environment, endpoint biodiversity

Operationalisation of land-use impacts on biodiversity means that, in Figures 2-2 and 2-3, the environmental quality Q on the vertical axis is replaced by a variable representing the magnitude of biodiversity. The land transformation process at time t_1 then leads to a sudden drop of biodiversity. The land occupation process between t_1 and t_2 keeps biodiversity at a level that is lower than the reference level. And land renaturalisation finally increases biodiversity again. Changes in biodiversity occur in a fast and concentrated way on the transformed or occupied area, but such change also takes place in the surrounding region, though at a slower rate and with lower intensity. Reasons for such impacts towards a larger surrounding region are, for example, worsening of pollination and of dissemination of seeds. As mentioned before, the extinction of species is a slow process that may take decades or even centuries from the first local reductions of populations to the final extinction, inclusive of disappearance of germinatable seeds. This means that biodiversity degradation can be increased not only by land-use types that change the relevant characteristics of the land but also by land-use types that maintain the relevant characteristics of the land at a low-quality level.

But how to express the magnitude of biodiversity? It is known that biodiversity contains 3 elements:
1) genetic diversity within a species,
2) diversity of species, and
3) diversity of ecosystems.

The knowledge and quantitative data on the first and third elements are much less developed than are those on species diversity. This is a practical reason why proposals for impact assessment of land use suggest expressing biodiversity degradation in terms of species diversity reduction only. Further simplification is necessary at the level of species diversity because the total number of species is on the order of 10 millions. A preference is

Table 2-6 Operationalisation characteristics for competition over land

Items according to Table 2-2	Properties of model
Baumann et al. 1992; Jolliet and Crettaz 1996	
Essentials[a]	Full quantification as $\Sigma^i A_i$, no regionalisation, no specification of land-use types
Sensitivity	Related to inventory data only
Mechanism and model	Direct link between m² used and land scarcity assumed. Flow character of land occupation (m²·y) not acknowledged.
Extent of representation	Only competition included
Choices or assumptions	All areas used cause same competition
Consistency	Units used inconsistent with framework
Applicability	Not applicable (lack of consistency)
Fava et al. 1993	
Essentials[a,b,c]	Full quantification as $\Sigma^i A_i/D \times P/D$, no regionalisation, no specification of land-use types
Sensitivity	Related to inventory data only
Mechanism and model	Direct link between m² used and land competition assumed. Flow character of land occupation (m²·y) not acknowledged.
Extent of representation	Only competition included
Choices or assumptions	All areas used cause same competition
Consistency	Not in inventory units for competition
Applicability	Not applicable due to lack of consistency
Van Ewijk 2000; Guinée et al. 2001	
Essentials[a,c,d]	Full quantification as $\Sigma^i A_i \times t/P$, no regionalisation, no specification of land-use types
Sensitivity	Related to inventory data only
Mechanism and model	Direct link between m²·y occupied and competition assumed
Extent of representation	Only competition included
Choices or assumptions	All areas used cause same competition
Consistency	Correct units and explicitly competition addressed
Applicability	Dependent on availability of land-use data

[a] A = Land area occupied (environmental intervention).
[b] D = Total land available.
[c] P = Yearly total land use.
[d] t = Time duration of occupation.

therefore given to determining species diversity on the basis of vascular plants only. This is due to the relatively good availability of current and historical data on vascular plants, and this choice can be justified further by the fact that vascular plants occupy a central position in terrestrial life as transmitters of solar energy, so that they are fairly good overall indicators for nonhuman life. Some evidence is available that, in practice, the diversity of vascular plants shows a reasonable correlation to the diversity of many other taxonomic groups of higher plants and animals (Duelli and Obrist 1998). Nevertheless, instead of using vascular plants, it would be preferable theoretically to use a set of species picked out from the whole taxonomic system in view of its representativeness for the totality of species. But such a set of species representing the totality of nonhuman life would probably be different from region to region, and data availability would cause additional problems. For this reason, it is an acceptable choice to start with vascular plant species diversity.

The expected number S of vascular plant species depends on the land-use type, on the area size A, and on biogeographical conditions. Under comparable conditions, the relationship between S and A may be approximated by $S = S_1 \times A^\forall$, that is to say, the species number is S_1 on a plot of 1 hectare, and it grows with the exponent $\forall$ if area A grows beyond 1 hectare. S_1 and $\forall$ are, under given biogeographical conditions, indicators characterising a given land-use type with respect to biodiversity. Practical figures for S_1 and $\forall$ per land-use type can be developed by statistical analysis of empirical floristic data from biogeographically similar plots. In this way, data on the slope factor α (see Lindeijer et al. 1998) and on the theoretical number of species, S_1, per 1 hectare (see Köllner and Jungbluth 2000; Köllner 2001) have been determined for a substantial number of land-use types.

If the biodiversity impact of any land-use type can now be represented by its specific indicator value, be it based on S_1, $\forall$, or a suitable function $f(S_1, \forall)$, it should be remembered that it is not the absolute value that counts but the difference from some reference situation. For example, if a 1-hectare cornfield area has an S_1 of 10 species only, and the more natural reference land cover would have an S_1 of 70 species, the biodiversity impact is then the difference between 'what could be' and 'what is', or $70 - 10 = 60$ species. In a further step, this result can be normalised by dividing through the 'what could be' number of species.

Various authors have selected different parameters (S_1, $\forall$, or others) for characterising vascular plant species diversity (or, to be more correct biologically, *species richness*) of a given land-use type, and they disagree also on the choice of the reference situation. Nevertheless, we illustrate the general concept here by giving the indicator for biodiversity damage, if an area A = 20 hectares is occupied during 1 year by a cornfield, whilst the reference situation would be an unfertilised grassland, and if the vascular plant species number S_1 (cornfield = 10, grassland = 70) is chosen for characterising the species diversity. The biodiversity loss in the area itself is then indicated by

$$[(S_{1\ \text{grassland}} - S_{1\ \text{cornfield}})/S_{1\ \text{grassland}}] \times A \times \text{time}_{\text{occupation}} = 0.86 \cdot 20 \text{ [hectar-years]} \quad (2\text{-}7).$$

In words: During the 1 year of occupation by the cornfield, the area of 20 hectares would not show 86% of its potential number of vascular plants.

Species diversity in the region surrounding an occupied area is influenced by species diversity inside the occupied land area. If the area occupied is very rich in species diversity, the influence on the regional species diversity will be relatively positive, and if the plot is poor in species or even exempt of species, the influence on the regional species diversity will go towards further degradation. This regional impact should ideally be included in the full impact assessment of land use, in addition to the local impact on the plot of land itself. The influence of land-use distribution on the biodiversity in a region has been analysed statistically in Köllner and Jungbluth (2000) and Köllner (2001). The interpretation of his work in the Eco-indicator 99 uses a factor of 20% on top of the local biodiversity impact of land occupation, in order to include the influence of the surrounding region on biodiversity.

The most important representatives of each type of approach to assessing biodiversity impacts of land use are shown below in Table 2-7; discussing all methods in detail here would not be feasible. A recent proposal operationalised for forestry in Sweden (Swan 2002) is not included here. One general statement that can be made is that vascular species diversity is at present the only proposed basis of an indicator within LCA on a species level. On an ecosystem level, only one

Table 2-7 Operationalisation characteristics for biodiversity

Items according to Table 2-2	Properties of model
Köllner 2000	
Essentials	Full quantification using several biodiversity indicators based on 7 land-use classes and 30 land-use types, no regional differentiation
Sensitivity	High; positive and negative impacts possible
Mechanism and model	Species loss: "Local" effects of 30 land-use types determined on 3706 field studies in Switzerland (CH) and Germany (D). "Regional" effects of 7 land-use classes based on regression analysis of historical change of species richness in 592 subregions in CH.
Extent of representation	Biodiversity impacts in terms of vascular plant species diversity
Choices or assumptions	Assuming transformations will be recoverable, and including occupation during recovery
Consistency	Reference (maximum Europe around 1850) not consistent with occupation interpretation?
Applicability	Good for mid-Europe
Goedkoop and Spriensma 1999	
Essentials	Full quantification based on 5 land-use types using a modification of the formulas of Köllner, no regional differentiation
Sensitivity	High; positive and negative impacts possible
Mechanism and model	Marginal analysis, including regional impacts, species-area curves fitted with formula from Arrhenius on a log-log scale
Extent of representation	Biodiversity impacts in terms of vascular plant species diversity
Choices or assumptions	Assumes transformations always will be recovered; expresses this by adding recovery time to occupation time
Consistency	Reference (average present mix) not consistent with occupation interpretation; consistent with Eco-indicator 99 damage approach framework.
Applicability	Good for mid-Europe
Lindeijer et al. 1998	
Essentials[a–c]	Full quantification via $\Sigma A_j \times t \times (\alpha_{ref} - \alpha_{act})/\alpha_{ref}$ for occupation, and $\Sigma A_j \times (\alpha_{ini} - \alpha_{fin})/\alpha_{ref}$ for transformation, rough regionalisation
Sensitivity	Lower for biodiversity (range 0–150) than for land area used (range 1–10.000)
Mechanism and model	Average analysis, excluding regional impacts when not counted, determined by scaling individual measurements with Fischer curve
Extent of representation	Biodiversity impacts in terms of vascular plant species diversity
Choices or assumptions	Rough biodiversity reduction factors for different land-use types
Consistency	Reference (present maximum) consistent with occupation interpretation
Applicability	Good for land-use types in mid-Europe; moderate outside Europe
Müller-Wenk 1978	
Essentials	Full quantification through damage function, no regionalisation
Sensitivity	Low: only 1 transformation type considered
Mechanism and model	Direct link between occupation and regsional species diversity loss, including rough damage assessment
Extent of representation	Focus on endangered species (intrinsic value of biodiversity, not life support function of it)
Choices or assumptions	Based on 2 datasets (CH, D); assumes full recovery after transformations, thus allows adding up occupation and transformation damages
Consistency	Damage assessment consistent with Eco-indicator 95
Applicability	Easy; range of a factor 5 in damage score for mid-Europe

Table 2-7 *continued*

Items according to Table 2-2	Properties of model
Sas et al. 1997	
Essentials	Full quantification ($A \times S \times t_{rec}$), no regionalisation but specification per species type
Sensitivity	Reasonable
Mechanism and model	Assesses only transformations; includes biodiversity recovery time to express vulnerability
Extent of representation	Focus on specific species
Choices or assumptions	Multiplicability of area, species density, and recovery time
Consistency	Consistent with related indicator for life support functions
Applicability	Depends on specific data availability
Weidema and Lindeijer 2001	
Essentials[d–k]	Full quantification by m²·y·$(S_{exi,nat}/S_{pot,nat}) \times [(A_{exi,b}/A_{pot,b})^{z-1} \times (A_{pot,max}/A_{pot,b}) \times (SD_{exi,b}/SD_{exi,min})]$, rough regional differentiation
Sensitivity	Large due to factors at ecosystem level
Mechanism and model	Species diversity + regionalisation through ecosystem characteristics (endangeredness, vulnerability, species richness)
Extent of representation	Includes scarcity on ecosystem level
Choices or assumptions	Choice and multiplication of factors and their ranges; value choice to assess impact on original (endemic) species only
Consistency	Assured among multiplication factors
Applicability	Potential vegetation type needs to be known; % of original species loss known for only a few land-use types
Felten and Glod 1995	
Essentials	Full quantification as scores for 9 land-use classes, no regionalisation
Sensitivity	Lower for biodiversity (range 0–1 with 2 decimals) than for land area used (range 1–10.000)
Mechanism and model	Diversity argued as stability principle; 9 Hemerobiestufen (land–use classes) distinguished
Extent of representation	Stability and scarcity aspect of biodiversity included
Choices or assumptions	Interventions based on ETH database (adding up occupation and transformation by assuming full recovery)
Consistency	Yes
Applicability	Easy; difference in scores for different weighting sets for the classes

continued

Table 2-7 *continued*

Items according to Table 2-2	Properties of model
Mattsson et al. 2000	
Essentials	Number of species quantified before land transformation on land-use type and for current land-use in 1 land-use type
Sensitivity	Low: only total number of different biodiversity groups
Mechanism and model	Species diversity and regionalisation through ecosystem characteristics
Extent of representation	Only number of species for a certain land-use type; no information on changes due to land transformations
Choices or assumptions	Higher number of species per area is of higher value
Consistency	Yes
Applicability	Potential vegetation type needs to be known; local data needs to be available

[a] A = area occupied with respect to area transformed (environmental intervention).
[b] t = time duration of occupation.
[c] α = measure of biodiversity (with indices: ref=reference, act=actual, ini=initial, fin=final).
[d] $A_{exi,b}$ = Existing area of biome b.
[e] $A_{pot,b}$ = Potential area of biome b.
[f] $A_{pot,max}$ = Area of biome with largest potential area.
[g] z = Steepness of area-species curve ($z = 0{,}15$ is reasonable for this).
[h] $SD_{exi,b}$ = Species density of biome (per 10.000 km^2).
[i] $SD_{exi,min}$ = Species density in biome with maximum species density.
[j] $S_{exi,nat}$ = Number of natural species still left on inventory spot.
[k] $S_{pot,nat}$ = Potential number of natural species on inventory spot.

proposal is made: Weidema and Lindeijer (2001) have suggested including, next to a species richness loss factor (% of original species lost), 3 additional factors:

1) the part of an ecosystem left (to express its endangeredness),
2) the relative potential ecosystem area (to express its vulnerability), and
3) the ecosystem value (to express the intrinsic value of the ecosystem).

This set of factors comes close to suggestions for global biodiversity indicators from ten Brink and Douma (1995); it is also the most area-related.

Several methods apply land-use classifications only, instead of continuous species measurements. Some of them suggest weighting sets between land-use types. Heijungs et al. (1992) suggest that only one transformation type (between agriculture and built-up land) is relevant in practice. However, the impact score was not interpreted explicitly, but there seem to be more links to life support (see Blonk and Lindeijer 1995). In Felten and Glod (1995), 3 diversity-based weighting sets for 9 land-use types were proposed (one based on the Simpson index, one on % Red List species, and one based on a combination of these). They are also mentioned in (Knoepfel 1995) and compared there with a German biotope weighting system based on a questionnaire on biotopes (Auhagen 1994).

Impacts on the natural environment, endpoint life-support functions

Operationalisation of land-use impacts on LSFs means that, in Figures 2-2 and 2-3, the environmental quality Q on the vertical axis is replaced by a variable representing the degree of fulfilment of LSFs. The land transformation process at time t_1 then leads to a sudden drop of LSF. The land occupation process between t_1 and t_2 keeps LSF at a level that is lower than the reference level. And land renaturalisation finally improves LSF again. Changes of LSF occur not only on the transformed or occupied area, but such changes may involve larger regions, depending on the specific LSF. For example, the regulation of climate LSF is obviously not changed on the occupied or transformed land area only, but rather at a regional or worldwide level, as a consequence of the long-term trend of all land-use changes.

Life-support functions concern the role that ecosystems play in maintaining natural processes that are important for supporting life, including nonhuman and human life. Important LSFs have already been mentioned at the beginning of this chapter: production of biomass, regulation of climate, (re)generation of soil, pollination of crops, mitigation of floods and droughts, and natural purification of air and water systems. But this enumeration is certainly not complete; there is no agreement on a criterion that separates functions of the components of ecosystems that merit being called LSFs from other ecosystem functions that are not LSFs. Even a limitation to the currently discussed LSFs would lead to a longer list. In consequence, we cannot expect to build a complete quantitative model of the links from land occupation and land transformation to the effects on all of the LSFs. The question is, rather, is it acceptable to select for modelling 1 or 2 LSFs that are fairly well understood and that could be taken as a proxy for the impact of land occupation or transformation on the unknown totality of LSFs?

It merits mentioning at this point that suitable indicators of biodiversity, as presented in the preceding section, could also be useful proxies for land occupation or transformation impacts on the totality of LSFs. An expert panel of 39 leading scientists came to the conclusion that there is a good correlation between changes in biodiversity and changes in ecosystem processes (Schläpfer and Schmid 1999).

Several authors, who have proposed methods for linking types of land use to the effects at the level of LSFs, have focused their model on the LSF of soil. Soil is not always clearly defined, but it appears that it is mostly meant as topsoil, the uppermost layer of soil, which is a mixture of mineral components and decay products of organic matter and which is interspersed with a large diversity of organisms. As a matter of fact, topsoil plays (together with freshwater availability) a central role in the functioning of terrestrial life, and several other LSFs depend on the quantity and quality of topsoil. A sufficient quantity and quality of topsoil provides for water buffering and filtering, for properly operating substance conversion processes, as well as for the growth of the vegetation cover on the soil surface. In turn, a strong vegetation cover leads to high biomass production, to soil stability, and to a partial compensation of the current destabilisation of the climate system. Baitz et al. (1998) have covered most of the physical impacts on soil-related life-support issues. Schweinle (1998) and Schenck (2001) have taken about the same route. Although here the importance of LSFs to life is not seen primarily from the anthropocentric perspective of human needs for food and shelter, it may be noted that a good state of topsoil quantity or quality is also decisive for food production.[10]

Various indicators have been proposed to express the degree of good functioning of the life-support system soil. Some try to indicate topsoil quantity and quality directly, and others try to give indirect indicators of topsoil quantity or quality. Below, we comment on these briefly.

Current biomass production on the occupied land area can be taken as an indicator of topsoil

---

[10] The soil is a crucial life-support system because the majority of food production is dependent on it (IUCN et al. 1980) and especially because its microorganisms regenerate necessary elements of it (Odum 1993).

quantity or quantity and other related LSFs (Knoepfel 1995; Sas 1997; Weidema and Lindeijer 2001). If the current biomass production on an occupied area is high, it can be reasonably expected that, at least in the short term, the soil quality is stable, the freshwater circuit is not out of order, and rainfall can be absorbed by the soil. However, the experience of the U.S. Midwest shows that high biomass production can be associated with high erosion and low-quality freshwater runoff. It may therefore be more appropriate to take into account only the part of the biomass that is not harvested by humans but also the part that is left on the plot for nonhuman life support. This may also give a better indicator of the long-term capability for life support of topsoil. A published proposal (Lindeijer et al. 1998) works with the free net primary production (fNPP, or so-called *biomass appropriateness*). The fNPP is the whole NPP grown on the occupied area A, minus the part of NPP that is taken away from area A as an agricultural or silvicultural product. The reference situation may be the NPP of a natural surface in the same region or the fNPP previous to the land-use change registered in the LCI. If the annual fNPP, caused by a certain type of land use, is smaller than the reference fNPP of the same area, this indicates that (Lindeijer et al. 1998) soil fertility is reduced because of quantity or quality loss of topsoil, that less biomass enters the carbon cycle through natural subsystems, and that the decomposition rate of carbon compounds is lower. On the other hand, fNPP expresses other LSFs less adequately than NPP (Weidema and Lindeijer 2001), for instance, the hydrological cycle, the nitrogen fixation, the concentration of atmospheric dust, and the energy balance. Thus, both indicators, fNPP and NPP, have some arguments in their favour.

Soil quantity and the related quality may be used directly as indicators for topsoil quantity or quality, provided that it is possible to obtain the necessary data. Soil erosion, soil organic content, and/or soil compaction of an area A have been proposed as indicators in Cowell (1998), Baitz et al. (1998), and Mattsson et al. (1998). If the soil volume and the organic matter content are high and the soil compaction is low, it can be reasonably expected that soil fertility or the potential for food production is high and that the rainfall absorption potential of the ground is high. A somewhat similar proposal for an indicator representing the impacts on life-support system topsoil, resulting from a land transformation, consists of calculating the time needed to regenerate the soil that was lost as a consequence of the transformation (Mila i Canals 2000). This indicator is to be multiplied with the area A. Actually, soil organic matter is proposed as the primary entity from which the soil recovery time is calculated, but other properties of the soil could be included.

Exergy is another indicator for expressing the impact on LSFs, proposed recently by Muys et al. (2000). According to thermodynamic theories, ecosystems are open systems dissipating solar energy but also increasing their own exergy or capability to perform work. By measuring the ecosystem surface temperature with thermal infrared airborne imagery, we can expect to obtain the necessary physical data for a model that expresses land-use impacts on ecosystems. However, this approach is not yet feasible.

As in the preceding section on biodiversity-oriented proposals, it is not possible here to discuss the LSF-oriented proposals in more detail. Therefore, the current list of proposals to operationalise impacts on the LSFs is concentrated in Table 2-8. A more elaborate discussion on these proposed indicators (soil organic matter, NPP, fNPP, and others) will be required in the future, in order to arrive at a guidance on which indicators to select, or how to deal with a set of indicators for LSFs.

Impacts on the manmade environment

The degradation of landscapes, consisting of a web of natural surfaces, traditionally cultivated surfaces, and manmade buildings, due to a continuous change of land use, is considered a serious problem by many. However, there is a lack of proposals to

Table 2-8 Operationalisation characteristics for life support

Items according to Table 2-2	Properties of model
Baitz et al. 1998	
Essentials	Full quantification of 11 different LSFs, fully regionalised, comprehensive
Sensitivity	High for individual functions but less for a weighted score; area efficiency differences result in large range
Mechanism and model	Complete list of land functions, without quantification of relationship with endpoints. Weighting suggested for aggregation of land-use functions.
Extent of representation	Full for human functionality; functions for nonhuman less dominant
Choices or assumptions	Equal weights for land functions is acceptable; net transformations are neglected.
Consistency	Reasoning on integrating occupation area neglects net transformation.
Applicability	Depends heavily on geographical information system (GIS) or local data availability
Schweinle 2000	
Essentials	Full quantification of several LSFs, fully regionalised, comprehensive
Sensitivity	High for individual functions but less for total set of scores
Mechanism and model	Complete list of land functions, without quantification of relationship with endpoints. Area is missing as intervention.
Extent of representation	Full for human functionality; functions for nonhuman less dominant
Choices or assumptions	Equal weights for land functions is acceptable; net transformations are neglected.
Consistency	Reasoning on integrating occupation area neglects net transformation; area is missing in argumentation.
Applicability	Depends heavily on GIS or local data availability
Mattsson et al. 2000	
Essentials	Full quantification possible (but not expressed in formulas) for soil erosion, soil organic matter, and soil compaction; fully regionalised; comprehensive
Sensitivity	High for abovementioned functions, lower for total set of scores
Mechanism and model	Complete list of land function is discussed initially. Due to data gaps and importance of land functions, erosion, organic matter, and soil compaction were chosen as 3 very crucial indicators of land-use quality
Extent of representation	Full for human functionality
Choices or assumptions	Equal weights for land-use functions; no weighting between indicators. Net transformations are not included.
Consistency	Yes
Applicability	Depends heavily on local data availability
Lindeijer et al. 1998	
Essentials[a,b,c]	Full quantification ($\Sigma A_i \times t \times (\text{NPP}_{ref} - \text{fNPP}_{act})$ for occupation, $\Sigma A_i \times (\text{NPP}_{ini} - \text{fNPP}_{fin})$ for transformation), rough regionalisation
Sensitivity	Low; area efficiency is more dominant.
Mechanism and model	Not a relative score. By subtracting the harvested C, the focus is on the natural processes.
Extent of representation	fNPP assumed to be an indicator for C-cycling and soil quality
Choices or assumptions	Assumes linear relationship between fNPP difference and LSFs.
Consistency	Consistent with related bidiversity indicator
Applicability	More detailed fNPP data are hard to get.

continued

Table 2-8 *continued*

Items according to Table 2-2	Properties of model
Sas et al. 1997	
Essentials[a,b,c]	Full quantification ($\Sigma A_i \times t \times (NPP_{ref}\, t_{rec})$) for transformations, specification per species type
Sensitivity	Reasonable
Mechanism and model	Not a relative score. By adding recovery time for biomass, vulnerability of species is included.
Extent of representation	NPP assumed to be indicator for all LSFs
Choices or assumptions	Assumes a linear relationship between NPP, recovery time, and LSFs.
Consistency	Consistent with related biodiversity indicator, including reproduction time
Applicability	More detailed NPP data are hard to get.
Knoepfel 1995	
Essentials	Quantification based on 5 land-use classes, no regionalisation
Sensitivity	Low; land-use efficiency dominates
Mechanism and model	3 weighting schemes based on stability (NPP), vulnerability (recovery time), and expert opinions, respectively.
Extent of representation	3 different links to different endpoints
Choices or assumptions	Various
Consistency	No decision taken.
Applicability	Easy

[a] A = area occupied.
[b] t = time duration of occupation.
[c] NPP, fNPP indices: act=actual, fin=final, ini=initial, ref=reference.

solve it (see the attempts in Lindeijer et al. 1998). The impact on crop productivity from erosion or other damages to soil fertility has been quantified in terms of monetary yield losses in the Environmental Priority Strategies (EPS) system (Steen 1999) and more recently by Swan (2002).

Weighting within impact category land use

From the preceding sections, it appears that a reasonably complete assessment of the relevant impacts of land use leads to at least 2 or 3 subcategories of impacts and possibly to more than one CI per subcategory. Similar situations arise in the use of abiotic and biotic resources. The interpretation of an assessed inventory therefore tends to become almost as difficult as the interpretation of a simple nonassessed inventory. This would mean that the impact assessment methodology is of limited use in practice, unless an additional support tool is made available for a subsequent weighting process (across CIs within a subcategory, across subcategories of a category, or across categories).

Consequently, there is a practical interest in proposals for weighting. At least for the simplest case of several indicators inside a subcategory, which address the same AoP, weighting methods with a low level of arbitrariness should be feasible on the basis of currently available knowledge. But regardless of this perspective, the number of indicators should be kept as small as possible, also with respect to the required data assessments to operationalise land-use impacts in generic LCAs.

Amongst newer methods, only the Eco-indicator 99 (Goedkoop and Spriensma 1999) and the EPS system (Steen 1999) provide for integrating land-use impacts with other impacts on the same AoP. Within the Eco-indicator 99, land-use impacts are added to ecotoxicity impacts and acidification or nutrification impacts, based on the common principle of the PAF. This leads to the quantitative result that land uses dominate other impact categories as contributors to ecosystem damage, especially in the case of forestry or agriculture products. Within the default method of the EPS system, the contribution of land use (m²·y) to the threat of extinction of species (expressed in normalised extinction of species [NEX]) is multiplied with the (Swedish) expenses for preservation of species ($1{,}1 \times 10^{11}$ Euro for avoiding 1 NEX). Littering land (m²) is monetised by calculating the costs of picking up litter and the costs of severe nuisance. This monetisation allows for adding up all types of monetised impacts.

Conclusions

The survey and qualification of impact assessment methods for resources and land use have proven to be difficult because the underlying assessment framework is not yet adequately stabilised. Because of a lack of agreed principles, definitions, and objectives, the proposed methods vary to an extent that makes them hardly comparable, so that they cannot be arranged in order of preference. We see them rather as quarries, from which useful building stones can be mined as soon as the LCA community moves towards a certain agreement on principles, definitions, and objectives.

The major methodological contribution of this chapter is therefore the proposal for principles, definitions, distinctions, and objectives, in order to improve the LCA framework. The following points have been mentioned explicitly here:

- Distinctions between natural resources extractions and products from man-controlled cultures
- Distinctions between abiotic and biotic resources
- Enumeration and precise description of the clearly relevant impacts on AoPs, as caused by abiotic extractions, biotic extractions, and land use
- Discussion of the competition for land use and resource extraction by current users or extractors, versus the damage to future generations from degradation of stock
- Definition of occupation and transformation as basic cases of land-use interventions
- Definition of environmental quality and of quality for the intended use of a land area
- Description of impacts of occupation and transformation on the various endpoints
- Proposed principles for attributing land transformation to functional units and for selecting reference states for land use.

In the future, the need to develop comprehensive, best available methods based on the above framework is the most important research recommendation we can make. The need to harmonise the indicators that point from different impact categories to the same endpoint should be considered. Further study on the consequences of choosing certain reference states in land use, and different levels of detail therein, is another important research topic. Broader efforts to study the assessment of impacts to LSFs are required because this endpoint has received little attention until now, in spite of its high importance. Finally, developers of impact assessment methods should bear in mind that a reasonably comprehensive impact assessment of land and resource use leads to a considerable number of CIs; a certain degree of aggregation will be required, in order to maintain acceptable practicability of the tool.

References

Angelstam PK. 2000. Assessing biodiversity in boreal forest—The Swedish experience. Presentation at the COST E9 meeting; 2000 Mar; Espoo, FS.

Auhagen A. 1994. Wissenschaftliche Grundlagen zure Berechnung einer Ausgleichsausgabe. Berlin, D: Auhagen & Partners GmbH. im Auftrag der Senatsverwaltung für Stadtentwicklung und Umweltschutz Abt. III.

Baitz M, Kreibig J, Schöch C. 1998. Methode zur Integration der Naturraum-Inanpruchnahme in Ökobilanzen. Stuttgart, D: IKP Universität.

Baumann H, Evall T, Svennson G, Rydberg T, Tillmann AM. 1992. Aggregation and operative units. In: Anonymous. p. 87–94.

Blonk TJ, Lindeijer EW. 1995. Towards a methodology for taking physical degradation of ecosystems into account in LCA. Delft, NL: IVAM Environmental Research. DWW Publication Series Nr. 1995/15. (in Dutch)

Cowell S. 1998. Environmental LCA of agricultural systems: Integration into decision-making [PhD dissertation]. Surrey, GB: Univ Surrey, Guildford Centre for Environmental Strategy.

de Groot RS. 1992. Functions of nature. Groningen, NL: Wolters Noordhoff. ISBN 9001 35594 3.

Duelli P, Obrist MK. 1998. In search of the best correlates for local organismal biodiversity in cultivated areas. *Biodivers Conserv* 7(3):297–309.

[EUROSTAT] Statistical Office of the European Communities. 1995. Europe's environment: Statistical compendium for the Dobris assessment. Luxembourg, L: Office for Official Publications of the European Communities.

[FAO] Food and Agriculture Organization of the United Nations. 1997. Review of the state of world fishery resources: Marine fisheries. Rome, I: FAO. FAO Fisheries Circular 920 FIRM/C920(En). ISSN 0429-9329.

[FAO] Food and Agriculture Organization of the United Nations. 2000. Global forest resources assessment 2000: Main paper. Rome, I: FAO. FAO Forestry Paper 140.

Fava J, Consoli F, Denison R, Dickson K, Mohin T, Vigon B, editors. 1993. A conceptual framework for life-cycle impact analysis. Pensacola FL, USA: Society of Environmental Toxicology and Chemistry (SETAC).

Felten P, Glod S. 1995. Weiterentwicklung ökologischer Indikatoren für die Flächenbeanpruchung und für Lärmwirkungen und Aufwendung auf Logistik-Konzepte einer Firma, Semesterarbeit. Zürich, CH: Eidgenossichen Technischen Hochschule (ETH).

Finnveden G. 1996. Resources and related impact categories, Part II. In: Udo de Haes (see under references section 4). Society of Environmental Toxicology and Chemistry (SETAC) Europe.

Finnveden G, Ostlund P. 1997. Exergies of natural resources in LCA and other applications. *Energy* 22(9):923–931.

Goedkoop M, Spriensma R. 1999. The Eco-indicator 99: Methodology report and annex. Amersfoort, NL: PRé Consultants. http://pre.nl/eco-indicator99/ei99-reports.html. Accessed 17 Apr 2002.

Guinée J, Gorrée M, Heijungs R, Huppes G, Kleijn R, de Koning A, van Oers L, Sleeswijk AW, Suh S, Udo de Haes HA, de Bruijn H, van Duin R, Huijbregts MAJ. 2001. Life cycle assessment, an operational guide to the ISO standard. Guide and scientific backgrounds. Leiden, NL: Leiden Univ Centre of Environmental Science (CML).

Guinée J, Heijungs R. 1995. A proposal for the definition of resource equivalency factors for use in product LCA. *Environ Toxicol Chem* 14(5):917–925.

Heijungs R, Guinée J, Huppes G. 1997. Impact categories for natural resources and land use. Leiden, NL: Leiden Univ Centre of Environmental Science (CML). CML Report 138.

Heijungs R, Guinée J, Huppes G, Lankreijer RM, Udo de Haes HA, Wegener Sleeswijk A, Ansems AMM, Eggels PE, van Duin R, de Goede HP. 1992. Environmental life cycle assessment of products: Guide and backgrounds. Leiden, NL: Leiden Univ Centre of Environmental Science (CML).

[IPCC] Intergovernmental Panel on Climate Change. 2001. A report of Working Group I of the Intergovernmental Panel on Climate Change (summary for policymakers). Geneva, CH: IPCC.

[ISO] International Organization for Standardization. 1999. Environmental management—Life cycle assessment—Part 3: Life cycle impact assessment. ISO Standard 14042 (TC 207/SC5).

[IUCN] International Union for Conservation of Nature. 2001. 2000 IUCN red list of threatened species. http://redlist.org. Accessed 15 Mar 2002.

[IUCN/UNEP/WWF] International Union for Conservation of Nature, United Nations Environment Programme, World Wildlife Fund. 1980. Gland, CH: World Conservation Strategy.

Jolliet O, Crettaz P. 1996. Critical surface-time 95, a life cycle assessment methodology including fate and exposure. Lausanne, CH: Ecole Polytechnique Fédérale de Lausanne (EPFL).

Jungbluth N, Frischknecht R. 2001. Qualitätsrichtlinien ECOINVENT 2000: Landnutzung und -umwandlung. Uster, CH: Zentrum für Ökoinventare im ETH-Bereich. Arbeitspapier No. 5.0. http://www.esu-services.ch/ecoinvent2000. Accessed 15 Mar 2002.

Klöpffer W, Renner I. 1995. Methodology of impact assessment within the framework of LCA. CAU, D, February, also in Methodik der produktbezogenen Ökobilanzen -Wirkungsbilanz und Bewertung - Texte 23/95, UBA Forschungsbericht 101 01 102. Berlin, D:

Umweltbundesamt (UBA). UBA-FB 94-095, ISSN 0722-186X.

Knoepfel I. 1995. Indicatorensystem für die ökologische Bewertung des Transports von Energie [dissertation]. Zürich, CH: Eidgenossichen Technischen Hochschule (ETH). ETH Nr. 11146.

Köllner T. 2000. Species-Pool Effect Potentials (SPEP) as a yardstick to evaluate land use impacts on biodiversity. *J Clean Prod* 8:293–312.

Köllner T. 2001. Land use in product life cycles and its consequences for ecosystem quality [dissertation PhD 2519]. Universitat St. Gallen, CH. Bamberg, D: Difo-Druck GmbH.

Köllner T, Jungbluth N. 2000. Life cycle impact assessment for land use: Assessing the impacts on the regional plat species pool diversity due to food products. Presentation at SETAC World Congress; 2000 May 21–25; Brighton, GB.

Lindeijer EW. 2000. Review of land use impact methodologies. *J Clean Prod* 8:273–281.

Lindeijer EW, van Kampen M, Fraanje P, van Dobben H, Nabuurs GJ, Schouwenberg E, Prins D, Dankers N, Leopold M. 1998. Biodiversity and life support indicators for land use impacts in LCA. Delft, NL: RWS Dienst Wegen Waterbouwkunde (DWW). IVAM ER, RWS DWW Publication series Raw Materials 1998/07.

Marks R, editor. 1999. Anleitung zur Bewertung des Leistungsvermögens des Landschaftshaushaltes. Forschungen zur Landeskunde, Band 229. Trier, D: Selbstverlag.

Mattsson B, Cederberg C, Blix L. 2000. Agricultural land use in life cycle assessments (LCA): Case studies of three vegetable oil crops. *J Clean Prod* 8:283–292.

Mattsson B, Cederberg C, Ljung M. 1998. Principles for environmental assessment of land use in LCA. Göteborg, S: Swedish Institute for Food and Biotechnology (SIK). SIK Report 642.

Mila i Canals L, Domènech X, Rieradevall J. 2000. Soil recovery time as a characterisation factor for impacts due to land use. Presentation at SETAC World Congress; 2000 May 21–25; Brighton, GB.

Müller-Wenk R. 1978. Die ökologische Buchhaltung. Frankfurt, D: Campus-Verlag. ISBN 3-593-32250-1.

Müller-Wenk R. 1999. Depletion of abiotic resources weighted on the base of 'virtual' impacts of lower grade deposits used in future. St. Gallen, CH: Institut für Wirtschaft und Ökologie (IWÖ). IWÖ Discussion Paper Nr 57. http://www.iwoe.unisg.ch/org/iwo/web.nsf/wwwPubDiskussionGer?openview&count=9999. Accessed 29 Mar 2002.

Muys B, Wagendorp T, Coppin P. 2000. Ecosystem exergy as indicator of land use impact in LCA. Presentation at the COST E9 Meeting; 2000 Mar 27–29; Espoo, FS.

Niemelä J. 2000. To what extent can biodiversity of forests really be measured? Presentation at the COST E9 Meeting; 2000 Mar 27–29; Espoo, FS.

Odum EP. 1993. Ecology and our endangered life-support systems, 2nd ed. Sunderland MA, USA: Sinauer Associates. ISBN 0 87893 634 3.

Pedersen B. 1991. Hvad er et baeredygtigt ressourceforbrug? Lyngby, DK: Tvaerfagligt center, Danmarks Tekniske Hojskole (DTU).

[RIVM] National Institute of Public Health and the Environment. 1992. The environment in Europe: A global perspective. Bilthoven, NL: RIVM. RIVM Report 481505001.

Sala OE, Chapin III FS, Armesto JJ, Berlow R, Bloomfield J, Dirzo R, Huber-Sanwald E, Huenneke LF, Jackson RB, Kinzig A, Leemans R, Lodge D, Mooney HA, Oesterheld M, Poff NL. 2000. Global biodiversity scenarios for the year 2100. *Science* 287(March 10):1770ff.

Sas H. 1997. Extraction of biotic resources: Development of a methodology for incorporation in LCAs, with case studies on timber and fish. Zoetermeer, NL: Ministry of Housing, Spatial Planning and the Environment (VROM). Publication Series Product Policy Nr. 1997/30.

Schenck RC. 2001. Land use and biodiversity indicators for life cycle impact assessment. *Int J LCA* 6(2):114–117.

Schläpfer F, Schmid B. 1999. Expert estimates about effects of biodiversity on ecosystems processes and services. *Oikos* 84(2):346–352.

Schweinle J. 1998. Methoden zur Integration des Aspectes der Flächennutzung in der Ökobilanzierung. Hamburg, D: BFH. BFH Mitteilung Nr. 202.

Steen B. 1999. A systematic approach to environmental priority strategies in product development (EPS). Version 2000—Models and data of the default method. Environmental systems analysis. Göteborg, S: Chalmers Univ of Technology. CPM Report 1999:5.

Swan G. 2002. A top-down approach to land use impacts. Submitted for publication in *J Clean Prod*.

ten Brink B, Douma W. 1995. Biodiversity indicators for integrated environmental assessments at the regional and global level, a discussion paper. Bilthoven, NL: National Institute of Public Health and the Environment (RIVM). July draft.

Udo de Haes HA, editor. 1996. Towards a methodology for life-cycle impact assessment. Brussels, B: Society of Environmental Toxicology and Chemistry (SETAC) Europe. Report of the SETAC Europe Working Group on Life-Cycle Impact Assessment (WIA).

Udo de Haes HA, Jolliet O, Finnveden G, Hauschild M, Krewitt W, Müller-Wenk R. 1999. Best available practice regarding impact categories and category indicators in LCIA. *Int J LCA* 4(2):66–74.

[UNEP] United Nations Environment Programme. 2000. UNEP-World Conservation Monitoring Centre: Global biodiversity, earth's living resources in the 21st century. Cambridge, GB: UNEP.

van Ewijk H, Saft RJ, Lindeijer EW. 2000. Manual IVAM LCA data, draft update. Amsterdam, NL: IVAM Environmental Research.

Vogtländer J, Lindeijer EW, Witte J-PM, Hendriks CF. 2002. Characterizing land use changes on the basis of diversity and rarity of vascular plants. *J Clean Prod*. Forthcoming.

[WCMC] World Conservation Monitoring Centre. 1992. Global biodiversity: Status of the earth's living resources. London GB: WCMC. http://www.unep-wcmc.org. Accessed 15 Mar 2002.

Weidema B, Lindeijer E. 2001. Physical impacts of land use in product life cycle assessment. Final report of the EURENVIRON-LCAGAPS sub-project on land use. Lyngby, DK: Technical Univ of Denmark.

Weidema BP. 2000. Can resource depletion be omitted from environmental impact assessments? Poster presented at SETAC World Congress; 2000 May 21–25; Brighton, GB.

[WRI] World Resources Institute. 1999. World resources 1998–99. http://www.wri.org/wr-98-99/pdf/wr98_oc.pdf. Accessed 15 Mar 2002.

Climate Change, Stratospheric Ozone Depletion, Photooxidant Formation, Acidification, and Eutrophication

José Potting, Walter Klöpffer, Jyri Seppälä, Greg Norris, Mark Goedkoop

Abstract — In this chapter, we review best available practice regarding category indicators (CIs) and sets of concomitant characterisation factors for climate change, stratospheric ozone depletion, ground-level ozone formation or photooxidant formation, acidification, and aquatic and terrestrial eutrophication. We approach best available practice in terms of uncertainty management, and we take the perspective that an optimal balance should exist between the several areas of uncertainty in life-cycle impact assessment (LCIA).

The chapter starts with a general review of several issues relevant for the nonglobal impact categories covered here (such as models and model domains, substances covered, spatial aspects, and temporal aspects). Although we review the state-of-the-art science generally, with regard to life-cycle assessment (LCA), we review each impact category separately.

The chapter ends with general recommendations and with specific recommendations for each impact category. The majority of our recommendations focus on the accuracy of characterisation results. Such focus is appropriate when comparison among impact categories is not essential; spatially resolved characterisation factors derived from relatively sophisticated environmental models should be used in that case (usually when data uncertainty is relatively low). Such methods are also fine for applications such as benchmarking and monitoring. When comparison across impact categories is essential, damage-based characterisation factors might be an alternative (at the present state of the art, data uncertainties are high, but interpretation uncertainties are lower). Spatial differentiation and other measures limit uncertainties and should therefore be used as much as possible.

Introduction

In this chapter, we give a state-of-the-art review of best available practices regarding category indicators (CIs) and lists of concomitant characterisation factors for life-cycle impact assessment (LCIA) of climate change, stratospheric ozone depletion, ground-level ozone formation[1] or photooxidant formation, acidification, and terrestrial and aquatic eutrophication. Eutrophication, often also indicated by the term *nutrification*, is divided here into aquatic and terrestrial impact subcategories, although the categories have clear links because of atmospheric nitrogen deposition. The division is in line with the terminology in current environmental science, and it has been made because there are different consequences and cause–effect relationships within the impact categories.

The starting point in this chapter is taken from the background document of the Society of Environmental Toxicology and Chemistry (SETAC) Europe Second Working Group on Life-

---

[1] Ground-level ozone formation is a concern in itself, but can also be seen as an indicator for the broader range of impacts caused by the photochemically formed substances, together called (including ozone) *photooxidants*.

Cycle Impact Assessment (WIA-2; Udo de Haes et al. 1999) and the earlier reports from Udo de Haes (1996) and Nichols et al. (1996) written for the SETAC Europe first working group (WIA-1). The next section, 'General issues', highlights issues of general importance for the impact categories covered in this chapter (see also Potting and Klöpffer 2002). Each impact category is next discussed in its own section (see Klöpffer and Potting 2002 for further backgrounds): climate change, stratospheric ozone depletion, photooxidant formation, acidification, terrestrial eutrophication, and aquatic eutrophication. The chapter ends with general and category-specific conclusions and recommendations.

General Issues

Best available practice and uncertainties

Udo de Haes et al. (1999) gives direction for a best available practice that is based partly on requirements as formulated in ISO 14042 (International Organization for Standardization [ISO] 2000). The majority of these requirements aim to increase the relevance and accuracy with which potential impacts are quantified, with the help of characterisation factors to reflect the actual impact. The quality of CIs and related characterisation factors can thus also be understood in terms of the uncertainty and relevance in the projected impact.

All impact CI results described here contain some uncertainty. An in-depth discussion of areas and sources of uncertainty in LCIA is given in the background document on general issues (Potting and Klöpffer 2002). Two types of uncertainty are highlighted here (there are other important types of uncertainty in LCIA that are not emphasised in this section):

1) Model uncertainty — Characterisation factors are typically derived from more sophisticated environmental models that describe the fate, exposure, and effect of a substance. Such an environmental model is, of course, a simplification of a complex system, and choices must be made in the process of simplifying reality assumptions. This means that each model will have uncertainties. Typical examples of such uncertainties are these:
   - Time frame: Do we integrate climate change over 20, 100, 500, or an indefinite number of periods?
   - Simplifications: Fate and transport of emissions from a given region are treated as equivalent.
   - Limitations: Only the effects on plants are taken into account.

2) Data uncertainties — All environmental models rely on data that are uncertain to some extent. The sensitivity of humans or species to environmental changes is difficult to measure, for instance, because it is related to many other parameters. The mean atmospheric lifetimes of global warming gasses are not known with complete precision. The timing of an emission is an important source of variability in effect per unit emission, which may contribute to uncertainty in the average expected effects. And, as a final example, aquatic eutrophication has different effects in summer and winter.

Life-cycle impact assessment calculates CI results in order to give an overview of all relevant environmental issues related to a product system. Based on this environmental profile, a decision must be made about whether to require weighting and/or aggregation of indicator results. An important aspect is how uncertainty from values used in weighting may influence the robustness of this decision. Uncertainty from weighting can be avoided largely by defining CIs as damage to a limited number of the same endpoints. Damage indicators are typically expressed in the same unit, which facilitates aggregation. The resulting measures in terms of damage may also be easier to interpret. However, such indicators are usually more difficult to model, and model and data uncertainties may therefore be significantly higher (Potting 2002). There is a clear tradeoff between certainty and ease of interpretation in indicators (Bare et al. 2000).

It is important to understand that only data uncertainties can be expressed statistically, for example, by a standard deviation (SD). Other uncertainties cannot be expressed in such a way. There are several ways to assess these, but one of the most practical is sensitivity analysis. In a sensitivity analysis, different model assumptions can be applied and different CIs can be used for given impact categories.

Models and model domains

Characterisation factors for impact assessment are typically derived as metamodels from more sophisticated models outside LCA. The more sophisticated models should ideally cover the whole world, but they typically are implemented for continents only, as is the case of the several nonglobal impact categories cited here. Whereas these continental models are sophisticated in being spatially resolved, they appear to differ between continents with regard to the environmental mechanisms addressed (i.e., the extent to which they cover the causality chain up to the endpoint).[2]

The differences in available models between continents pose problems for deriving globally applicable characterisation factors because the processes that make up a product system may take place all over the world.[3] One solution to this problem is to choose the CI as early in the causality chain as required by the least-sophisticated continental model. Calculating characterisation factors for other continents could then be adapted accordingly, to arrive at a globally consistent set based on one and the same CI. Category indicators become less environmentally relevant when they are defined so early in the causality chain, however. It therefore seems better to define impact indicators and calculate characterisation factors as close to the endpoint as sophisticated models for a continent allow. The geographical centre of the processes in the product system under study, typically the region commissioning the LCA, will then determine CIs and choose characterisation factors.

Definition of the category indicator

The availability of a sophisticated model does not put an end to the discussion about the choice of a CI and related characterisation factors. Some available model sophistication can deliberately be put aside, and the same information (or model components) can be used in several ways to arrive at different types of CIs. The section on acidification clearly illustrates this with a review of the different types of acidification factors from Lindfors et al. (1998), from Potting et al. (1998a, 1998b) as recommended by Hauschild and Potting (2002), and from Huijbregts et al. (2000) as adopted by Guinée (2002). All of these authors used the same modelling framework to calculate characterisation factors, but each employed a definition of the CI that differs fundamentally from the others. Only Lindfors et al. (1998) took an average approach. All of the other CIs are based

---

[2] Thus, North American methods employ sophisticated emission inventories and atmospheric transport models but do not employ continent-wide effect models analogous to the critical loads in Europe. The European critical loads, together with sophisticated emission inventories and atmospheric transport models, are established and maintained under the several protocols negotiated in the context of the United Nations Economic Commission for Europe Convention on Long-Range Transboundary Air Pollution (UNECE CLRTAP). The differences in sophistication between continents may be due to the varying importance of these impact categories on given continents and to cultural differences in problem solving and environmental management.

[3] A product mainly manufactured and marketed in Europe may require importing materials directly (e.g., wool or soy) or indirectly (e.g., coke coal used in electricity production) from faraway continents.

on a marginal approach, but they are defined at different places in the causality chain. The chosen definition of the CI seems to be based on different values regarding environmental management (emission reduction as such, or impact reduction through emission reduction). The consequences of the several impact definitions are yet poorly understood, however.

The choice of an average or marginal approach is context and application dependent. A marginal approach is appropriate for a change-oriented LCA, while an average approach is seen as suitable for an information-oriented approach.

Most of the CIs mentioned in this chapter are operationalised by means of spatially resolved characterisation factors for acidification, terrestrial eutrophication, aquatic eutrophication, and photochemical ozone creation (see 'Substances covered', below). Such factors relate an emission in a given region to the impact on its receiving area.

Substances covered

Life-cycle impact assessment typically aims to characterise and aggregate the impact from all processes in a given product system. The principal pollutants in a product system may be of minor importance from a regional perspective (i.e., they may not be covered by the regional models to calculate characterisation factors from). For example, hydrogen chloride from waste incineration often dominates the total acidifying impact from polyvinylchloride (PVC) products, whereas it is negligible in total regional acidifying emissions that consist almost fully of nitrogen and sulphur compounds (the only substances covered by regional acidification models). However, if the acidifying contribution of hydrogen chloride is ignored, misplaced optimisations of PVC product systems could result. It is therefore important that characterisation factors be available for all potentially relevant substances. Approximations can often be made for substances not covered by regional models, as illustrated by Potting et al. (1998a, 1998b).

Spatial aspects

The relatively long atmospheric lifetime and the related transport over large distances of relevant pollutants are mutual characteristics of the impact categories discussed in this chapter. The long-range transport results in a large area being affected by an emission at a specified point. The gradients of exposure increase, and subsequent impacts from an emission at that point therefore strongly overlap with those from a geographically nearby emission. This results, for example, in rather similar acidification factors for the 3 small and neighbouring Benelux countries. As the distance becomes larger between 2 points of emissions, the gradients of exposure increase and the subsequent impact starts to deviate substantially. So, Dutch and Slovenian acidification factors will differ considerably. The required spatial resolution differs between the nonglobal impact categories in this chapter, but it is expected to be optimal for acidification, terrestrial eutrophication, and ground-level ozone formation or photooxidant formation for regions that are several hundred kilometres square. This resolution typically coincides with that of full countries, larger administrative regions, or federal states within large countries such as the U.S.

The values spanning the range of spatially resolved characterisation factors can differ by orders of magnitude (cf. Potting et al. 1998a, 1998b; Huijbregts et al. 2000; Krewitt et al. 2001; Bare et al. 2002; Hauschild and Potting 2002). The SD of and the range in site-dependent characterisation factors increase considerably as the indicator is defined closer to the endpoint (Potting 2002). The SD takes a value similar to that of the mean (which could be defined as the *site-generic characterisation factor*). Given the large uncertainty in site-generic characterisation, this chapter advocates a site-dependent approach over a site-generic one (this deviates from the preference stated by

Udo de Haes et al. [1999]). Several reasons[4] may exist to deviate from a site-dependent approach. All these reasons may be valid enough, and it is in the end up to the practitioner whether or not to perform site-dependent characterisation. However, these reasons then should be reported explicitly, and the accompanying spatially determined uncertainty must be accepted explicitly as well.

Site-dependent impact assessment is advocated here when the years in which and the geographical locations where an extraction or releases take place are known. However, LCA practitioners may lack information about geographical location or may deliberately choose to refrain from using it. Lack of this information can arise when emissions take place in an unknown and far future. Also, purchase of resources on the spot-market may cause uncertainty about the geographical location of processes (though an indication of annual average supplier composition is often available). European ecolabelling is often mentioned as an application that deliberately refrains from site-dependent assessment to avoid potential economic favouring of member states. In all these cases, site-generic characterisation factors should be available to perform an impact assessment. The uncertainty posed by refraining from site-dependent assessment should be given, however. Site-generic factors can be calculated as the emissions-weighted mean of the available site-dependent ones. However, Potting and Hauschild (2002) showed that the mean value depends highly on the selected regions on which the mean is based. The 'European mean' (44 regions) is very different from the 'EU15 mean', and both are very different from the 'East-European mean'. This once again underlines that site-generic impact assessment should be possible, but it is not advocated where site-dependent factors are available.

The above discussions about site dependency do not apply to climate change and stratospheric ozone depletion because the pollutants contributing to these global impact categories disperse globally and because no relationship exists between the geographical location of the emission and the resulting impact.

Temporal resolution

According to Udo de Haes et al. (1999) impact assessment in LCA not only integrates over space (see preceding section and last section), but also over time. It is not clear whether this integration relates to the residence times of the relevant substances or to the duration of their impact. The latter theoretically requires characterisation of first- to last-order impacts. This would come down to estimating the course of evolution. We, therefore, understand integration over time as the period over which the relevant substances directly exert their impact (i.e., their residence time in the relevant environmental compartments). This is in line with CIs and characterisation factors for global change (global warming potentials [GWPs]) and stratospheric ozone depletion (ozone depletion potentials [ODPs]). Integration over a shorter period is often taken for very long-lived substances (i.e., those contributing to climate change), however.

The choice of the time period is clearly a value choice that cannot be solved here. However, a long time integration should be chosen if future generations are considered important. A short time integration is possible if the basic assumption is that future generations will find new solutions.

---

[4] Such reasons may include the absence of spatially resolved characterisation factors for the relevant regions or the requirements of a given decision-making context (European ecolabelling is often mentioned in this context). The present lack of software supporting spatial differentiation may also hinder a site-dependent approach. Hauschild and Potting (2002) describe a procedure in which the large uncertainty (SD 100%) in the site-generic characterised impact is reduced step by step by a site-dependent characterisation of the dominant contributors. Another reason can be prioritising uncertainty reduction in weighting by choosing CIs close to the endpoint in order to express them in the same units and thereby facilitate their aggregation (spatially resolved endpoint characterisation is not possible with present state-of-the-art modelling).

Goedkoop and Spriensma (1999) developed alternatives for most impact categories, using different time perspectives.

Typical CIs and the alternatives for acidification, terrestrial and aquatic eutrophication, and ground-level ozone formation or photooxidant formation do not integrate over time. The revised critical load concept for acidification and terrestrial eutrophication will account for soil deterioration and restoration time as a result of prolonged exposure over the critical load. This would provide a way to integrate impacts over time.

The impact of an emission on its receiving area depends on the operative background exposure, if exceedance of a threshold is covered by the CI (Lindfors et al. 1998; Hauschild and Potting 2002). The operative background exposure is the sum of all sources in a large region[5] and therewith depends on total economic activity in that region. Potting et al. (1998a, 1998b) and Krewitt et al. (2001)[6] calculated site-dependent characterisation factors for actual 1990 European emissions levels, and they forecast 2010 emissions. The 1990 and 2010 characterisation factors differ considerably, as expected, on the basis of the large difference in economic activity between the 2 years. Nevertheless Potting et al. (1998a, 1998b) show the characterisation factors to be rather robust for moderate changes in the overall emission situation. This indicates that characterisation factors will not change dramatically between successive years when changes in economic activity are usually moderate. Potting (2000) therefore suggests, similarly to the site-dependent approach, that time-dependent characterisation factors relate to a time frame of several years. This would facilitate impact assessment for a product system with processes that take place relatively far in the future.

Interrelations between impact categories

Presently, LCA typically assesses the environmental performance of product systems in terms of their contribution to a number of environmental themes or impact categories. These themes are in many ways related directly or indirectly through damage to the same set of endpoints, as is exemplified here for terrestrial ecosystems.

Nitrogen changes the availability of nutrients as well as the acidity of the receiving soils, which both may cause ecosystems to shift in their species composition. However, the shift caused by nutrient enrichment may be different from the one caused by acidity. The same terrestrial systems are influenced by toxic tropospheric ozone levels (as presently occur in some regions), as well as by changes in temperature and precipitation, which may result from greenhouse gas-evoked changes in climate.

Interactions between impact categories currently are not covered in LCIAs that treat each impact category separately and usually define the CI earlier in the causality chain. This is largely a consequence of the state-of-the-art science not yet being able to jointly model all those impact categories at endpoint level in a reliable way. A few models integrate some impact categories, although at different levels in the causality chain. Goedkoop and Spriensma (1999) used the Simulation Model for Acidification's Regional Trends/Multistress Model voor de Vegetatie (SMART/MOVE) models of Latour et al. (1997) to compute characterisation factors for a combined acidification–eutrophica-

---

[5] To which each source contributes only little or marginally (except sometimes in the first 100 m from that source).

[6] They both define the CI as the change in area of ecosystems exposed above their critical loads, caused by the change in emission in a given region. Their characterisation factors thus basically represent the slope of the dose–response function (acidifying load versus unprotected ecosystem) in the point of the actual background load.

tion impact category at endpoint level.[7] The Regional Air Pollution Information Simulation (RAINS) model[8] was used by a number of LCA method developers to derive spatially resolved characterisation factors that have their CI defined at midpoint (Potting et al. 1998a, 1998b; Huijbregts et al. 2000). The model integrates acidification, terrestrial eutrophication, and ozone formation. These 3 problems are related to each other through the common stressor NO_X.

Climate Change

Description of the impact category

The physical basis of the impact category 'climate change' is the so-called *greenhouse effect* or, more specifically, the enhanced greenhouse effect attributable to human influence. The mechanism that causes the warming effect is called *radiative forcing* and consists essentially of infrared absorption in the spectral region between 10 and 15 µm, the spectral window of the atmosphere. The enhanced radiative forcing and thereby enhanced global warming can be seen as the primary effects caused by the increase of greenhouse gases in the atmosphere. Several secondary and tertiary effects have been identified, which may follow the primary radiative forcing and warming, such as climatic instabilities (e.g., an increase in the number and intensity of storms), increasing sea level, and changing of oceanic streams. Therefore, the more general category name *climate change* replaced the formerly used *greenhouse effect* and *global warming*.

Within the community of atmospheric scientists, there is consensus that the midpoint called *enhanced radiative forcing* (i.e., the absorption of infrared radiation in the spectral window from about 10 to 15 µm) is the common and global primary effect. This primary effect may cause several serious secondary and tertiary effects whose actual regional consequences seriously endanger the future of human society. Enhanced radiative forcing is linked to global warming, that is, the increase in average temperature near the surface of the earth. For this reason, it is advisable to select enhanced radiative forcing as the midpoint to be modelled and characterised as the indicator for the renamed impact category climate change.

The name *climate change*, which implies a broader definition, gives the option of defining further indicators closer to the endpoints. These indicators are based on tentative models about possible secondary and tertiary effects and do not have the same degree of scientific confidence as the midpoint indicator (which, in turn, cannot predict human health effects). The following discussion is based on the midpoint approach.

Relevant substances

The most important gases contributing to climate change are carbon dioxide (CO_2), methane (CH_4), nitrous oxide (N_2O), and synthetic, persistent, and

---

[7] The change in species diversity is computed on a 250 × 250-m grid for the Netherlands. Empirical data from Ellenberg (1992) is used for each plant species, which correlates their chance of occurrence with the combined acidity and nutrient availability. Because the model has a fate part that links the deposition of a sulphur or nitrogen emission to a change in these parameters, an overall assessment of the change in species diversity can be computed. A particularly difficult aspect is the incorporation of positive effects. Many species flourish if nutrient availability and acidity increase. However, these are useful species, not the desired species. To cope with this problem, the concept of target species must be used. Only species that were assumed to be desirable for a particular ecosystem on a grid cell are assessed.

[8] RAINS is used in and established under the UNECE convention on CLRTAP. Calculations with the RAINS model have been the basis for several protocols, among which is the recently adopted Göteborg protocol to abate acidification, terrestrial eutrophication, and ground-level ozone in Europe. The model estimates deposition of nitrogen and sulphur and formation of ozone spatially resolved over the full European domain by taking into account the pattern of regional emissions and spatial differences in meteorology and tropospheric chemistry. Next, the spatially resolved deposition patterns and ozone levels are compared with critical acidifying and eutrophying loads and critical ozone values (accumulated ozone over threshold [AOT]60 for human beings and AOT40 for natural environment).

volatile chemicals (e.g., tetrachloro- and tetrafluoromethane, sulfurhexafluoride, hydrofluoroalkanes). The GWPs of some compounds are several orders of magnitude (10^3 to 10^4) higher, compared to CO_2. Because the absolute amounts of these other compounds in the atmosphere are much lower, the main overall contribution to global warming is nevertheless due to CO_2. This is true from a global perspective; for certain product systems, other greenhouse gases may be more important.

State-of-the-art science

According to ISO 14042 (ISO 2000), the indicator model chosen for an impact category ideally should be based on scientific evidence and supported by an international organisation. The Intergovernmental Panel on Climate Change (IPCC), working under the auspices of the United Nations Environmental Programme (UNEP) and the World Meteorological Organisation (WMO), fulfils both conditions. The role of the IPCC is to periodically assess the scientific, technical, and socioeconomic information relevant to understanding the risk of climate change. IPCC does not perform new research, nor does it monitor climate-related data. Its assessments are based on state-of-the-art published and peer-reviewed scientific and technical literature. The reports published by IPCC are written and peer-reviewed in an open process by a panel consisting of several hundred leading experts from around the world. The models and the results obtained are based on the best scientific evidence available.

The models used by IPCC serve different purposes. One purpose is the prediction of future and the explanation of past temperature increases that are due to the anthropogenic greenhouse effect. The known development since the beginning of industrialisation is used to calibrate the models. A second purpose, which is of paramount importance for environmental policy, aims at proposing reduction rates and political and technical measures for the most important emissions of greenhouse gases. For this purpose, different scenarios are calculated, showing the temperature increase after a given time horizon. A third purpose, which is the most important for LCIA, is to calculate the relative contribution of the different gases at the basis of equal weight. The question to be answered here is this: How much more (or less) does 1 mass unit of gas A relative to 1 mass unit of gas B contribute to the global warming at a given time horizon?

The ratios calculated for different time horizons are GWPs normalised with regard to 1 mass unit of CO_2. For example, a GWP_i of 100 says that 1 kg of the substance i has the same global warming effect (at a given time horizon, e.g., 100 y) as does 100 kg CO_2.

The form chosen by IPCC for quantifying these ratios is ideally suited for the characterisation of climate change. It is actually even the other way round. Because the form given by IPCC is so well suited to LCIA, this form has been adopted as the general model for the characterisation of basically all midpoint indicators of the impact categories (Heijungs et al. 1992; ISO 2000).

Other indicators, based on tentative models about possible secondary and tertiary effects, do not have the same degree of scientific confidence as the midpoint indicator (which, in turn, cannot predict human health effects). IPCC deliberately employs several models in calculating those indicators as a way to explore the considerable uncertainties in modelling indicators closer to the endpoint than to the global warming midpoint indicator. Such an approach is possible but not practicable in LCIA. However, endpoint indicators do serve another purpose in LCIA because they facilitate aggregation of different impact categories (see 'Other aspects' p 73).

Category indicator and characterisation factors

Given the above discussion, the enhanced radiative forcing or global warming is still the obvious CI for climate change. Lists of GWP_i values have been published by IPCC (Houghton et al. 1996, 2001), by WMO (1999), and in the background document (Klöpffer and Potting 2002). It is immediately clear that a list of all known GWP_i, including those greenhouse gases quantified in the inventory (ISO 14041 [ISO 1998]), allows the aggregation of the masses per functional unit into one figure, the GWP.

The formula for calculating the indicator result from the inventory data classified for the impact category climate change is given in Equation 3-1:

$$GWP = \Sigma_i (m_i \times GWP_i) \qquad (3\text{-}1),$$

where GWP is the indicator result for the impact category climate change (kg CO_2 equivalents per functional unit), m_i is the mass of greenhouse gases i assigned to the impact category climate change during classification (kg i per functional unit), and GWP_i is the global warming potential of gas i (kg CO_2 equivalents per kg i) for a time horizon of 100 years, if not otherwise requested.

Nitrogen oxides (NO_X) and volatile organic compounds (VOCs) contribute indirectly to climate change through the potential forming of tropospheric ozone. Their GWP_i is very uncertain, however. They are therefore rarely taken into account in LCIA.

Spatial aspects

Because the average tropospheric lifetime of all greenhouse gases, even of the relatively short-lived ones (CH_4, about 10 y), exceeds the tropospheric mixing time (about 1 y), it is not important where the emissions occur. Climate change is therefore truly a global impact category. The secondary and tertiary effects modelled in endpoint models may vary in space, but the GWP midpoint indicator does not.

Temporal aspects

The time horizon chosen for the calculation influences the GWP values significantly only in the case of relatively short lifetimes (up to about 100 y). The typical uncertainty of the GWP_i of individual greenhouse gases i is in general about ±35%. Since GWP_i of individual gases are relative numbers, this uncertainty does not contain the absolute uncertainty of climate modelling of CO_2. However, this is of no concern to LCIA because only relative values are necessary in the indicator model chosen.

A long time horizon should be chosen in LCIA (Udo de Haes et al. 1999) in order to take into account possible negative effects for coming generations. For this reason, the longest horizon used in the GWP calculations (500 y) would be appropriate. However, scenarios defined today are very unlikely to hold true for the future and seem to be most accurate for the present. It may therefore be a good compromise to use GWP data calculated for a time horizon of 100 years, as has been done in most LCIAs in the past.

Other aspects

The problem of damage modelling of climate change is that the impact category has many different effects on human health and ecosystems. Some, like the increase of agricultural production in at least some areas, can even be considered positive. Other problems are the large uncertainty in the climate models to date and the possible difficulties in predicting consequences such as a change in ocean currents or the release of huge amounts of methane from the Russian and Canadian tundras. IPCC deliberately employs several models as a way to explore the considerable uncertainties in modelling these consequences. Such an approach is possible but not practicable in LCIA.

Because of these difficulties, the use of such endpoint models in LCA is not without danger. However, if the midpoint level is used, the stakeholder or the panel that needs to weight or assess the midpoints has exactly the same problems as the modellers who developed the endpoint model. The endpoint models add all available (but uncertain) information, which the midpoint models usually lack.

Recommendations

We recommend that LCA practitioners characterise climate change according to Equation 3-1 with help from the most recent list of GWP_i values published by IPCC. The present GWP_i values for indirect greenhouse gases are very uncertain (see 'State-of-the-art science', p 72). We therefore advise characterising the climate change effect from direct greenhouse gases only until more reliable GWP_i values are available for indirect greenhouse gases.

Researchers in the field of climate change should provide reliable GWP_i values for all chemicals that are persistent in the troposphere. This asks for closer research on the effects of indirect greenhouse gases (including particles) and modified GWP_i values where necessary.

We recommend that LCA researchers investigate the effect of changing time horizons in real-life LCAs. They also should consider requirements for the life-cycle inventory (LCI) because some greenhouse gases are not generally contained in present-day inventories (e.g., CH_4 from anaerobic processes).

The characterisation of climate change should be developed further and adjusted to the rapid scientific progress in this area. A promising development is seen in damage factors proposed by Goedkoop and Spriensma (1999) and Steen (1999) to characterise climate change at the endpoints. However, such modelling is as yet very uncertain, and only partially complete. For this reason, we recommend using midpoint modelling in this impact category.

Stratospheric Ozone Depletion

Description of the impact category

The environmental concern about the impact category 'stratospheric ozone depletion' is based on the ultraviolet (UV) absorption capacity of the ozone present in the stratosphere. The UV absorption hinders radiation below 300 nm from reaching the troposphere and the surface of the earth. The ozone is present in the stratosphere in very low concentration, but the layer thickness to be passed by the photons is very large (about 25 km). The absorption of short-wavelength radiation is therefore complete despite the small concentration of ozone (O_3). This absorption capacity, among other beneficial properties of the stratospheric ozone layer, is at stake if ozone is depleted by anthropogenic emissions.

As early as 1970, the possible depletion of the ozone layer by supersonic jets emitting NO_X during flights in the lower stratosphere was discussed. An extension of this work to the ClO_X cycle, which was new at that time, was presented by Molina and Rowland (1974) and Rowland and Molina (1975). In this work, a possible connection between the emission of chlorofluorocarbons (CFCs) and the postulated ozone depletion was outlined and supported with reasonable data and assumptions.

In addition to this mechanism of homogeneous catalysis, which leads to a more or less equal and slow depletion of stratospheric ozone around the globe, the so-called *ozone hole* was detected over Antarctica in 1985 (Farman et al. 1985) and later was found to be due to heterogeneous catalysis. It is important to know that this effect was not predicted by Roland and Molina's theory, although the chemicals causing it are the same. This shows the importance of the precautionary principle and our limited knowledge of complex reactions in the environment.

Relevant substances

The relevant gaseous emissions, quantified as mass per functional unit in the inventory, originate from many human activities.[9] Most of these uses have in the past been performed with CFCs and similar chlorinated solvents. The majority of these chemicals are now forbidden in industrialised countries as a consequence of the Protocol of Montreal (1987) and the subsequent adjustments and amendments of London (1990), Copenhagen (1992), Vienna (1995), and again Montreal (1997) (WMO 1999).

Nitrogen oxides entering into the stratosphere (e.g., N_2O) may also deplete stratospheric ozone. No simple quantitative relationship exists at present, however, which could be used as an indicator model in LCIA.

State-of-the-art science

As in the case of climate change, modelling plays a major role in predicting the further development of the ozone layer as a function of the further development of the critical emissions and in identifying and quantifying the contributions of the individual substances that cause the adverse effects. The condition of acceptance by an international scientific organisation is fulfilled by the Global Ozone Research and Monitoring Project of the WMO, UNEP, and other national and international bodies (WMO 1991, 1999). The scientific evidence accumulated within this program endorses the causal relationships outlined above, that is, the halogen input by manmade persistent chlorine- and bromine-containing chemicals and the catalytic destruction of stratospheric ozone.

Two basic models were developed in order to quantify the ozone depletion capacity of chemicals (WMO 1991) that are applicable to LCIA: the chlorine loading potential (CLP) model and the ODP model.

The CLP is the simplest model and considers only the tropospheric lifetimes of the compound (relative to trichlorofluoromethane [CFC-11]), the molar mass, and the number of chlorine atoms in the molecule considered. Since bromine is also — and even more — effective in degrading ozone, a bromine loading potential (BLP) was defined in an analogous manner. The ozone depletion efficiency in the stratosphere depends not only on the factors included in the calculation of CLP and BLP, however, but also on the stratospheric lifetime, which is controlled mainly by photolysis and not by the OH reaction that dominates the tropospheric degradation of organic chemicals. Hence, the ODP has been defined as a relative measure of the ozone depletion capacity, which avoids the deficiencies of CLP and BLP and allows the description of chlorine- and bromine-containing molecules in one parameter.

The ODP is, in analogy to the older CLP and BLP, a relative number and uses the ozone depletion capacity of CFC-11 as a reference. The ODP is defined (WMO 1991) by Equation 3-2:

$$ODP_i = (\text{global } \Delta O_3 \text{ due to } i) / (\text{global } \Delta O_3 \text{ due to CFC-11}) \quad (3\text{-}2),$$

where ODP_i is the ozone depletion potential of compound i.

ODP is ideally suited as a midpoint indicator. Because some effects of increased short-wavelength UV radiation are well known, especially with regard to skin cancer formation, this knowledge can be used for endpoint modelling (See Itsubo 2000). Possible adverse effects towards the terres-

---

[9] This concerns uses as aerosol sprays, polymer foam production (e.g., polyurethanes), cooling agents for refrigerators and small air conditioners (cars), cleaning agents (e.g., in the electronics industry), smaller applications in medicine (asthma sprays), analytical chemistry (extraction agents, solvents for IR-spectroscopy), fire extinguishing (halons), agriculture, and pesticides (CH_3Br).

trial and marine ecosystems are difficult to estimate, however. It is argued, therefore, that the midpoint modelling, taking ozone depletion as the indicator, better takes into account the precautionary principle, that is, our ignorance about the causal relationships between increased short-wavelength UV radiation and ecosystem damages.

Category indicator and characterisation factors

Ozone depletion potential data have been published by WMO (1991, 1999) and in the background paper (Klöpffer and Potting 2002). The formula for calculating the indicator result from the inventory data classified for the impact category stratospheric ozone depletion is given as

$$\text{ODP} = \Sigma_i \, (m_i \times \text{ODP}_i) \qquad (3\text{-}3),$$

where ODP is the indicator result for the impact category stratospheric ozone depletion (kg CFC-11 equivalents per functional unit), m_i is the mass of ozone-depleting gas i assigned to the impact category stratospheric ozone depletion during classification (kg i per functional unit), and ODP_i is the ozone depletion potential of gas i (kg CFC-11 equivalents per kg i) for steady state, if not requested otherwise.

The highest values of ODP_i are those of the halons because of the nearly 10-fold catalytic activity of bromine, compared to chlorine. The other perhalogenated compounds (Cl, F) are in the range of ODP_i 0.5 to 1.1. Compounds containing H atoms that can react with OH in the troposphere, and are therefore less persistent, show much smaller ODP_i values. The ODP_i values of compounds containing only F as halogen are 0 by definition, since F does not catalyse the ozone destruction. Iodine acts as a catalyst, but I-containing compounds have a very short tropospheric lifetime (hours to a few days [WMO 1999]) due to photolysis, and therefore only a very small fraction enters the stratosphere. Several atmospheric models were used for the calculation of ODP values, resulting in similar but not identical ODPs. The uncertainty is in the range of 20% to 50%.

Spatial aspects

Because of the long lifetime of most ozone-depleting gases, a good mixing in the troposphere can be expected. Therefore, as an approximation for LCIA, no regional dependence of emissions has to be taken into account. This is also true if secondary effects (e.g., UV-B exposure of humans) should be considered. Whereas the effects are spatially differentiated (e.g., the ozone hole), the responsible agents still distribute globally after emission, and the spatial differentiated effects thus cannot be traced back to a specific region of emission. In the case of quantifying secondary effects, other indicators must be defined.

Temporal aspects

The temporal effects of ozone depletion are less straightforward. The question of time dependency is discussed in a WMO (1994) document. According to this discussion, ODP (stationary) values of such relatively short-lived compounds as hydrochlorofluorocarbons (HCFCs) are small because they are derived with regard to CFC-11, whose steady state — assuming constant emissions — will be reached in centuries. The short-term impact of less persistent ozone-depleting substances is therefore underestimated. If ODP is calculated for a short time horizon of a few years, the ODP of relatively short-lived compounds may be higher by an order of magnitude (but still smaller than that of CFC-11 because of their shorter tropospheric lifetimes).

Other aspects

The common primary effect of stratospheric ozone depletion is the increased ozone destruction in itself. Increased UV-B radiation at the surface of the earth is already a secondary effect, and not the only one. Because absorption of solar radiation (not only UV) by stratospheric ozone contributes significantly to the warming of the stratosphere (the tropopause being the thermocline between the stratosphere and the colder upper troposphere), ozone depletion may also cool the stratosphere and possibly change the stratification of the atmosphere. Possible tertiary effects from this are not calculable and belong in the topic of climate change.

If the ODP is to be expressed in damages, warnings similar to those for climate change apply. Modelling the damage to ecosystems is particularly difficult, as there are indicators that increased UV radiation has some positive effects on parts of ecosystems, while it is damaging for other parts. The effects on human health are not extremely difficult to model because the number of diseases is relatively limited.

Recommendations

We recommend that LCA practitioners characterise stratospheric ozone depletion according to Equation 3-3. The ODP_i values should be taken from the most recent WMO (1999) report. We propose that practitioners use an infinite time horizon (steady-state model), if not required otherwise, in the goal definition of the LCA. The model used for the calculation should be given if time-dependent data are used.

We urge atmospheric or environmental researchers to investigate the effects leading to ozone depletion by other mechanisms, especially by nitrogen oxides. We encourage them to provide ODP_i values for N_2O and NO_X (for supersonic aircraft) to be used in addition to the established values of the halogen-containing gases. Alternatively, a subcategory and an indicator dealing with these effects may be created.

Photooxidant Formation

Description of the impact category

The photochemical smog, also known as *Los Angeles smog*, has been known for about 50 years. Its popular name is derived from the air quality problems in the metropolitan area of Los Angeles, California, USA. Those problems are due to the high density of car traffic in the area, in combination with the strong solar irradiance and a high frequency of meteorological situations that inhibit the exchange of air. These factors form the basis for a sequence of chemical reactions in the lower troposphere, which leads to the formation of ozone and other reactive toxic and ecotoxic reaction products. The mixture of products formed in this photochemical oxidation process, including ozone and peroxyacetyl nitrate (PAN), is called *photooxidants*, hence the name of this impact category.

The reaction sequence leading to ground-level ozone and the other photooxidants depends on the presence of nitrogen oxides (NO_X), OH-reactive hydrocarbons (HCs) and CO, and solar radiation in the visible and near-UV spectral region (Finlayson-Pitts and Pitts 1986). The reaction can be modelled in computer programmes that use several hundred chemical and photochemical reactions characterised by rate constants, quantum efficiencies, and meteorological conditions. These, however, vary strongly in time and space (as do the emissions that cause the smog) so that photooxidant formation differs basically from the impact categories of climate change and stratospheric ozone depletion. One molecule of NO_X, CO, or ethene may have quite different consequences with regard to smog formation, depending on the site and time of emission.

Photooxidants do not occur only in typical smog events; there is evidence of a general increase in tropospheric ozone concentrations, especially in the Northern Hemisphere. The pre-industrial level of about 10 ppbv has increased to between 30 and

50 ppbv, now found globally in the lower atmosphere (Finlayson-Pitts and Pitts 1993). Peak values observed in severe smog events amount to mixing ratios of 450 to 500 ppbv, clearly above the World Health Organization (WHO) published guideline values of 100 to 120 ppbv (8-h average; 1 ppbv ≈ 2 µg/m^3 concentration at ground level).

The positioning and defining of the indicator for this impact category are difficult because the effects range from local to regional and from acute toxicity in humans during smog episodes to chronic ecotoxic effects such as a possible contribution to forest die-back. The general increase in ozone concentrations has not yet reached a level that poses a concern with regard to known human toxic effects.

Ozone levels experienced at a particular location are difficult to estimate because they are influenced by the following (Syri et al. 2001):
1) Hemispheric concentrations of ozone in the free troposphere (resulting from emissions in the Northern Hemisphere)
2) Ozone generated by long-range transport of the precursor emissions over a scale of several hundreds to thousands of kilometres
3) Locally increased ozone production downwind of sources of precursor emissions (in sunny weather)
4) Local destruction of ozone (titration) due to nearby NO_X emissions (particularly important at sites close to high NO_X emissions, i.e., urban areas)
5) Deposition of ozone to the ground.

Thus, mechanisms of ozone formation at the regional scale differ from those at the local scale. Ozone levels are usually somewhat lower in urban areas than in the surrounding countryside, since ozone is consumed locally by vehicle emissions of nitric oxide (NO) (Pleijel 1999). On the other hand, some cities suffer from high ozone concentrations within photochemical smog. This phenomenon is found particularly in inland basins, and in coastal areas where air pollutants may be contained in a land–sea breeze circulation system, as in Barcelona, Spain, or Athens, Greece (European Environmental Agency 1995).

Unlike the 2 truly global categories, ozone precursors emitted in different geographical regions or at different times may have either a strong effect or none at all. That means that tropospheric chemistry will more or less transform precursors in ozone and other photooxidants, and the number of possible adverse effects will be very different between regions because of spatial differences. This is elaborated in 'Spatial aspects and site-dependent characterisation factors', p 80.

Relevant substances

The relevant gaseous emissions, quantified as mass per functional unit in the LCI, originate from many human activities, for example,
1) traffic by cars, trucks, ships, and aeroplanes (VOC, NO_X, and CO);
2) use of solvents in industry (VOC);
3) energy-related processes based on fossil fuels (NO_X, CO); and
4) heating of homes.

The most important single source for local smog events, and the most likely to cause adverse effects to human health, is road traffic.

State-of-the-art science

The earliest approach to classifying reactive HCs with regard to their ozone-forming potential is the one by Darnall et al. (1976). These authors defined groups of different reactivity (reactivity scale) based on k_{OH}, the most OH-reactive substances belonging to the most efficient ozone-forming HCs and vice versa. This is only partly true, however, because secondary processes also contribute to the overall ozone efficiency of a particular compound.

More advanced methods try to calculate the contribution of individual compounds to the formation of ozone (or PAN). The approach developed in the U.S. is based on incremental reactivities (IRs), as defined in Finlayson-Pitts and Pitts (1993):

$$IR = \Delta O_3/\Delta C \text{ atom of VOC added} \qquad (3\text{-}4).$$

This index, in contrast to the k_{OH}-reactivity scale, may even be negative if the secondary reactions hinder the development of the smog, for example, for benzaldehyde that removes NO_X without radical formation. The IR values still depend on the special characteristics of a smog event and thus cannot be used for a general ranking of VOC with regard to ozone formation. Therefore, a peak IR value can be defined, called *maximum incremental reactivity* (MIR, mg O_3 formed per mg VOC added). MIR values can be converted to relative values, taking the MIR_i of one characteristic compound equal to 1. This would be in complete accordance with the procedure observed in aggregating the indicator results in the global impact categories (GWP and ODP) and in photochemical ozone creation potential (POCP) (Equation 3-4).

The European counterpart of MIR is the POCP according to Derwent and Jenkins (1991) and Derwent et al. (1996, 1998). These relative reactivities have been calculated for ozone formation in an air parcel travelling through western Europe, that is, in a moderate climate and without peak ozone formation (smog formation in the local sense). Relative $POCP_i$ values were derived from the model calculations. The values given by Derwent and Jenkins (1991) were proposed for the characterisation step of LCIA by Heijungs et al. (1992) and subsequently used in many LCAs. The photochemical ozone creation of ethylene, a strong ozone-forming compound, was defined as the reference ($POCP_{ethene}$ (1 kg) = 1). The $POCP_i$ value for a given HC assesses its ability to form ozone relative to ethylene for an identical atmospheric mass emission.

A drawback of both MIR and early POCP data is the restriction to organic compounds as the only candidates for assessing the photooxidant formation. The assumption that composition and amount of the VOCs are the only chemical smog-controlling factors is misleading, however. In addition to VOC + CO, NO_X is the second controlling factor both in polluted and in rural or background atmospheres (Finlayson-Pitts and Pitts 1993). A minimum requirement for the indicator chosen is therefore the inclusion of both VOC + CO and NO_X. This requirement is fulfilled by recent recalculations of $POCP_i$ values by Derwent et al. (1996, 1998), including values for NO_2, CO, and SO_2. Derwent et al. (1996) also checked for the robustness of POCP values to changes in the NO emission densities across Europe, accounting for the general increase of ozone-formation efficiency of VOCs if background NO_X increases. It is not clear at the moment whether MIR values can be adjusted to NO_X.

Ideally, the inventory should list individual VOCs, so that the different reactivities can be taken into account. Unfortunately, emissions of HCs and other VOCs are often only poorly defined and given as the sum of VOCs, the sum of HCs, or the sum of nonmethane hydrocarbons (NMHCs) or organic gases (NMOGs). NO_X is part of each inventory, but this information has not been used in most LCIAs that have been published.

Category indicator and site-generic characterisation factors

A list of selected recent $POCP_i$ values (ethylene equivalents) and a list of MIR_i values (recalculated relative to ethylene, so that the comparability with POCP is given) is shown in the background document (Klöpffer and Potting 2002). The total range of values does not exceed 2 to 3 orders of magnitude, and most of the reactive compounds are even in a range of about 0.1 to 1 (POCP) and 0.1 to 1.5 (MIR) ethylene equivalents. This means that the exact composition of the mixtures 'VOC', etc., is not very important for the results.[10] However, ozone precursors emitted in different geographical regions or at different times may have either a strong effect or none at all on ozone formation and subsequent effects (see the next section, 'Spatial aspects and site-dependent characterisation factors').

The formula used for calculating the indicator result from the inventory data classified for the impact category 'photooxidant formation' is given in Equation 3-5.

$$POCP_{reg} = \sum_i (m_i \times POCP_i) \qquad (3\text{-}5),$$

where $POCP_{reg}$ is the photooxidant formation potential (regional, e.g., western Europe) and m_i is the mass of ozone-forming gas i assigned (classified) to the impact category photooxidant formation during classification (kg i per functional unit).

Alternatively, the relative MIR_i values may be used if local ozone formation is defined as the indicator in the goal definition phase:

$$POCP_{loc} = \sum_i (m_i \times MIR_i) \qquad (3\text{-}6),$$

where $POCP_{loc}$ is the photooxidant formation potential, local (e.g., a metropolitan area of the Los Angeles type).

Choosing the indicator in the same way as for the global categories requires a stricter interpretation of the precautionary principle because the adverse effect from precursors released in one region may be larger than that released from precursors in another region. In a precautionary approach, therefore, the most severe situations should be prioritised above others to be improved because general pollution prevention would lead to suboptimisation. Such spatially resolved assessment would require an indicator differentiated towards human beings and the natural environment. This is further elaborated in the following section.

Spatial aspects and site-dependent characterisation factors

The extent to which precursors lead to adverse effects on human beings and the natural environment depends on spatial differences in meteorological conditions, source density and resulting background concentrations of relevant precursors, and in receptor density. The effect from precursors released in one region may therefore be larger than the effect from precursors released in another region. The RAINS model, as used in and established under the UNECE CLRTAP, estimates ozone formation spatially resolved over the full European domain by taking into account the pattern of regional emissions and spatial differences in meteorology and tropospheric chemistry. The spatially resolved ozone levels are compared with critical ozone values (accumulated ozone over threshold [AOT]60 for human beings and AOT40 for the natural environment). The area where critical values are exceeded is multiplied with receptor density to arrive at the total number of receptors exposed above critical ozone levels. Following the approach of Potting (1998a, 1998b) for spatially resolved acidification factors, the RAINS model is used by Hauschild and Potting (2002) to arrive at simple factors that relate the emission of a given ozone precursor in a given region to its effect on the full impact area. Spatially resolved factors are available to calculate the total number of human beings exposed above specified levels. Similar factors are available to quantify ecosystems exposed above critical levels. Likewise, for the U.S., Norris (Bare et al. 2002) combined modelling of atmospheric reactions and transport with population density information to arrive at spatially resolved factors analogous to those of Hauschild et al. (see Bare et al. 2002).

These spatially resolved factors can be used to modify the ozone formation as calculated with

---

[10] Methane should not be included in this basket used for defining the VOC, however, because this compound is, owing to its small reactivity toward OH, much less reactive according to both models.

help of the POCPs into its actual effects on human beings and the natural environment. The results show that the uncertainty in the calculated effect by refraining from spatially resolved modelling of the actual effect (factor 100 between smallest and largest factor) is far larger than the uncertainty from ignoring the ozone-forming potential of different precursors (factor 2 difference for majority of VOCs). The results also show that NO_X as a precursor of ozone and its effects is far more important than VOCs. This underlines the importance in LCA of taking NO_X into account as a precursor of ozone and the other photooxidants.

Temporal aspects

The formation of ozone follows seasonal and even diurnal variations. Under conditions of high isolation and low winds, concentrations of ozone and photochemical oxidants can build up to high levels. Usually such situations take several hours to develop (European Environmental Agency 1995). In northern Europe, ozone episodes occur during most springs and summers, though to a highly variable extent. An episode is a relatively short period, anything from a few hours to a few days, of greatly elevated ozone concentrations, sometimes in excess of 100 ppb (Pleijel 1999).

Other aspects

The general increase in ozone concentrations has not yet reached a level that poses a concern with regard to known human toxic effects. Tropospheric ozone is a greenhouse gas, however, and interferes with the tropospheric chemistry in a complex manner (Finlayson-Pitts and Pitts 1986, 1993). It may be considered, therefore, as a global effect, although not yet for human health outside densely populated areas.

If a damage approach is required, it is important to incorporate a unit of damage that includes nonfatal health effects. Very few people die from summer smog, but many short-term hospital admissions result from it. The work of Steen (1999) and Goedkoop and Spriensma (1999) seems to suggest that the damage caused by smog can be about 2 orders of magnitude lower than that of other impact categories for many LCAs. This means it possibly gets more attention than it deserves. The effects on ecosystems have so far not been modelled very precisely.

Currently, the integrated assessment modelling of ozone formation in Europe uses the results of large-scale models intended for estimating rural background ozone levels. For this reason, modelling of population exposure to high ozone concentrations is subject to many uncertainties because small-scale phenomena in urban areas can significantly change ozone levels from those of their surroundings (Syri et al. 2001).

Recommendations

We recommend that LCA practitioners use the most recent set of $POCP_i$ or MIR_i values, depending on the aim and scope of the study, considering also the main geographical locations covered by the study. We strongly advise them to include NO_X in the calculation of POCP and MIR values. Depending on the aim and scope of the study, we recommend using the spatially resolved modifiers from Hauschild and Potting (2002), Krewitt et al. (2001), or Norris (Bare et al. 2002) to arrive at a more accurate estimate of ozone formation and its actual effect. However, there is a special need for research and development work in the field of accurate modelling of the small-scale and short-term effects posing risk to human health.

We further recommend that researchers should check the relevance of MIR values for European (and other) conditions, as well as the relevance of POCP values for North American (and other) conditions. Results obtained with both reactivity scales in real-life LCAs should be compared. We advise reconsidering the influence of different definitions and measurement practices of sum parameters such as VOC, HC, and NM-VOC.

Also, inclusion of increased ozone (local) in the impact category 'human health' and inclusion of increased ozone (regional to global) in the impact category climate change should be considered. Spatially resolved characterisation factors for regions other than those in Europe and the U.S. should be made available (i.e., a global set of spatially resolved characterisation factors).

Acidification

Description of the impact category

Acidification refers literally to processes that increase the acidity (hydrogen ion concentration) of water and soil systems. The common mechanism for acidification is deposition of negatively charged ions (anions) that are then removed by leaching or biochemical processes, leaving excess (positive) hydrogen ion (H^+) concentrations in the system.

Emissions of potentially acidifying substances lead to deposition (through a complex set of atmospheric transport and chemistry processes), which in turn can lead to damages to plant and animal populations (through a complex set of chemical and ecological processes). Deposition occurs through 3 routes:
1) Wet (rain, snow, sleet, etc.)
2) Dry (direct deposition of particles and gasses onto leaves, soil, surface water, etc.)
3) Cloud water deposition (from cloud and fog droplets onto leaves, soil, etc.).

Acidification of soils can promote leaching of nutrients, which reduces forest and plant health, and leaching of toxic aluminium, which leads to ecotoxicological impacts. Further, increases in lake and stream acidity and loss of acid-neutralising capacity are associated with loss of aquatic life.

Relevant substances

From a regional perspective, the major acidifying emissions are oxides of nitrogen (NO_X) and of sulphur (SO_2) and ammonia emissions (NH_3), which lead to nitrogen and sulphate depositions. However, LCIA typically aims to characterise and aggregate the impact from all processes taking place in a given product system, and the above pollutants may be minor ones in a given product system. For example, hydrogen chloride from waste incineration often dominates the total acidifying impact from PVC products, whereas it is negligible in total regional acidifying emissions that almost fully consist of nitrogen and sulphur compounds. Ignoring the acidifying contribution of hydrogen chloride could lead to misplaced optimisations in the case of PVC product systems. It is therefore important that characterisation factors cover the full range of substances potentially relevant from a product perspective. This condition is fulfilled by the 'old' hydrogen release potentials (see 'Category indicator and site-generic characterisation factors', p 84) but, except for the factors from Hauschild and Potting (2002) and Potting et al. (1998a, 1998b), not by the newer and sophisticated acidification factors (see 'Spatial aspects and site-dependent characterisation factors', p 84). In the factors of Norris (in Bare et al. 2002), the full range of substances is treated with national average factors, but only the nationally major contributing pollutants are treated in a regionalised way.

State-of-the-art science

Acidifying emissions are often added together in LCAs and other emission inventories without extensive modelling. The starting point for the most basic acidification equivalency factors is the number of hydrogen ions that theoretically can be formed per mass unit of pollutant X released. This number is given by the stoichiometric coefficient v in the chemical reaction shown in Equation 3-7:

$$X + \ldots \rightarrow vH^+ + \ldots \qquad (3\text{-}7).$$

From this basis, the equivalency factors assume that 1 mole of SO_2 will produce 2 moles of H^+; 1 mole nitrogen oxide compounds (NO_X) will produce 1 mole of H^+; and 1 mole of reduced nitrogen compound (NH$_X$) will produce 1 mole H+ equivalent. Because pollutant releases are specified in mass of emissions rather than moles, the coefficient v must be divided by the molecular weight of the pollutant.

The hydrogen release potentials overestimate the importance of nitrogen relative to sulphur, since nitrogen can be assimilated by ecosystems and then does not contribute to acidification. This basic acidification characterisation method also lacks attention to spatial variability in the likelihood of emissions leading to deposition on land or freshwater ecosystems rather than on the ocean, and regional variability in the sensitivity of the receiving environment. Both spatial variability and the behaviour of acidifying compounds in ecosystems are addressed in the more sophisticated models developed and maintained under CLRTAP. The modelling framework provides the main input to international negotiations about national emission ceilings in Europe. A number of protocols already have been established on this basis, the latest being the multiple pollutant–multiple effect protocol recently signed in Göteborg, Sweden.

Atmospheric emission inventories and transport processes are addressed in the European CLRTAP modelling framework by using region-to-grid atmospheric transfer matrices derived from simulations using the EMEP single-layer trajectory model (e.g., Barret and Burge 1996; EMEP is a cooperative programme for monitoring and evaluating the long-term transmission of air pollutants). The EMEP-based transfer matrices relate emissions from 44 European regions to deposition in 612 grid elements (150 × 150 km) superimposed on the same 44 regions. Atmospheric transport and process models have also been used to derive region-to-grid transfer matrices for North America. One such model is the Advanced Statistical Trajectory Regional Air Pollution (ASTRAP) model (Shannon 1991, 1992, 1996), originally developed to support the U.S. National Acid Precipitation Assessment Program (NAPAP). ASTRAP treats vertical diffusion, dry deposition, and chemical transformation to calculate normalised long-term average surface air concentrations, total airborne loading, and dry deposition increments as functions of effective emission height and time since release. A 2-dimensional program calculates seasonal mean horizontal trajectories and wet removal occurrences for a grid of source areas consisting of the 48 contiguous U.S. states and Washington DC, the Canadian provinces, and northern Mexico. Finally, a concentration and deposition calculation combines the statistics from the first 2 programs with an emission field to produce source–receptor coefficients that relate seasonal emissions by source area to deposition (wet and dry) per hectare in a grid of receiving areas. The resulting transfer matrices have been used in North America in the development of characterisation factors for acidification and eutrophication by Norris (Bare et al. 2002).

Spatially dependent sensitivity of the receiving environment was characterised by the methods and spatial databases developed under the ongoing European critical loads research program (e.g., Posch et al. 1995, 1997). The critical loads framework takes into account both the acidifying influence and the environmental sensitivity associated with sulphur and nitrogen deposition for a given ecosystem. One EMEP grid element may contain multiple ecosystems with differing critical load functions (some elements contain more than 30,000 ecosystems). These critical load functions can be superimposed and used to derive protection percentage isolines (e.g., Posch et al. 1997), which bound the deposition levels below which a given percentage of the contained ecosystems will not exceed critical loads.

Two heuristic methods for addressing the influence of release location upon expected impact were proposed by LCA practitioners in the U.S. during the mid-1990s. These methods weight emissions sources based on qualitative estimates of the sensitivity of the source region itself, rather than modelling fate and transport. Hogan and colleagues (1996) presented a Threshold Inventory Interpretation Methodology (TIIM), which neglected process emissions from the inventory analysis for which the emissions sources were located in areas not sensitive for the subject impact category. In the case of acidification, emissions originating from western U.S. states were neglected entirely, while those from eastern states were included in full. An approach similar to TIIM was demonstrated by Tolle (1997), who developed and applied qualitative scaling factors (using a 1 to 9 scale) on a state-by-state basis for the U.S. For acidification, he developed qualitative scores for states, using maps of acid-sensitive soil types and of regions likely to have acid-sensitive freshwater lakes based on bedrock geology, combined with information on the existence within each state of very large point-source emitters of acidification precursors. Then, as in TIIM, emission sources within the LCI were weighted on the basis of the source being located within a sensitive region. A modelling framework similar to the above one described for Europe is available for Asia (Hettelingh et al. 1995), though it addresses only the acidifying effect from sulphur in more detail. The state-of-the-art modelling for other continents is unclear (see also 'Models and model domains', p 67).

Category indicator and site-generic characterisation factors

Analogous to GWPs or ODPs, characterisation factors are developed relative to one of the acidifying substances on the basis of their potential to release hydrogen ions (see 'State-of-the-art science', p 82). Sulphur dioxide was arbitrarily selected as a reference compound, so acidification characterisation as presently typical in LCIA expresses acidification potential in terms of SO_2 equivalents. On this basis, Heijungs et al. (1992), Hauschild and Wenzel (1998), and Wenzel et al. (1998) calculated site-generic acidification potentials that can be used as follows:

$$AP_{reg} = \sum_i (m_i \times AP_i) \qquad (3-8),$$

where AP_{reg} is the acidification potential, regional (e.g., western Europe); m_i is the mass of acidifying emission i assigned (classified) to the impact category 'acidification' (kg i per functional unit), and AP_i is the characterisation factor of emission i within acidification.

As discussed in 'State-of-the-art science' (p 82), hydrogen release potentials overestimate the importance of nitrogen relative to sulphur because of the assimilation of nitrogen by ecosystems. This may lead to false optimisations for acidifying emission reductions. Recently available spatially resolved and site-generic characterisation factors do account for assimilation of nitrogen because the CI is defined further along the causality chain (see 'Spatial aspects and site-dependent characterisation factors' below for a discussion). Table 3-1 lists the site-generic acidification factors for the several indicators.

Spatial aspects and site-dependent characterisation factors

The most significant limitation in the basic acidification characterisation method is its lack of attention to

- spatial variability in the likelihood that emissions lead to deposition on land or freshwater ecosystems, rather than the ocean, and
- regional variability in the sensitivity of the receiving environment.

A series of advances in ability to account for one or both of these regional aspects have been proposed and applied in case studies. Norris (1998) provides a discussion that also can be found in the background document (Klöpffer and Potting 2002).

Potting et al. (1998a, 1998b) first used integrated assessment models to develop regionalised acidification factors, taking account of the influence of atmospheric transport on expected spatial distributions of deposition and the regional differ-

Table 3-1 Site-generic acidification factors typically used in LCA and new site-generic factors based on sophisticated spatially resolved models (New factors calculated as weighted mean from 1990 spatially resolved factors for EU15 + Norway + Switzerland)

Unit	SO_2 equivalent[a]	Swiss SO_2 equivalent[b]		0.01 m² unprotected ecosystem/g			Moles H+ equivalent/kg[c]
		Above and below	Only above	Mean (SD)[d]	Aquatic[e]	Terrestrial[e]	
SO_2	1.00	0.79	0.33	1.77 (2.29)	3.2	9.9	50.79
SO_3	0.80			1.41 (1.83)			
H_2SO_4	0.65			1.15 (1.49)			
H_2S	1.88			3.32 (4.29)			
NO_2	0.70			0.86 (0.72)			
NO_x	0.70	0.41	0.12	0.86 (0.72)	1.8	5.5	40.04
NO	1.07			1.31 (1.11)			61.26
HNO_3	0.51			0.63 (0.53)			
NH_3	1.88	1.3	0.48	2.31 (3.04)			95.5
HCl	0.88			6.20 (9.53)			44.7
HF	1.60			11.30 (17.36)			81.26
H_3PO_4	0.98			0.00			
Spatially resolved factors available?	No	Yes	Yes	Yes	Yes	Yes	Yes
Factors available for other base years?	No	Yes	Yes	Yes	Yes	Yes	No

[a] Wenzel et al. 1998.
[b] Huijbregts et al. 2000.
[c] Norris 1999 in Bare et al. 2002.
[d] Potting et al. 1998a, 1998b; Hauschild and Potting 2002.
[e] Krewitt et al. 2001.

ences in sensitivity of the receiving environment (see 'Relevant substances', p 82). Potting et al. (1998a, 1998b) used the superimposed curves for a grid element that allow calculation of an estimated change in 'unprotected' ecosystem areas (where critical loads are exceeded) in response to specified changes in sulphur and/or nitrogen deposition rates.

The critical load functions spatial database and the atmospheric transfer matrices jointly allow calculation of spatial acidification factors, in units of hectares (of ecosystem) per tonne (of emission), relating emission changes within one of the 44 regions to changes in total acreage of protected ecosystems across the 44-region area. As with the more simplistic acidification regionalisation approaches described earlier, the only additional information needed from the LCI analysis is the location of the emissions source (at roughly the European country level, in this case). Differences in the acidification factors among regional sources obtained by Potting et al. (1998a, 1998b) were commonly on the order of a factor 5 to 10 and ranged up to a factor of 1000.

Potting et al.'s (1998a, 1998b) method weights emissions based on estimated total acreage pushed beyond critical load thresholds by a marginal change in emission, which the authors term *a change in risk*. Krewitt et al. (2001) used the EcoSense integrated assessment model to calculate monetised characterisation factors, which like those of Potting et al. (1998a, 1998b), are tied to the change in ecosystem area protected against acidification. An initial comparison of Krewitt et al.'s (2001) country-specific factors with those developed by Potting et al. (1998a, 1998b) found considerable differences. These differences were explained by noting differences in emissions data for different scenario years, different critical loads datasets used (1995 versus 1997), and different air quality modelling approaches. Krewitt et al. noted the need for a more detailed comparison of the models, basic assumptions, and results among the 3 methods.

Heijungs and Huijbregts (1999) contend that a problem with the approaches of Potting et al. (1998a, 1998b) and Krewitt et al. (2001) is that the characterisation factors they provide are sensitive to the selection of an emissions change scenario. Other issues with using change in unprotected areas as the basis for characterisation include its implicit assumption that loadings in regions away from the critical load threshold are of no environmental consequence. Like Potting et al. (1998a, 1998b), Huijbregts et al. (2000) used an adaptation of the RAINS model to calculate acidification and eutrophication potentials, for NH_3, NO_X, and SO_2 air emissions for Europe and a number of European regions, taking both fate and effects into account. Analogous with toxicity potentials in use for LCIA, the characterisation factors of Huijbregts are based on marginal changes in what he calls *accumulated relative risk* in Huijbregts (1999) or *hazard index* in Huijbregts et al. (2000). Huijbregts calculated 2 sets of acidification factors, one set based on marginal changes in all ecosystems and another set based on marginal changes in ecosystems exposed above their critical loads only. The implicit assumption of these potentials (both sets) is that a loading much less than critical load is equally severe as a loading at or above critical load. Hence, the concept of relative risk as used by Huijbregts is different from the one used in risk assessment. The risk ratio in risk assessment evaluates the margin of safety from a full exposure instead of focusing on exposure increases as with Huijbregts (the latter one does not account for background exposures). A comparison found low correlations (r^2 generally below 0.5) between the resulting characterisation factors and those of Potting et al. (1998a, 1998b).

Norris developed regionalised acidification characterisation factors for the U.S., which use the results of atmospheric fate and transport modelling (Shannon 1996) to account for expected source-location–dependent differences in wet and dry deposition on regions other than oceans (Bare et al. 2002). Spatially comprehensive, receiving environment modelling capabilities were not available to enable inclusion of effects modelling in the development of these factors. The resulting fate-based regional characterisation factors range from roughly 20% of the U.S. average to 160% of the U.S. average, and deviation from the U.S. average is variable between SO_2 and NO_X.

Temporal aspects

Terrestrial ecosystem research has demonstrated that the nitrogen holding capacity of soil is not only spatially variable but also a dynamic property of the ecosystem, depending upon prior land-use practices and deposition histories spanning decades and even centuries (e.g., Aber et al. 1997). Ecosystem impacts as well as nitrogen leaching rates from soils change dramatically before and after this holding capacity is exceeded. A temporally related question is whether the impacts of deposition to environments with remaining holding capacity should be counted as depleting holding capacity.

Other time-oriented aspects include the influence of emissions-scenario timing upon results. For example, both Potting et al. (1998a, 1998b) and Krewitt et al. (2001) found considerable changes (e.g., for some source countries more than a doubling and for others more than a halving) in the acidification factors computed using a 1990 emissions scenario versus a 2010 emissions scenario. These differences are fully related to the sensitivity-of-area-beyond-threshold methods for emissions change scenarios highlighted by Huijbregts. The acidification factors of Huijbregts are scenario independent because they do not account for background exposure.

Other aspects

Some other aspects relevant to this impact category are the following:
- Benefits of nitrogen in some environments, for example, agriculture
- Linkages with eutrophication
- Non-ecosystem impacts of acidification, such as upon monuments and historic artefacts (reported as very important in countries such as Italy and Greece) and upon buildings more generally.

Recommendations

We recommend that LCA practitioners characterise acidification according to Equation 3-8. The presently typical hydrogen release potentials overestimate the importance of nitrogen relative to sulphur. We therefore advise practitioners to start using one of the sets of characterisation factors from Table 3-1 in order to avoid false optimisations. Practitioners should take care not to ignore the acidifying contributions from compounds other than nitrogen and sulphur compounds (since these are not covered by some of the improved factors).

We urge researchers from the field of acidification to develop global models or to harmonise continental models with regard to the environmental mechanisms covered. This requires extension of non-European continental models with effect information. Acidification should be developed further (e.g., by improved endpoint modelling) and adjusted to the scientific progress.

We advise LCA researchers to investigate the effect of using the alternative characterisation factors from Table 3-1. Special attention should be given to evaluating the feasibility and additional information gained from spatially resolved LCIA. Additional requirements are to be considered for the inventory (spatial information is not generally contained in present-day inventories).

Terrestrial Eutrophication

Description of the impact category

The impact category 'terrestrial eutrophication' covers adverse effects of excess nutrients on plant functioning and on species composition in natural or seminatural terrestrial ecosystems. The ecosystems include those of potential interest for nature conservation, such as forests, heathlands and shrubs, permanent grasslands, and bogs and marshes (Grouzet et al. 1999).

Under natural or seminatural conditions, plant nutrient requirements are met mainly from soil reserves, and the growth and competition of plants are commonly controlled by the limited availability of nitrogen. However, the present atmospheric input of nitrogen from human activity also contributes considerably and, in many ecosystems, already has caused the adverse effects of excess nutrients. Also, land use and management (e.g., cultivation, drainage, and burning) can affect nitrogen transformations and mobility in soils (Hornung et al. 1994; Grouzet et al. 1999).

When nitrogen-limited ecosystems are exposed to increased loads of nitrogen, their structure and functioning may change in a number of ways. For example, there is an increased competition from nitrogen-adapted species at the expense of less adapted species and an altered tolerance towards diseases, drought, frost, or herbivory (e.g.,

Hornung et al. 1994; Grouzet et al. 1999). At the community level, nutrient imbalances can cause changes in species composition. In most cases, a limited number of species benefit more than the average from the increased availability of nitrogen (Grouzet et al. 1999).

At present, terrestrial eutrophication is not typically characterised in LCIA. However, some scientists emphasise that the problem may be at least as serious as the eutrophication phenomena in European aquatic ecosystems (Grouzet et al. 1999). A recent global assessment of the nitrogen loads to terrestrial ecosystems (Busch et al. 2001) shows that the eutrophication problem is not limited to Europe and North America. Also, tropical and subtropical regions receive considerably increasing loads of nitrogen.

Relevant substances

In LCIA, only ammonia (NH_3) and nitrogen oxide (NO_X) emitted to air are classified under terrestrial eutrophication. This is in line with environmental science. Though phosphorus is a key nutrient in eutrophication of inland waters (rivers and lakes), it has only minor relevance for terrestrial ecosystems because, under natural conditions, their growth is rarely limited by phosphorus (Chardon 2000).

State-of-the-art science

The critical load concept for terrestrial eutrophication (e.g., Posch et al. 1995), developed in the frame of the UNECE CLRTAP (1996), offers a more sophisticated characterisation by defining the CI closer to the endpoint. The spatially resolved critical loads of nitrogen used in integrated assessment models, as discussed in 'Relevant substances' (p 82), allow defining a CI closer to the endpoint than does the one from Lindfors et al. (1995). Potting, Schöpp, Hauschild (2002) and Krewitt et al. (2001) both used such models to arrive at spatially resolved characterisation factors for terrestrial eutrophication. They based their impact indicator, similar to the one Potting et al. (1998a, 1998b) proposed for acidification, on a change in the area of ecosystems above critical loads from a marginal change in emissions in a given region and year. This leads to a characterisation approach in which spatial differentiation can be included. Analogous to toxicity potentials in use for LCIA, the characterisation factors of Huijbregts are based on marginal changes in 'accumulated relative risk'.

Category indicators and site-generic characterisation factors

Terrestrial eutrophication usually is not covered by current LCIA. The simplest CI for terrestrial eutrophication, as proposed by Lindfors et al. (1995), is the amount of nitrogen emitted to air because this indicates the increased biomass production in terrestrial ecosystems. The characterisation factors for atmospheric nitrogen compounds can be determined according to the nitrogen contents of compounds.

However, both spatially resolved and site-generic characterisation factors have recently become available to characterise this impact category. Table 3-2 lists the site-generic factors for the several indicators. They can be used following Equation 3-9:

$$TEP_{reg} = \sum_i (m_i \times EP_i) \qquad (3\text{-}9),$$

where TEP_{reg} is the terrestrial eutrophication potential, regional (e.g., western Europe), m_i is the mass of terrestrial eutrophying emission i assigned (classified) to the impact category terrestrial eutrophication during classification (kg i per functional unit), and EP_i is the characterisation factor of emission i within terrestrial eutrophication.

Spatial aspects and site-dependent characterisation factors

Nitrogen deposition and ecosystem sensitivity to nitrogen input vary between geographic regions.

Table 3-2 Site-generic factors to assess terrestrial eutrophication based on sophisticated spatially resolved models (Site-generic factor calculated as weighted mean from 1990 spatially resolved factors for EU15 + Norway + Switzerland)

Unit	Swiss NO_x equivalents (base year 1995)[a]		0.01 m² unprotected ecosystem/g (base year 1990)	
	Above and below	Only above	Mean (SD)[b]	Mean[c]
NO_2			2.48 (2.65)	
NO_x	3.7	2.9	2.48 (2.65)	18.8
NO			3.79 (4.05)	
HNO_3			1.79 (1.93)	
NH_3	0.99	0.62	14.24 (18.76)	
Spatially resolved factors available?	Yes	Yes	Yes	Yes
Factors available for other base years?	Yes	Yes	Yes	Yes

[a] Huijbregts et al. 2000.
[b] Hauschild and Potting 2002.
[c] Krewitt et al. 2001.

This spatial differentiation can be taken into account in determining characterisation factors for emissions that cause terrestrial eutrophication in a similar way to the determination for acidifying emissions. In this case, the effect can be defined using a marginal approach based on the critical load mapping and deposition values calculated by integrated assessment models like RAINS (Alcamo et al. 1990; Schöpp et al. 1999). The area of ecosystem that becomes unprotected as a result of reduced emission reveals tradeoffs between characterisation factors of different areas (see Krewitt et al. 2001; Hauschild and Potting 2002; Potting, Schöpp, Hauschild 2002). Huijbregts et al. (2000) also used the RAINS model to calculate characterisation factors for another type of CI. Analogous with toxicity potentials in use for LCIA, the characterisation factors of Huijbregts are based on marginal changes in what is called *accumulated relative risk* in Huijbregts (1999) or *hazard index* in Huijbregts et al. (2000). Table 3-2 lists the site-generic acidification factors for the several indicators.

Temporal aspects

Nitrogen accumulation in terrestrial ecosystems is a long-term process, and it can take decades of exceedance of the critical load before effects in the ecosystem can be observed. The same concerns recovery of impacted terrestrial ecosystems. At present, it is not clear how the time lags can be included in the characterisation.

Hauschild and Potting (2002), Potting, Schöpp, and Hauschild (2002), and Krewitt et al. (2001) presented the integrated assessment model calculation in which changes of emissions and background concentrations over time were taken into account. In their method, country-specific characterisation factors for European airborne nitrogen emissions of 2 different base years, namely actual emission levels in 1990 and forecasted emission levels in 2010, were calculated by the marginal approach (the 10% change of nitrogen emissions from the individual European Union [EU] countries).

Other aspects

At present, critical loads of nutrient nitrogen are calculated primarily for forest ecosystems. Other potential nature types may have different critical loads, compared to those loads derived from forest ecosystems.

The N:P ratios in terrestrial plants differ from those found in aquatic plants. The average ratio of terrestrial plants is twice as big as the Redfield ratio of 16, which describes the ratio N:P of phytoplankton in the aquatic environment (Koerselman and Meuleman 1996). For this reason, there is no basis to aggregate nutrients into one eutrophication category according to the Redfield ratio (see 'Aquatic eutrophication', below).

Nitrogen deposition is also involved in acidification and aquatic eutrophication. If nitrogen deposition is higher than critical loads of nutrient nitrogen, surplus nitrogen is leached. The leaching may increase aquatic eutrophication. If the nitrogen deposition is higher than the minimum critical load (sum of nitrogen uptake and nitrogen immobilisation), acidification occurs if the rate of denitrification is too low. Damage can also occur if the total load of nitrogen and sulphur exceeds the critical load of acidity.

Recommendations

We recommend that LCA practitioners add eutrophication of natural and seminatural terrestrial ecosystems as an impact category to the presently typical ones in LCIA. This impact category should be taken into account if emissions of ammonia and nitrogen oxide to air exist in the inventory. Spatial differentiation should be quantified where possible. We recommend that LCA practitioners characterise terrestrial eutrophication according to Equation 3-9 and with help of one of the sets of characterisation factors from Table 3-2.

We urge researchers from the field of terrestrial eutrophication to develop global models or to harmonise continental models with regard to the environmental mechanisms covered. This requires extension of non-European continental models with effect information (as in the form of critical loads). Different terrestrial ecosystems must be taken into account. This category should be developed further (e.g., by improved endpoint modelling) and adjusted to the scientific progress. A worldwide overall picture of the magnitude and location of terrestrial eutrophication is needed.

We advise LCA researchers to investigate the effect of using the alternative characterisation factors from Table 3-2. Special attention should be given to evaluating the feasibility and additional information gained from spatially resolved LCIA. Additional requirements for the inventory are necessary (spatial information is not generally contained in present-day inventories).

Aquatic Eutrophication

Description of the impact category

Aquatic eutrophication can be defined as nutrient enrichment of the aquatic environment. The supply of nutrients to water, under natural circumstances, is in balance with the subsequent growth of biomass such that a stable though variable society of plants and animals is ascertained. Such society is to some extent able to cope with variations in nutrient supplies. A too-large nutrient increase pushes this stable society out of balance, however, and through a chain of ecological effects, may provoke a shift of the biological structure (Kristensen and Hansen 1994).

Water becomes dominated by phytoplankton after a high nutrient input that makes the water turbid. The turbidity prevents the light from reaching the water bottom, which makes submerged plants disappear. The fish community also changes because predatory fish are unable to see and catch smaller fish. The fish community becomes dominated by zooplankton-eating species that are more tolerant of turbidity. Zooplankton eats phytoplankton, and the decrease in zooplankton therefore results in a further increase of phytoplankton. The excess phytoplankton dies and sinks to the bottom, where the decay of phytoplankton uses oxygen. The decrease of oxygen leads to further fish kills and disappearance of bottom fauna, and may also evoke a release of phosphorus from the lake bottom. The released phosphorus may be used for a new increase of phytoplankton, after which a new cycle of phytoplankton dying and sinking starts (Kristensen and Hansen 1994).

The aquatic environment consists of fresh waters (lakes, reservoirs, rivers), coastal waters, and marine waters. The effects of eutrophication are generally more apparent in standing or slow-moving waters than in (small) fast-flowing rivers.

Relevant substances

Typically, under aquatic eutrophication in LCIA, only those nutrients that are the most important for limiting the growth rate or biomass yield of aquatic plants are taken into account. They are, in practice, phosphorus and nitrogen compounds, including NH_3 and NO_X emitted to air.

Both decomposition of biomass and emissions of organic material into water consume oxygen. Due to the effect of consuming oxygen, emissions of organic material measured as biological oxygen demand (BOD) or chemical oxygen demand (COD) are often handled together with nutrients under aquatic eutrophication.

State-of-the-art science

The concentration of nutrients above which eutrophication becomes an environmental problem depends on many factors (e.g., topography, physical and chemical nature of water bodies). The same amount of nutrient can cause significantly different responses in various aquatic environments. This feature concerns the issue of thresholds. In addition, biomass growth may be limited by different nutrients in different aquatic ecosystems. For example, in Europe, most freshwater is phosphorous limited, whereas in marine waters, it is nitrogen that limits production of algal biomass (e.g., Grouzet et al. 1999). However, the concept of the limiting nutrient is not always clear because in the same water body, the limiting nutrients can change over seasons and years.

The state-of-the-art modelling of aquatic eutrophication is yet poor compared to modelling of atmospheric environmental problems. Sophisticated spatially resolved models that characterise runoff and leaching of nutrients to surface waters are available for most continents. However, further transport through rivers and lakes to marine waters is not yet modelled in detail, and appropriate effect measures for water quality are lacking. Characterisation of aquatic eutrophication is therefore often based on loading assessment as also used in LCIA.

Category indicators and site-generic characterisation factors

All aquatic eutrophication impact methods in LCIA base their characterisation factors for emissions on the Redfield ratio, which describes the overall composition of carbon, nitrogen, and phosphorus (106:16:1) in phytoplankton (Finnveden and Potting 1999). According to this ratio, 1 mol of nitrogen emitted will produce 6.6 mol organic carbon in biomass, whereas in the case of phosphorus, the corresponding amount is as much as 106 mol. In fact, although the ratio depends on many factors (e.g., species, aquatic circumstances) and varies case by case, it is commonly used as a baseline for the nutrient utilisation of phytoplankton in limnology (Grouzet et al. 1999). Thus, the basic approach to characterise different nutrients is

$$\text{AEP} = \sum_i (m_i \times \text{EP}_i) \qquad (3\text{-}10),$$

where AEP is the aquatic eutrophication potential, m_i is the mass of aquatic eutrophying emission i

assigned (classified) to the impact category terrestrial eutrophication during classification (kg i per functional unit), and EP_i is the characterisation factor of emission i based on the Redfield ratio (Redfield et al. 1993; see, e.g., Heijungs et al. 1992 and Wenzel et al. 1998 for actual factors).

The impact indicator of aquatic eutrophication is taken to be algal growth when characterisation factors of phosphorus and nitrogen compounds are directly based on the Redfield ratio. An alternative impact indicator used in aquatic eutrophication is oxygen consumption that results from the degradation of organic material (Finnveden et al. 1992; Samuelsson 1993; Lindfors et al. 1995). This approach is based on the assumption that all organic material is eventually degraded, including the biomass produced from nitrogen and phosphorus. The corresponding amount of organic material needs a certain amount of oxygen in order to degrade completely (e.g., Samuelsson 1993). According to this knowledge and the Redfield ratio, the emissions of phosphorus and nitrogen compounds can be expressed in a unit of mg O_2. This is the same unit in which BOD or COD is also measured, and emissions of organic material, phosphorus, and nitrogen can thus be aggregated into a single score by characterisation factors related to the impact indicator of oxygen consumption.

The disadvantage of the aggregation of COD or BOD and nutrients with the help of oxygen consumption is that COD or BOD has different cause–effect relationships compared to phosphorus and nitrogen compounds (see 'Description of the impact category', p 90). The primary effect of COD or BOD discharges is oxygen consumption that does not result in biomass growth, as nitrogen and phosphorus compounds do. So, double counting is another disadvantage of aggregation of COD or BOD and nutrients. Nitrogen or phosphorus discharged to water is often bound to organic matter. In that case, it is reported twice in LCA inventory analysis, as organic matter (expressed in BOD or COD) and as nutrient emission. For this reason, organic material is excluded from the eutrophication category in the method proposed by Hauschild and Wenzel (1998), Wenzel et al. (1998), and Hauschild and Potting (2002). There also is an approach in which oxygen depletion and eutrophication are handled as separate impact categories (Seppälä 1997, 1999). However, this approach is difficult to conduct because contributions of organic matter (often expressed in BOD or COD) and nutrients to oxygen depletion are difficult to separate in the water bodies. In addition, organic materials can cause indirect phosphorus releases through anoxic conditions at the sediment–water interface (e.g., Wetzel 1983).

Spatial aspects and site-dependent characterisation factors

In the earlier approaches, no spatial differentiation is included and the characterisation factors describe only biomass production derived from the Redfield ratio. Lindfors et al. (1995) suggested a more advanced approach, in which the limiting nutrient issues are handled according to scenarios.

Seppälä (1997, 1999), Lindfors et al. (1998), Pleijel et al. (1998), and Norris (in Bare et al. 2002) have used characterisation methods in which the fate of waterborne emissions is taken into account. In this approach, site-generic characterisation factors (see 'Category indicators and site-generic characterisation factors', p 91) are multiplied by the corresponding site-dependent transport factors (or fate factors) in order to obtain site-dependent characterisation factors. The transport factor represents the proportion of a nutrient emission reaching a water body in which the nutrient is the limiting nutrient. The transport factor varies between 0 and 1 depending on recipient, and it can be determined on the basis of scientific models or expert judgements.

In fact, only a fraction of airborne nutrient will be transported to the aquatic environment, causing eutrophication. Integrated assessment models for airborne nitrogen emissions such as the EMEP models for Europe or ASTRAP for North America offer a good starting point to estimate 'effective' air emissions under aquatic eutrophication (Seppälä 1997, 1999; Bare et al. 2002; Potting, Beusen et al. 2002; Potting, Schöpp, Hauschild 2002). Huijbregts and Seppälä (2000) presented country-specific fate factors for direct deposition in the European marine environment of ammonia and nitrogen oxide emitted to air. However, the factors do not cover the amounts of runoff and leaching to the marine environment. In their recent paper, Huijbregts and Seppälä (2001) take into account aspects of leaching from agricultural soil and calculate characterisation as well as normalisation factors for the Netherlands, Europe, and the world. Potting, Beusen et al. (2002) combined modelling of deposition on water and soil with detailed, spatially resolved modelling of leaching and runoff of nutrients to surface water. They thus provide factors for emissions to air that fully integrate deposition after atmospheric transport, and leaching and runoff to surface water. Like Potting, Norris combined atmospheric transport and deposition models with hydrologic modelling of runoff to fresh- and saltwater environments to derive aquatic eutrophication factors for airborne sources in each of the 48 continental U.S. states (Bare et al. 2002).

The concept of critical load is not applied to aquatic eutrophication. However, there have been approaches in which spatial differentiation with respect to damage levels of aquatic ecosystems was taken into account in determining characterisation factors. Norris suggested using the damage factors (Bare et al. 2002). Seppälä (1997, 1999) used an approach in which air emissions of nitrogen falling on marine areas that represent the 'natural state' are not considered. In this case, transport (or fate) factors describe the proportion of the total atmospheric emissions of NH_3 and NO_X falling on marine areas where adverse effects of eutrophication exist.

Temporal aspects

Adverse effects of aquatic eutrophication have seasonal dimensions that vary according to geographic region. For example, in northern Europe, algal growth is greatly reduced or negligible during the winter period of low light and temperature. In the regions where seasonal climatic changes are minimal, the length of the active growing season is extended to the entire year (Wetzel 1983). It is also important to note that, because of the internal load, the effects of eutrophication can continue years after external nutrient loads have been finished. Despite the temporal aspects of aquatic eutrophication, there have been no attempts to account for these aspects in LCIA.

Other aspects

The biological availability of total nitrogen and phosphorus emitted from various anthropogenic sources can vary. *Biological availability of a nutrient* can be defined as the sum of nutrient forms that can be transformed into forms that algae are able to use. In general, all major forms of nitrogen (ammonium, nitrite, nitrate, urea, etc.) are ultimately bioavailable. Phosphorus is more or less bioavailable from different particulate or dissolved compounds (e.g., Grouzet et al. 1999). The biological availability of total nitrogen and phosphorus emitted from various sector sources (e.g., pulp and paper industry) is included in the solution presented by Seppälä (1997, 1999). The characterisation factor for total nitrogen or phosphorus based on the Redfield ratio and the transport factor is multiplied by the 'effect factor', indicating what quantity of the transported emission can cause increased production of biomass in the area. In practice, bioavailability means that source-dependent characterisation factors are needed for total nitrogen and phosphorus emitted from a certain area.

Recommendations

We recommend that LCA practitioners characterise aquatic eutrophication according to Equation 3-10 and with help of the Redfield ratio, which for the time being offers an acceptable starting point as factors to characterise emissions of nitrogen and phosphorus that cause aquatic eutrophication. However, these factors produce the maximum contributions of emissions to aquatic eutrophication. To obtain an impact indicator that is closer to the expected occurrence of actual impact, we recommend using site-dependent characterisation factors that include an appropriate estimate of the fate and/or marginal effect of adding substances to the region.

Discussion is ongoing whether BOD and COD should be part of this impact category. We advise practitioners, therefore, to decide whether to address these inputs. However, we ask them to make their reasoning explicit.

Much work is still to be done for researchers in the field of aquatic eutrophication. There is a need for better knowledge about the contribution of soil emissions and air emissions to aquatic eutrophication. Models for assessing runoff and leaching of atmospheric nitrogen should be developed and/or modified in this direction. Spatial information about the sensitivity of aquatic systems, including the limiting nutrient and damage issues, is needed. The definition of the CIs, the seasonal variation in utilisation of nutrients, and the biological availability may need some further discussion. The issue of including nutrients such as silicon also should be investigated.

Overall Recommendations

Recommendations on a general level

Characterisation in LCA quantifies the potential environmental impact accumulated over the short and long range of emissions from processes in a product system over the full receiving area. Emissions and their subsequent impact may take place in a near as well as a distant future. The impact or CI, in principle, can be defined at any level of the causality chain. The CI may calculate exposures and evaluate this against threshold values or choose to refrain from that.

Quality requirements for impact characterisation in LCIA should be clarified, analogous with quality requirements for inventory data. Uncertainty management in impact characterisation in LCA should be guided by optimal reduction of uncertainty through careful balancing between different types of uncertainty. Choices that most reduce uncertainty may be context and application dependent. Uncertainty from data uncertainty in LCA outcomes should be quantified as far as possible (by mean, median, and SD). Uncertainty from model uncertainty, value choices (like definition of CI), and assumptions (e.g., about background exposures) should be avoided. However, under all circumstances, value choices and technical assumptions should be explicitly and transparently reported (and preferably, their influence on LCA results should be quantified by means of sensitivity analysis).

The choice of CI and related characterisation factors for nonglobal impact categories should be guided by the geographical emphasis of the processes in the product system under study (determined often by the region where the LCA is commissioned). For example, North American studies typically employ indicators and factors based on North American models, while European countries typically use European ones. However, work in the future should move towards harmonising these approaches to yield globally consistent characterisation factors. Models from authoritative bodies should be employed where possible.

The definition of the CI and related characterisation factors can differ with respect to
- position in the causality chain,

- spatial and temporal resolution (for the nonglobal categories), and
- marginal or average approach.

With regard to the nonglobal impact categories, systematic comparison of CIs and related sets of characterisation factors is needed to enhance understanding of consequences on the LCIA outcome from the chosen position in the causality chain and whether to follow a marginal or average approach.

Characterisation factors should be available for all substances that potentially contribute in a nonnegligible amount to a given impact category, in order to facilitate characterisation of the full impact of the product system under study.

Site-generic characterisation factors are sufficient for the global impact categories, but spatial differentiation in impact assessment is advocated for the nonglobal categories, to reduce the considerable spatially determined uncertainty in LCIA outcomes. Site-dependent characterisation factors should be resolved at the optimal spatial level for the given impact categories (which is several hundred kilometres square for the nonglobal impact categories covered here; this usually compares to the land surface of countries or larger administrative areas within large countries like the U.S.). Site-generic characterisation factors should be available to facilitate site-generic assessment for applications requiring this, or for processes for which information about geographic location is lacking. The reasons for and uncertainty of refraining from spatial differentiation should be quantified where possible.

As a general guiding principle, CIs and characterisation factors in LCA should integrate over the period in which the relevant substances directly exert their impact (i.e., their residence time in the relevant environmental compartments). In line with site-dependent assessment, we advocate temporal differentiation in impact assessment to reduce temporally determined uncertainty in LCIA outcomes. To date, time-dependent characterisation has received little research attention, but such attention is warranted.

The majority of the above recommendations focus on the accuracy of characterisation results. These recommendations are appropriate when comparison with other impact categories is not essential. Spatially resolved characterisation factors derived from relatively sophisticated environmental models should be used here (usually relatively low data uncertainties). Such methods are also fine for applications such as benchmarking and monitoring. In applications where comparison across impact categories is essential, damage-based characterisation factors might be an alternative (high data uncertainties at present state-of-the-art, but lower interpretation uncertainties; see 'Recommendations by impact category', below). Also, spatial differentiation and all other measures to limit the uncertainties should be used as much as possible.

Recommendations by impact category

The above recommendations are formulated in a general fashion. The following recommendations to LCA practitioners apply to each impact category.

- Characterise climate change according to Equation 3-1 with help from the most recent list of GWP_i values as published by IPCC. We advise characterising direct greenhouse gases only until more reliable GWP_i values for the indirect greenhouse gases are available.
- Characterise stratospheric ozone depletion according to Equation 3-2 with the ODP_i values from the most recent WMO/UNEP report (1999). We propose that practitioners use an infinite time horizon (steady-state model) if the goal definition of the LCA does not require otherwise. The model used for the calculation should be given if time-dependent data are used.

- Characterise photochemical ozone creation according to Equations 3-5 and 3-6 with the most recent set of respectively $POCP_i$ or MIR_i values, depending on the aim and scope of the study and considering the main geographical locations covered by the study. Practitioners should include NO_X in the calculation of POCP and MIR values. Depending on the aim and scope of the study, we recommend using the spatially resolved modifiers from Hauschild and Potting (2002) or Krewitt et al. (2001) to arrive at a more accurate estimate of ozone formation and its actual effect.
- Characterise acidification according to Equation 3-8 with the characterisation factors from Table 3-1 (in order to avoid false optimisations because, presently, typical hydrogen release potentials overestimate the importance of nitrogen relative to sulphur). Practitioners should take care not to ignore the acidifying contributions from compounds other than nitrogen and sulphur compounds (because these are not covered by some of the improved factors).
- Characterise eutrophication of natural and seminatural terrestrial ecosystems as an additional impact category in LCIA, if emissions of ammonia and nitrogen oxide to air exist in the inventory. Spatial differentiation should be quantified where possible. We recommend that LCA practitioners characterise terrestrial eutrophication according to Equation 3-9, with the help of one of the sets of characterisation factors from Table 3-2.
- Characterise aquatic eutrophication according to Equation 3-10 and with help of the Redfield ratio, which for the time being offers an acceptable starting point for characterisation factors for emissions of nitrogen and phosphorus that cause aquatic eutrophication. These factors produce the maximum contributions of emissions to aquatic eutrophication. To obtain an impact indicator that is closer to the expected occurrence of actual impacts, we recommend using site-dependent characterisation factors that include an appropriate estimate of the fate and/or marginal effect of adding substances to the region.

If a damage approach is preferred, and the larger model uncertainties are accepted because of the less uncertain interpretation, we suggest the approaches of Steen (1999) and Goedkoop and Spriensma (1999).

References

Aber JD, Ollinger S, Driscoll C. 1997. Modeling nitrogen saturation in forest ecosystems in response to land use and atmospheric deposition. *Ecol Model* 101:61–78.

Alcamo J, Shaw RW, Hordijk L, editors. 1990. The RAINS model of acidification. Science and strategies in Europe. Dordrecht, G: Kluwer. ISBN 0-7923-0781-X (HB); 0-7923-0782-8 (PB).

Bare JC, Hofstetter P, Pennington DW, Udo de Haes HA. 2000. Life cycle impact assessment workshop summary; Midpoints vs endpoints: The sacrifices and benefits. *Int J LCA* 5(6):319–326.

Bare JC, Norris GA, Pennington DW. 2002. TRACI: The US EPA's tool for the reduction and assessment of chemical and other environmental impacts. *J Ind Ecol*. In review.

Barrett K, Berge E, editors. 1996. Transboundary air pollution in Europe. Part 1: Estimated dispersion of acidifying agents and of near surface ozone. Oslo, N: Norwegian Meteorological Institute. EMEP MSC-W status report. Research Report Nr. 321996.

Busch G, Lammel G, Beese FO, Feichter J, Dentener FJ, Roelofs G-J. 2001. Forest ecosystems and the changing patterns of nitrogen input and acid deposition today and in the future based on a scenario. *Environ Sci Pollut Res* 8(2):95–102.

Chardon WJ. 2000. The role of the soil in the phosphorous cycling. In: Weidema BP, Meusen MJG, editors. Agricultural data for life cycle assessment. Volumes 1 and 2. The Hague, NL: Agricultural Economics Research Institute (LEI). p 13–24.

Darnall KR, Lloyd AC, Winer AM, Pitts Jr JN. 1976. Reactivity scale for atmospheric hydrocarbons based on reaction with hydroxyl radical. *Environ Sci Technol* 10:692–696.

Derwent RG, Jenkin ME. 1991. Hydrocarbons and the long-range transport of ozone and PAN across Europe. *Atmos Environ* (Part A: General Topics)25:1661–1678.

Derwent RG, Jenkin ME, Saunders SM. 1996. Photochemical ozone creation potentials for a large number of reactive hydrocarbons under European conditions. *Atmos Environ* 30:181–199.

Derwent RG, Jenkin ME, Saunders SM, Piling MJ. 1998. Photochemical ozone creation potentials for organic compounds in north-west Europe calculated with a master chemical mechanism. *Atmos Environ* 32:2429–2441.

Ellenberg H, Weber HE, Düll R, Wirth V, Werner W, Paulissen D. 1992. Zeigerwerte von Pflanzen in Mitteleuropa. Göttingen, D: Verlag Erich Groltze. Scripta Botanica XVVIII.

European Environmental Agency. 1995. Europe's environment. The Dobris assessment. Copenhagen, DK: European Environmental Agency.

Farman JC, Gardiner BG, Shanklin JD. 1985. Large losses of total ozone in Antarctica reveal seasonal ClO_X/NO_X interaction. *Nature* 315:207.

Finlayson-Pitts BJ, Pitts Jr JN. 1986. Atmospheric chemistry. Fundamentals and experimental techniques. New York NY, USA: J Wiley.

Finlayson-Pitts BJ, Pitts Jr JN. 1993. Atmospheric chemistry of tropospheric ozone formation: Scientific and regulatory implications. *Air Waste* 43:1091–1100.

Finnveden G, Andersson-Sköld Y, Samuelsson M-O, Zetterberg L, Lindfors L-G. 1992. Classification (impact analysis) in connection with life cycle assessment: A preliminary study. In: Lindfors L-G, editor. Product life cycle assessment: Principles and methodology. Copenhagen, DK: Nordic Council of Ministers. Nord 1992:9. p 172–231.

Finnveden G, Potting J. 1999. Eutrophication as an impact category. *Int J LCA* 4(6):311–314.

Goedkoop M, Spriensma R. 1999. The Eco-indicator 99. A damage oriented method for life cycle impact assessment. The Hague, NL: Ministry of Housing, Spatial Planning and the Environment (VROM).

Grouzet P, Leonard J, Nixon S, Rees Y, Parr W, Laffon L, Bogestrand J, Kristensen P, Lallana C, Izzo G, Bokn T, Bak J, Lack TJ, Thyssen N, editors. 1999. Nutrients in European ecosystems. Copenhagen, DK: European Environment Agency. Environmental Assessment Report Nr. 4.

Guinée J, editor. 2002. Handbook on life cycle assessment. Operational guide to the ISO standard. Dordrecht, NL: Kluwer Academic. ISBN 1-4020-02289.

Hauschild M, Potting J. 2002. Spatial differentiation in life cycle impact assessment: Guidance document. Copenhagen, DK: Danish Environmental Protection Agency. Forthcoming.

Hauschild M, Wenzel H. 1998. Nutrient enrichment as a criterion in the environmental assessment of products. In: Hauschild M, Wenzel H, editors. Environmental assessment of products. Volume 2, Scientific background. London, GB: Chapman and Hall. p 179–202.

Heijungs R, Guinée J, Huppes G, Lankreijer RM, Udo de Haes HA, Wegener Sleeswijk A, Ansems AMM, Eggels PG, van Duin R, de Goede HP. 1992. Environmental life cycle assessment of products. Guide and background. Leiden, NL: Leiden Univ Centre of Environmental Science (CML).

Heijungs R, Huijbregts MAJ. 1999. Threshold-based life cycle impact assessment and marginal change: Incompatible? Leiden, NL: Centre of Environmental Science-Section Substances and Products (CML-SSP). Working Paper 99.002. http://www.leidenuniv.nl/interfac/cml/ssp/publications. Accessed 15 Mar 2002.

Hettelingh J-P, Chadwick M, Sverdrup H, Zhao D. 1995. RAINS-Asia, an assessment model for acid rain in Asia: Impact module. In: Foell W, Amann M, Carmichael G,

Chadwick M, Hettelingh J-P, Hordijk L, Zhao D, editors. Acid rain and emissions in Asia. Bilthoven, NL: National Institute for Public Health and the Environment (RIVM). Report to the World Bank.

Hogan LM, Beal RT, Hunt RG. 1996. Threshold inventory interpretation methodology: A case study of three juice container systems. *Int J LCA* 1(3):159–167.

Hornung M, Ineson P, Bull KR, Cresser M, Davison A, Fowler D, Harriman R, Irwin JG, Wilson RB, Pitcairn CER, editors. 1994. Impacts of nitrogen deposition in terrestrial ecosystems. London, GB: United Kingdom Review Group, Department of the Environment.

Houghton JT, Ding Y, Griggs DJ, Noguer M, van der Linden PJ, Dai X, Maskell K, Johnson CA, editors. Climate change 2001: The scientific basis. Cambridge, GB: Cambridge Univ Pr.

Houghton JT, Meira Filho LG, Callander BA, Harris N, Kattenberg A, Maskell K, editors. 1996. Climate change 1995: The science of climate change. Cambridge, GB: Univ Pr.

Huijbregts MAJ. 1999. Life-cycle impact assessment of acidifying and eutrophying air pollutants. Calculation of characterisation factors with RAINS-LCA. Amsterdam, NL: Interfaculty Dept Environmental Science.

Huijbregts MA, Schöpp W, Verkuijlen E, Heijungs R, Reijnders L. 2000. Spatially explicit characterization of acidifying and eutrophying air pollution in life-cycle assessment. *J Ind Ecol* 4(3):75–92.

Huijbregts MA, Seppälä J. 2000. Towards region-specific, European fate factors for airborne nitrogen compounds causing aquatic eutrophication. *Int J LCA* 5(2):65–67.

Huijbregts MA, Seppälä J. 2001. Life cycle impact assessment of pollutants causing aquatic eutrophication. *Int J LCA* 6(6):339–344.

[ISO] International Organization for Standardization. 1998. Life cycle assessment: Goal and scope definition and inventory analysis. Geneva, CH: ISO. ISO 14041.

[ISO] International Organization for Standardization. 2000. Life cycle assessment: Life cycle impact assessment. Geneva, CH: ISO. ISO 14042.

Itsubo N. 2000. Screening life cycle impact assessment with weighting methodology based on simplified damage functions. *Int J LCA* 5(5):273–280.

Klöpffer W, Potting J, editors. 2002. Best available practice in life cycle impact assessment of climate change, stratospheric ozone depletion, acidification, eutrophication and tropospheric ozone formation. Backgrounds on impact categories. Bilthoven, NL: National Institute of Public Health and the Environment (RIVM). RIVM Report 408660 002. Report by SETAC Europe Scientific Task Group on Global and Regional Impact Categories (STG-GARLIC).

Koerselman W, Meuleman AFM. 1996. On N:P ratios and the nature of nutrient limitation. *J Appl Ecol* 33:1441–1450.

Krewitt W, Trukenmüller A, Bachmann TM, Heck T. 2001. Country-specific damage factors for air pollutants: A step towards site dependent life cycle impact assessment. *Int J LCA* 6(4):199–210.

Kristensen P, Hansen HO. 1994. European rivers and lakes. Assessment of their environmental state. Copenhagen, DK: European Environmental Agency. EEA Environmental Monographs 1.

Latour JB, Staritski JG, Alkemade JRG, Wiertz J. 1997. De natuurplanner. Decision support system. Bilthoven, NL: Natuur en Milieu. RIVM Report 711901019.

Lindfors L-G, Alkemark M, Oscarsson C, Spännar C. 1998. A manual for the calculation of ecoprofiles intended for third party certified environmental product performance declarations. Stockholm, S: Swedish Environmental Research Institute (IVL).

Lindfors L-G, Christiansen K, Hoffman L, Virtanen Y, Juntilla V, Hanssen O-J, Rønning A, Ekvall T, Finnveden G. 1995. Nordic guidelines on life cycle assessment. Copenhagen, DK: Nordic Council of Ministers. Nord 1995:20.

Molina MJ, Rowland FS. 1974. Stratospheric sink for chlorofluoro-methanes: Chlorine atom catalyzed destruction of ozone. *Nature* 249:810–814.

Nichols P, Hauschild M, Potting J, White P. 1996. Impact assessment of non toxic pollution in life cycle assessment. In: Udo de Haes HA, editor. Towards a methodology for life-cycle impact assessment. Brussels, B: Society of Environmental Toxicology and Chemistry (SETAC) Europe.

Norris G. 1998. Background report on life cycle impact assessment methods for acidification, eutrophication and photochemical oxidant formation. North Berwick ME, USA: Sylvatica.

Pleijel H. 1999. Ozone and the environment: A background. In: Pleijel H, editor. Ground-level ozone: A threat to vegetation. Stockholm, S: Swedish Environmental Protection Agency. Report 4970.

Pleijel K, Altenstedt J, Pleijel H, Lövblad G, Grenfelt P, Zetterberg L, Fejes J, Lindfors L-G. 1998. A tentative methodology for the calculation of global and regional impact indicators in type-III ecolabels used in Swedish case studies. Stockholm, S: Swedish Environmental Research Institute (IVL).

Posch M, de Smet PAM, Hettelingh JP, Downing RJ. 1995. Calculation and mapping of critical thresholds in Europe. Bilthoven, NL: Coordination Centre for Effects (CCE), National Institute of Public Health and the Environment (RIVM).

Posch M, Hettelingh JP, Downing RJ, de Smet PAM, editors. 1997. Calculation and mapping of critical thresholds in Europe. CCE status report 1997. Bilthoven, NL: Coordination Center for Effects (CCE), National Institute of Public Health and the Environment (RIVM).

Potting J. 2000. Spatial differentiation in life cycle impact assessment [PhD thesis]. Utrecht, D: Utrecht Univ.

Potting J. 2002. Acidification and terrestrial eutrophication. Comparison of different levels of sophistication. *Int J LCA*. Forthcoming.

Potting J, Beusen AHW, Øllgaard H, Hansen OC, de Haan B, Hauschild M. 2002. Aquatic eutrophication. In: Potting J, Hauschild M. 2001. Spatial differentiation in life cycle impact assessment: Technical backgrounds. Copenhagen: DK: Danish Environmental Protection Agency. Forthcoming.

Potting J, Hauschild M. 2002. Spatial differentiation in life cycle impact assessment: Technical backgrounds. Copenhagen: DK: Danish Environmental Protection Agency. Forthcoming.

Potting J, Klöpffer W, editors. 2002. Best available practice in life cycle impact assessment of climate change, stratospheric ozone depletion, acidification, eutrophication and tropospheric ozone formation. General issues. Bilthoven, NL: National Institute of Public Health and the Environment (RIVM). RIVM Report 408660 001. Report by SETAC Europe Scientific Task Group on Global and Regional Impact Categories (STG-GAR-LIC).

Potting J, Schöpp W, Blok K, Hauschild M. 1998a. Comparison of the acidifying impact from emissions with different regional origin in life-cycle assessment. *J Hazard Mater* 1998(61):155–162.

Potting J, Schöpp W, Blok K, Hauschild M. 1998b. Life cycle impact assessment of acidification. *J Ind Ecol* 2(2):63–87.

Potting J, Schöpp W, Hauschild M. 2002. Terrestrial eutrophication. In: Potting J, Hauschild M. 2001. Spatial differentiation in life cycle impact assessment: Technical backgrounds. Copenhagen, DK: Danish Environmental Protection Agency. Forthcoming.

Redfield AC, Ketchum BH, Richards FA. 1993. The influence of organisms on the composition of sea water. Proceedings of the 2nd International Water Pollution Conference; Tokyo, Japan. Oxford, GB: Permagon. p 215–243.

Rowland FS, Molina MJ. 1975. Chlorofluoromethanes in the environment. *Rev Geophys Space Phys* 13:1–35.

Samuelsson M-O. 1993. Life cycle assessment and eutrophication: A concept for calculation of the potential effects of nitrogen and phosphorus. Stockholm, S: Swedish Environmental Research Institute (IVL). IVL report B1119.

Schöpp W, Amann M, Cofola J, Heyes C, Klimont Z. 1999. Integrated assessment of European air pollution emission control strategies. *Environ Model Software* 14(1):1–9. ISSN 1364-8152. Reprinted as IIASA RR-99-002.

Seppälä J. 1997. Decision analysis as a tool for life cycle assessment. Helsinki, FS: Finnish Environment Institute. The Finnish Environment 123.

Seppälä J. 1999. Decision analysis as a tool for life cycle impact assessment. In: Klöpffer W, Hutzinger O, editors. LCA documents, Volume 4. Bayreuth, D: Eco-Informa Pr.

Shannon JD. 1991. Modeled sulfur deposition trends since 1900 in North America. In: van Dop H, Kallos G, editors. Air pollution modeling and its application VIII. New York NY, USA: Plenum Pr.

Shannon JD. 1992. Regional analysis of S emission-deposition trends in North America from 1979 through 1988. In: van Dop H, Kallos G, editors. Air pollution modeling and its application VIII. New York NY, USA: Plenum Pr.

Shannon JD. 1996. Atmospheric pathways module. In: Tracking and analysis framework (TAF) model documentation and user's guide. Argonne IL, USA: Argonne National Laboratory. ANL/DIS/TM-36.

Steen B. 1999. A systematic approach to environmental priority strategies in product development (EPS). Version 2000: General system characteristics. Göteborg, S: Chalmers Univ of Technology, Technical Environmental Planning. CPM Report 1999:4.

Syri S, Amann M, Schöpp W, Heyes C. 2001. Estimating long-term population to ozone in urban areas of Europe. *Environ Pollut* 113(1):59–69.

Tolle D. 1997. Regional scaling and normalization in LCIA. *Int J LCA* 2(4):197–208.

Udo de Haes HA, editor. 1996. Towards a methodology for life-cycle impact assessment. Brussels, B: Society of Environmental Toxicology and Chemistry (SETAC) Europe. Report of the SETAC Europe First Working Group on Life-Cycle Impact Assessment.

Udo de Haes HA, Jolliet O, Finnveden G, Hauschild M, Krewitt W, Müller-Wenk R. 1999. Best available practice regarding impact categories and category indicators in life cycle impact assessment. *Int J LCA* 4(2):66–74 and 4(3):167–174.

[UNECE] United Nations Economic Commission for Europe. 1996. Manual on methodologies and criteria for mapping critical levels/loads and geographical areas where they are exceeded. Berlin, D: Umweltbundesamt.

Wenzel H, Hauschild M, Alting L. 1998. Environmental assessment of products. Methodology, tool and techniques, and case studies in product development. London, GB: Chapman and Hall.

Wetzel RG. 1983. Limnology. 2nd ed. Philadelphia PA, USA: CBS College Publ.

[WMO] World Meteorological Organization. 1991. Scientific assessment of ozone depletion. Global ozone research and monitoring project. Geneva, CH: WMO. Report Nr. 25.

[WMO] World Meteorological Organization. 1994. Scientific assessment of ozone depletion: 1994. Geneva, CH: WMO. Report Nr. 37.

[WMO] World Meteorological Organization. 1999. Scientific assessment of ozone depletion: 1998. Global ozone research and monitoring project. Geneva, CH: WMO. Report Nr. 44. ISBN 9280717227.

Fate and Exposure Assessment in the Life-Cycle Impact Assessment of Toxic Chemicals

Edgar G. Hertwich, Olivier Jolliet, David W. Pennington, Michael Hauschild, Carsten Schulze, Wolfram Krewitt, Mark Huijbregts

Abstract — The Society of Environmental Toxicology and Chemistry (SETAC) Europe First Working Group on Life-Cycle Impact Assessment (WIA-1) laid out a framework in which an assessment of the human or ecological exposure to a chemical is combined with an indicator of the chemical's toxicity to characterise toxic releases in life-cycle assessment (LCA). Here, we outline some of the key issues in the assessment of chemical exposure in LCA, covering topics such as the degree of sophistication and detail necessary to modelling a chemical's behaviour in the environment and the extent to which location-dependent variations in exposure can be taken into account. We recommend a procedure that uses generic assessments to identify the most important releases; site-specific assessment is conducted for the most important releases only if location-dependent variations are judged to be important for the overall assessment. We review a range of models that help assess the distribution of a chemical in the environment and its availability to organisms of concern. Available models are grouped into 2 categories:
1) Multimedia fate and exposure models take into account the fate of pollutants across medium boundaries and model multipathway exposure routes.
2) Spatially explicit single-medium models can take into account chemical reactions such as oxidation and photooxidant generation.

Limiting the assessment to a single modelling approach would result in the omission of important environmental impacts. We recommend the combination of results from complementary models.

Introduction

The Society of Environmental Toxicology and Chemistry (SETAC) Europe First Working Group on Life-Cycle Impact Assessment (WIA-1) distinguished between 4 types of information potentially useful for the characterisation step in life-cycle impact assessment (LCIA):
1) effect information,
2) fate information,
3) background level information, and
4) spatial information (Udo de Haes 1996).

Effect information describes the sensitivity of the target system in Figure 4-1. Fate information concerns the distribution and chemical transformation of the chemical. Background-level information may be relevant if the sensitivity to the toxicant is nonlinear. Spatial information is relevant for the modelling of the fate and for determining the presence of the target system.

For the toxicological impact categories, the group laid out the following general characterisation framework (Jolliet 1996). The effect score S is the product of an effect factor E, a fate and expo-

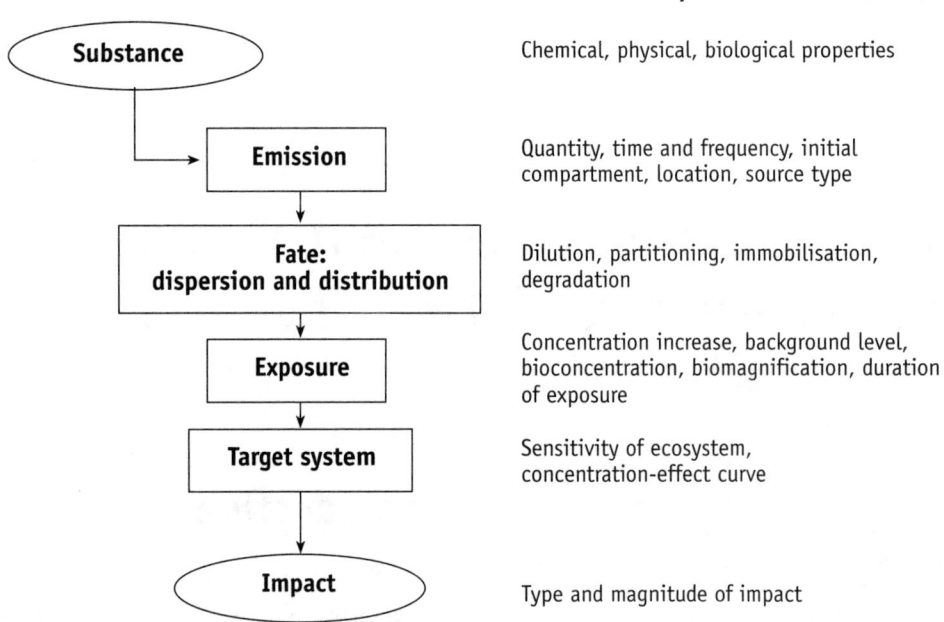

Figure 4-1 Causality chain from emission to impact. At each link of the chain, the consideration of different descriptors determines the level of sophistication of the model.

sure factor F, and the total mass loading or emissions M:

$$S_i^{mn} = E_i^m F_i^{nm} M_i^n \quad (4\text{-}1),$$

where i denotes the chemical, n the release compartment, and m the exposure route. The total effect score (or category indicator [CI]) for all emissions in the life cycle, or life-cycle stage, is obtained by summing across chemicals, release compartments, and exposure routes.

$$S = \sum_{i,m,n} S_i^{nm} \quad (4\text{-}2)$$

The effect factor E may consist of a simple indicator of toxicity, such as the inverse of the predicted no-effect concentration (PNEC) for ecosystems or the allowable daily intake (ADI) for humans, or it may be a more detailed description of the dose–response curve. It may even consist of a detailed assessment of the specific toxic effect and a valuation of its adversity. The effect factors for human and ecological toxicity are discussed in Chapters 5 and 6.

Given potential confounding factors, such as location and background concentration, a number of options for the terms in Equation 4-1 have been suggested. These options make explicit the dependency of fate and/or effect on the background concentration, the timing, and the location, either through modifying or correction terms or by expressing E and F as a function of these factors (Jolliet 1996; Potting and Hauschild 1997; Heijungs and Wegener Sleeswijk 1999). Another suggestion involves breaking up the fate factor and separately describing the distribution and fate of the chemical in the environment and its transfer through different exposure pathways (Heijungs and Wegener Sleeswijk 1999), to account for the modelling practice for human toxicity (Guinée and Heijungs 1993).

In this chapter, we first identify and discuss choices that must made to develop characterisation methods based on the WIA-1 framework. These choices concern the depth of analysis, the degree of sophistication, and the degree of accuracy. There currently is no consensus regarding many of these choices (Bare, Hofstetter et al. 2000). In the second part of the chapter, we describe and investigate 2 different approaches to fate and exposure modelling in life-cycle assessment (LCA) toxicity assessments. One uses multimedia environmental fate and exposure models ('Multimedia Models in LCA', p 106). The second approach uses spatially explicit single-medium models ('Medium-Specific Approaches', p 109). These model results could be combined. It may also be possible to use, for example, a modular approach to build an integrated fate and exposure model. The modelling approaches are briefly compared in 'Use of the Models in LCIA' (p 112). In 'Evaluation of Fate and Exposure Methods' (p 113), we use a set of criteria, initially developed by the task force on ecotoxicity (Chapter 6), to present an evaluation of modelling approaches and model capabilities. In 'Recommendations' (p 116), we offer key opinions. The main weight in the presentation is given to the discussion of the multimedia models, which constitute the most elaborated approach in LCA at present.

This chapter concentrates on the fate and exposure of chemicals for human toxicity and ecotoxicity but does not account directly for photooxidation, eutrophication, acidification, or radioactivity. Photooxidants, such as ozone, whose formation is assessed using sophisticated atmospheric chemistry models, can be addressed using the model described in 'Application of air quality modelling: The EcoSense Model' (p 110), but in this book it is treated as a separate category, as are acidification and eutrophication. For damage-oriented approaches, the effects of these impact categories could be included in assessments of human health and ecosystem health, together with the impacts resulting from ozone depletion, climate change, noise, and land use (European Commission 1999; Goedkoop and Spriensma 1999; Steen 1999).

Methodological Choices in the Assessment of Toxicity

Desired level of sophistication

The inclusion of toxicity in LCIA is complicated by 2 factors:
1) the large number of stressors (chemicals) and
2) the fact that these chemicals act according to different mechanisms of action.

Some categories are homogeneous, that is, each pollutant has the same mechanism of action (global warming, ozone depletion). Other categories are necessarily heterogeneous, that is, they contain pollutants that act according to different mechanisms of action (eutrophication, toxicity). The justification for creating heterogeneous categories is that they have the same areas of protection (AoPs) as well as similar environmental mechanisms. Hence, the same scientific tools are used to study and assess the action of pollutants within a category (Hertwich and Hammitt 2001). Chemical risk analysis and risk management, for example, were developed to address toxic chemicals.

The problem of heterogeneous categories is that we do not have the same degree of information on, and understanding of, the action of every pollutant grouped in such a category. When we compare a well-studied pollutant with one that is poorly understood, we must discard the information available for just one pollutant (Figure 4-2), if only a single model is to be used universally. An important consideration for the development of an impact assessment method for toxic chemicals is, hence, the amount of information required by the method.

Different methods may be adopted for different chemicals, depending on the relevance of the model and the availability of data. Methods that require little information can include many chemicals but disregard much of what we know about

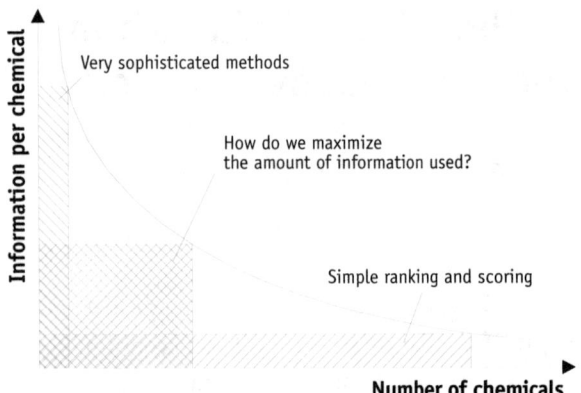

Figure 4-2 Limits in the amount of information that can be used in comparing chemicals to each other

some of them, whereas methods that require a lot of information necessarily include estimates and guesses for many others. The available information is symbolised by the area under the curve in Figure 4-2. The amount of information that is used by a method is symbolised by the rectangle under the curve. An optimal impact assessment method maximises the area of this rectangle, that is, the amount of information used.[1]

The amount of information required by an impact assessment method depends on the depth of the analysis[2] and the degree of sophistication. The *depth of the analysis* (or comprehensiveness) indicates which elements along the impact chain are taken into account in the assessment (Hertwich et al. 1997), whereas the *degree of sophistication* indicates the detail with which each of the elements is assessed (Udo de Haes et al. 1999). Depth and sophistication determine the environmental relevance of the characterisation indicator as well as the level of accuracy (or uncertainty) with which this indicator can be determined.

Figure 4-3 represents the impact chain and indicates the depth of analysis of 2 methods used for the assessment of human toxicity:
1) the human toxicity potential (HTP) and
2) the disability-adjusted life years (DALYs) (Guinée and Heijungs 1993; Hofstetter 1998).

Note that the depth of analysis of both methods is higher, that is, the investigated chain is longer, than that of the global warming potential (GWP; Hertwich, Hammitt, Pease 2000). Approaches to human and ecological toxicity assessment vary in their depth of the analysis of the assessment and valuation of effects. This affects the toxicological models and the valuation of different health or species endpoints, but not necessarily the fate and exposure modelling.

Fate and exposure models range in the degree of sophistication along a continuum from simple scoring methods to detailed, site-specific risk analysis (Swanson and Socha 1997; Hertwich et al. 1998; Bare, Udo de Haes, Pennington 2000; Pennington and Yue 2000; Pennington and Bare 2001). To illustrate the differences along this continuum, we outline and compare 3 of the different options. All 3 options have the same depth of analysis — they include environmental concentration, exposure, and toxicity — but they differ in the detail with which each of these factors is assessed.

Option 1: Simple ranking and scoring

Chemical ranking and scoring is commonly based on ordinal scoring of attributes that are judged to be important indicators of the implicit concern posed by a chemical (Swanson and Socha 1997): persistence, bioaccumulation, and toxicity (PBT).

---

[1] In principle, there are 2 ways out of this problem, which both depend on the shape of the curve in Figure 4-1. The classical decision analysis approach would be to use a quantitative uncertainty analysis for the entire assessment and to represent the unknown quantities for little-studied chemicals by probabilistic distributions that reflect the variation of this parameter for better-studied chemicals. The second option is to use multiple assessment steps of increasing sophistication. Each step assesses in more detail those emissions that have been identified as more hazardous in the previous step. This second approach depends on there being a correlation between the hazard posed by chemicals and the amount of information available about them.

[2] The term *depth of analysis* is also used in life-cycle inventory (LCI) modelling, where it describes how much of the supply chain (how many tiers of suppliers) has been considered.

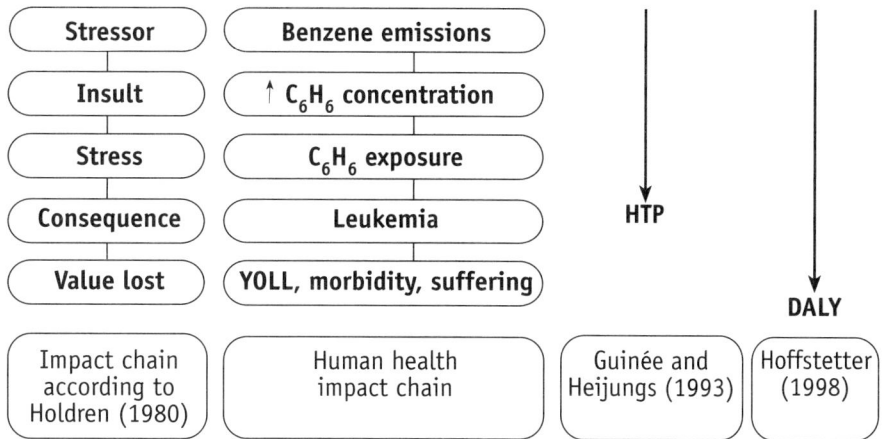

Figure 4-3 Human toxicity impact chain, with benzene as an example. (The HTP method assesses exposure and effect; the DALY also assesses the value lost, or the consequences of the effect [Chapter 5]. YOLL = years of life lost.)

The U.S. Environmental Protection Agency (USEPA) Waste Minimization Prioritization Tool (WMPT), for example, adopts 3 attributes to reflect the implicit concern, or hazard, of a chemical (USEPA 1997). Each attribute is assigned a score of 1, 2, or 3 to reflect low to high concern. These attributes are subsequently added to generate a total score on a scale from 3 to 9. A hierarchy of different sources, ranging from well-documented measurements to expert judgement, is used to determine each attribute score. This reflects the basis of many chemical screening approaches and allows the USEPA to assess an unprecedented number of chemicals (thousands). Preliminary studies suggest that the approach provides a reasonable indication of a chemical's implicit concern when compared to some more sophisticated methodologies, such as the potentials described in the next paragraph (Pennington and Bare 2001).

Option 2: Toxic equivalency potentials

Toxic equivalency potentials (TEPs) are cardinal indices calculated from model representations of source-to-dose relationships in the environment. The models can vary in terms of sophistication and comprehensiveness (Bare, Hofstetter et al. 2000). Both measured and estimated data are used in the calculations. While the calculated potentials still have a physical interpretation (e.g., the increase in concentration resulting from specific emission flow rates, or the time-integrated exposure associated with the emission of a given mass of chemical), they usually represent stylised conditions. The TEPs developed by Guinée and Heijungs (1993), for example, are based on a multimedia environmental fate model that assumes uniformly mixed environmental compartments. The multimedia models in this case represent the partitioning and intermedia transfer of chemicals, but not their spatial distribution. They represent the behaviour of chemicals in a 'unit world' model environment, not in an observable setting.[3]

Option 3: Detailed risk assessment

In principle, we could conduct a detailed risk assessment for each release and then use the result scaled down to the functional unit of the study in terms of the predicted damages in an LCA. This would mean that for each release, the increase in

---

[3] The GWP is based on a similar, deliberate simplification of reality: The atmospheric lifetime of greenhouse gases is calculated under steady-state conditions and does not use the predicted concentration increases. This is relevant because of the nonlinear processes that control the fate of CO_2.

concentration in a specified location and the associated exposed populations are taken into account. In this way, the location of emissions, which determines the number of people exposed, their health status, as well as the background concentration and meteorological conditions, can be taken into account. Such an assessment has been conducted, for example, for criteria air pollutants released by power plants in Europe, that is, with a functional unit being the full output of the power plant (Krewitt et al. 1999; Spadaro and Rabl 1999).

There is agreement that chemical ranking and scoring should be used only for a screening-level assessment, to identify substances that require more detailed evaluation in an LCA study. There is also agreement that a full-scale risk analysis is beyond the scope of many LCA studies (Udo de Haes et al. 1999) because emission sources associated with a given product or service are often widely distributed around the world and at multiple sites. (This does not mean that risk analysis should not be used in decisions, particularly because LCA provides useful insights into where more detailed analyses may be beneficial.) The most commonly adopted level of sophistication, therefore, corresponds to the level of the TEP.

Even with this basic decision taken, in the determination of fate factors and effect factors, there are many details that further define the level of sophistication. These concern, for example, the number of exposure pathways taken into account and the detail with which these pathways are modelled.

The interface between fate and effect assessment

The assessment of chemical fate and exposure, captured by the multiplication factor F in the framework described in the 'Introduction' (p 101), must match the effect assessment, captured by E. It must describe the exposure concentration or dose on the same basis as the effect factor (Table 4-1), it needs to provide background dose information if a nonlinear dose–response curve is used, and it needs to describe exposure or dose as a function of space and/or time if the effect assessment is temporally or spatially specific.

The choice of the interface between toxic effects and fate and exposure is a tradeoff between the availability of toxic data (e.g., in terms of ingested doses, air concentration, blood concentration) and the feasibility of the fate and exposure modelling. For most indicators, the toxic effect is defined in terms of media concentration–effect or dose–effect characteristics. Toxicodynamics, or the study of uptake rates, excretion rates, and residence times within an organism, is usually addressed as an integral component of the toxicological dose–response assessment, hence by the task force for toxicological effects (see Chapters 5 and 6).

In agreement with the human toxicity and the ecotoxicity task forces, our task force decided that fate and exposure modelling should provide a concentration increase in different media for ecotoxicity, as well as ingested or inhaled doses through different pathways for human toxicity. These variables will serve as the interface between effect on the one hand and fate and exposure on the other hand.

Multimedia Models in LCA

The application of multimedia models for the calculation of TEPs is well established in LCA (Guinée and Heijungs 1993; Pratt et al. 1993; Jia et al. 1996; Jolliet and Crettaz 1997; Hertwich et al. 1998; Huijbregts, Thissen, Guinée et al. 2000). Detailed discussions of multimedia fate models have been published (Cowan et al. 1995; Mackay, Di Guardo, Paterson, Kicsi et al. 1996; Wania and Mackay 1999), and these discussions are not replicated here. Rather, we focus on issues in multimedia modelling that are relevant for LCA.

Table 4-1 Characteristics of a consistent fate and exposure analysis, for different types of effect coefficients[a]

Characteristic	Deposition	Media concentration	Dose	Organ or internal concentration
Relevant characteristics for effect	Critical deposition rate	Concentration–effect–response relationship	Dose–effect relationship	Concentration in biota or target organ (e.g., body burden)
Fate factor required for a consistent fate analysis	Links emissions to an increase of deposition rate	Links emissions to a concentration increase	Links emissions to an increase of inhaled, ingested, or absorbed doses	Links emission to an increase in organ or internal body concentration

[a] Note that this concerns the interface between fate and effect factors; further depth of analysis can be added in effect modelling, e.g., through assessing the potentially affected fraction (PAF) of species resulting from a specific deposition rate, or the years of life lost (YOLLs) resulting from a specific human dose.

Integrated multimedia fate and exposure models represent the distribution of a chemical among different environmental compartments and the transfer of chemicals through various exposure routes to a species of interest (Figure 4-4). The models usually treat environmental compartments such as air, surface water, and sediments as uniformly mixed. Transport processes are described by simple elementary models that use measured or, more commonly, estimated partitioning constants and intermedia transport rates for processes such as rainfall and runoff. The transformation rate (or reaction half-life, in most cases) of a chemical is an empirical or estimated quantity entered into the model. For human toxicity, the models calculate a potential dose (also called *individual dose fraction*), which is indicative of the level of impact expected (Guinée and Heijungs 1993; Hertwich et al. 1998). For ecological toxicity, the models calculate either environmental compartment concentrations or potential doses for animals at different levels of the food chain (Jolliet 1996; Campfens and Mackay 1997; Huijbregts 1999).

The environmental fate model, as represented by the left part of Figure 4-4, determines the concentration in different compartments (air, surface water, sediments, etc.) through the solution of mass balance equations that describe the release, transformation, and intercompartmental transfer of a pollutant. Mathematical solutions for simultaneous differential equations are well established in the literature, particularly for linear first-order systems. In most cases, the transport and transformation rates are assumed to be linear, which simplifies the mathematical solution and avoids the need to account for background information. Steady state is usually assumed because it further simplifies modelling and provides an estimate of the time-integrated exposure of the mass of chemical emitted (Heijungs 1995). The equations can be solved explicitly (Mackay and Paterson 1981).

Multimedia fate models were developed using semivolatile organic chemicals and radionucleids as case studies. They have proven to be powerful tools for gaining insights into the behaviour of such substances. Their use has been extended to nonvolatile organic chemicals (Mackay and Diamond 1989) and inorganic compounds (McKone 1993; Crommentuijn et al. 1997). Multimedia modelling approaches are not suited, without appropriate changes, to address dissociating (e.g., organic acids) or speciating chemicals (e.g., mercury), although such efforts have been pursued (USEPA 1999; Woodfine et al. 2000).

The representation of the environment in multimedia models ranges from a simple 'uniform world' (Mackay 1991), that is, a single box for each medium, to a high degree of spatial detail (Wania and Mackay 1995; USEPA 1999). The

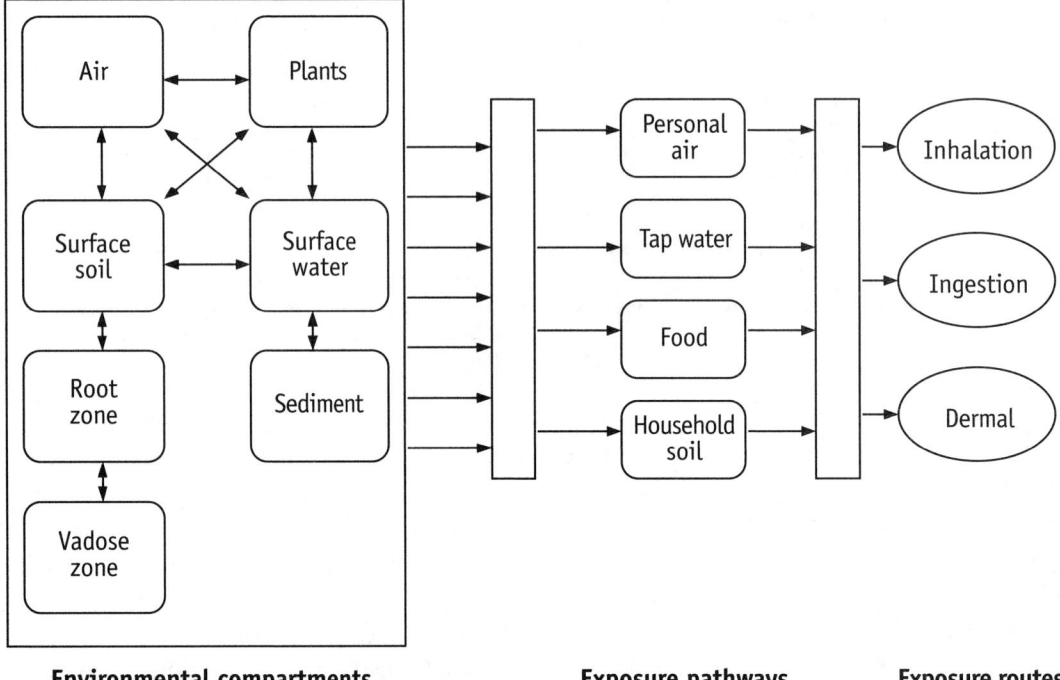

Figure 4-4 The general integration of source, dispersion, and exposure in integrated environmental fate and exposure models, such as the Uniform System for the Evaluation of Substances (USES) and CalTOX. (Credit: J.S. Eisenberg)

spatial detail is achieved through the use of multiple boxes for each compartment, considering advective exchange between adjacent boxes similar to climate models. To date, the models used for LCA have been either uniform world models (Jia et al. 1996; Hertwich et al. 2001) or a simple nested box model, which consists of continental Europe and 3 climatic zones of the Northern Hemisphere (Huijbregts, Thissen, Guinée et al. 2000). Jolliet and co-workers (Jolliet and Crettaz 1997, 2000; Jolliet 1999) suggested a modular approach in which the mass balance equations are solved first in each compartment. Intermedia transfer fractions are then used to couple the different compartments. This facilitates the possibility of coupling medium-specific models, still providing multimedia solutions. The importance of so-called *feedback* or *looping* between compartments is quantified, providing a measure of the required degree of integration of the single-medium models (Margni et al. 2001).

Exposure pathway models calculate the exposure of an organism through a stated pathway, resulting from a given environmental concentration. These factors take into account transfer factors, uptake rates such as the rate of inhalation, partitioning or bioconcentration factors, and environmental concentrations. Individual components of an exposure pathway can be estimated on the basis of measured or predicted data. For example, the bioconcentration of organic pollutants in fish is often based on empirical bioconcentration factors or on an estimation equation based on observed correlations between the bioconcentration factor and the octanol–water partition coefficient (Mackay and Boethling 2000). The bioaccumulation factor (accounts for both bioconcentration as well as biomagnification; increased uptake by ingestion), however, also can be modelled, taking into account factors such as the trophic level, growth dilution, and biomagnification (Campfens and Mackay 1997).

For metals, exposure pathway models require the input of measured transfer and partitioning factors. For organic chemicals, these transfer and partitioning factors are often predicted by simple correlations. The use of such estimation methods is convenient because it allows for the assessment of many chemicals for which detailed empirical data do not exist. However, the estimation methods can result in errors when the chemical–physical properties of an assessed chemical lie outside the domain for which the property estimation method was developed. For example, in a systematic assessment of a number of model components, Hertwich, McKone, and Pease (2000) noted that the estimation methods used to determine crop concentrations in the exposure component of CalTOX produced flawed results for hydrophilic chemicals with a low Henry's Law constant. These methods, commonly used in exposure analysis, had to be replaced to extend the model domain to this class of chemicals (Hertwich et al. 2001).

When multimedia models are used to characterise toxic chemicals, a number of assumptions are commonly adopted:
1) The model is applicable for all chemicals.
2) The output of the model has the same meaning over a wide range of parameters.
3) The chemical is of concern in its emitted form, not because of its action in the environment (e.g., as a catalyst) or its transformation products.

A number of other impact categories address the action of chemicals in the environment (e.g., acidification) and the effects of transformation products (photochemical oxidant formation). In principle, transformation products can be included in multimedia models, and some case studies use linked models to consider 'ultimate degradation'. The inclusion of transformation products is considered desirable but is not common practice because of the limited availability of transformation rate data and knowledge of environmental chemistry.

Table 4-2 includes a list of multimedia environmental fate models and their references. This is not intended to be an exhaustive list. The models are grouped according to certain similarities. Column 2 indicates special properties of the model.

Model parameters, intermedia transport correlations, and the number of compartments can differ, which may lead to differences in the results. A SETAC workshop evaluated and compared a number of multimedia fate models, not considering exposure pathways (Cowan et al. 1995). It compared the outcomes of ChemCAN, SimpleBOX, and CalTOX and found that they are very similar if the models are parameterised in the same way. Original differences in model setup often reflect differences in purpose and local conditions. The Netherlands, for example, has no soil erosion, while California has a high erosion rate, leading to the transport of pollutants from soil to surface water. The environment modelled also differs, ranging from a regional environment without an ocean (CalTOX) to one that considers different climate zones that represent the entire Northern Hemisphere (SimpleBOX 2.0).

All models serve as a reasonable basis for fate and exposure modelling in LCA, but none is yet ideal. In our evaluation, a model should include coastal and oceanic environments and offer a detailed description of human exposure pathways, including dermal exposure. We recommend the development of a new model that combines the best aspects of the existing models under initiatives such as the United Nations Environment Programme (UNEP)–SETAC Life-Cycle Initiative.

Medium-Specific Approaches

This section presents some medium-specific approaches and a discussion of their potential application in LCA.

The use of medium-specific models within LCA is possible only if compartments can indeed

Table 4-2 Prominent multimedia models

Model	Remarks	Reference
EQC ChemCAN	Steady-state multimedia models that describe the regional partitioning of chemicals. Currently used by DuPont and the USEPA in some scoring applications, as well as in some LCA case studies where exposure factors were added.	http://www.trentu.ca/envmodel; Mackay, Di Guardo, Paterson, Cowan 1996
SimpleBOX 1.0, 2.0; Uniform System for the Evaluation of Substances (USES) 1.0, 2.0; European Union System for the Evaluation of Substances (EUSES); USES-LCA	SimpleBOX is a steady-state multimedia model; recent release is an evaluative model with nested boxes (local, regional, continental, global). USES and EUSES combine SimpleBOX with an exposure model both for humans, ecosystems, and a limited number of animals. EUSES was developed for risk screening in the European regulation of hazardous chemicals. USES-LCA is a modified version of USES 2.0, which has been adapted for calculating human and ecological toxic equivalency factors and includes sea compartments.	Natl Institute of Public Health and the Environment (RIVM) et al. 1994, 1998; Brandes et al. 1996; Huijbregts, Thissen, Guinée et al. 2000; http://www.leidenuniv.nl/interfac/cml/lca2/
CalTOX	Integrated fate and exposure model for humans, used by Cal-EPA for hazardous waste site assessment and in some USEPA tools for HTP calculations. Steady-state model version is used typically, although a time-dependent representation of the soil compartment exists.	http://eetd.lbl.gov/ied/era/; McKone 1993
Total Risk Integrated Methodology (TRIM)	Describes the spatial distribution of pollutants by using a large number of adjacent boxes for each compartment. Exposure assessment for both humans and ecosystems. Complicated by the level of spatial and temporal representation.	USEPA 1999
Modular approach	The mass balance is developed in 2 separate steps: 1) Exposure is calculated for each medium in an independent module based on the steady-state mass balance within the medium, accounting for transfer to other media as a sink but assuming that there is no feedback transfer. 2) Compartments are connected on the basis of intermedia transfer factors to determine the first-order increases in concentration. It thus becomes possible to determine the level of coupling between compartments or to couple single-medium models.	Jolliet and Crettaz 1997; Jolliet 2000; Margni et al. 2001

be considered independently, that is, if the coupling (intermedia exchange) is not important. Therefore, it is necessary to determine the degree of intermedia coupling and establish whether a single-medium approach can be retained or how it could be combined with other models.

Application of air quality modelling: The EcoSense Model

The EcoSense Model was designed to support the assessment and valuation of environmental and health impacts caused by the emission of airborne pollutants from combustion processes, as part of the ExternE project (European Commission 1999). The model supports the quantification of environmental impacts by following a detailed site-specific 'impact pathway' approach. It models the causal relationships from the release of pollutants through their interactions with the environment to a physical measure of impact and, where possible, a monetary valuation of the resulting welfare losses (Krewitt, Hurley et al. 1998; Krewitt, Mayerhofer et al. 1998; Krewitt et al. 1999).

The current version of EcoSense covers about 20 substances, including SO_2, NO_X particles, and

CO, as well as a number of heavy metals and organic substances. Although EcoSense was primarily designed to assess impacts from power plant emissions, the scope of the model was broadened to cover other industrial activities. EcoSense has been extensively tested and reviewed, and so far has been used mainly for the assessment of energy- and transport-related externalities, and in studies analysing environmental policy measures aimed at the reduction of airborne pollutants (Krewitt, Friedrich et al. 1998; Krewitt et al. 1999). Within the LCA context, EcoSense could contribute to a further investigation of site-dependent effects, as well as generic characterisation factors for criteria air pollutants (such as SO_X, NO_X, etc.). The model was developed first for Europe, and versions have now been completed for Brazil and Latin America and for China and Asia.

A schematic flowchart of EcoSense is shown in Figure 4-5. EcoSense provides harmonised air quality and impact assessment models together with a comprehensive set of relevant input data for Europe and South America, which allows a site-specific impact assessment. A link to the European CORINAIR air emission database allows the definition of emission scenarios by taking into account emission reduction measures in specific countries and industry sectors.

To cover different pollutants and different scales, EcoSense provides 3 air quality models completely integrated into the one system:

1) The Industrial Source Complex (ISC) Model is a Gaussian plume model developed by the USEPA (Brode and Wang 1992). The ISC is used for transport modelling of primary air pollutants on a local scale.
2) The Windrose Trajectory Model (WTM; Trukenmüller and Friedrich 1995) is a user-configurable trajectory model based on the windrose approach of the Harwell Trajectory Model developed by Derwent and Nodop (1986). The WTM is used to estimate the concentration and deposition of nonreactive

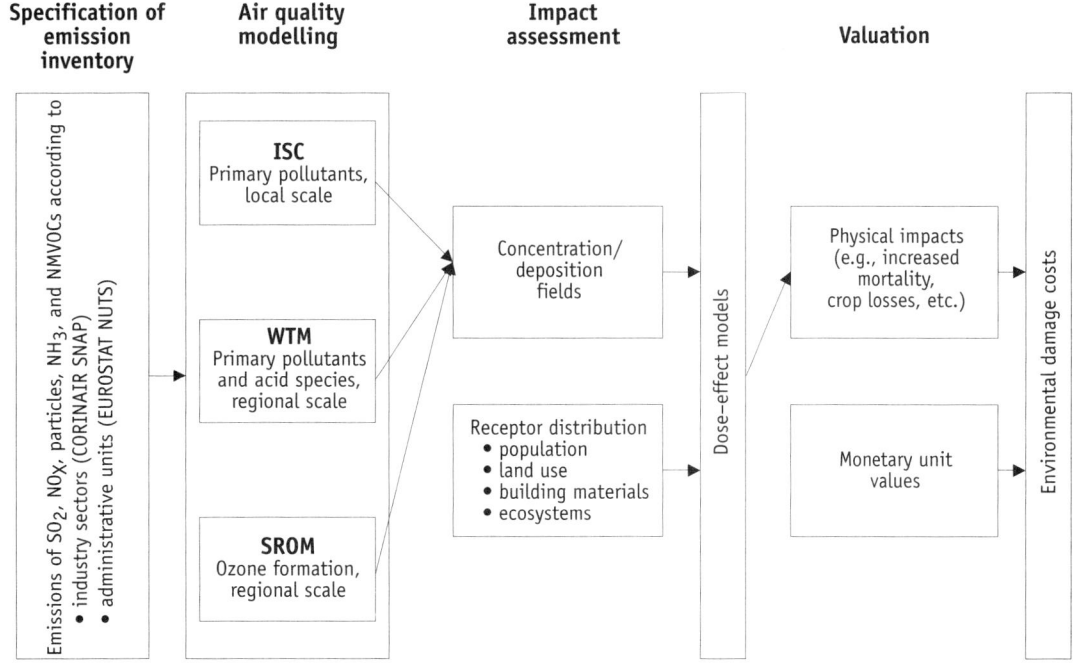

Figure 4-5 Flowchart of the EcoSense Model
(NMVOC = nonmethane volatile organic compound, SNAP = Selected Nomenclature for Air Pollution, EUROSTAT NUTS = European Statistical Office Nomenclature of Territorial Units for Statistics)

substances and acid species on a Europe-wide scale.

3) The Source–Receptor Ozone Model (SROM), based on the EMEP country-to-grid matrices (Simpson et al. 1997), is used to estimate ozone concentrations on a European scale.

The current version of the EcoSense package includes a large number of exposure–response functions and monetary values that were compiled and thoroughly reviewed within the ExternE projects (European Commission 1999). The impact assessment modules calculate the physical impacts by applying the relevant exposure–response functions to each individual grid cell of a Europe-wide grid (EMEP 50 × 50 km), taking into account the information on receptor distribution and concentration levels of air pollutants from the database. Because EcoSense uses a full European emission inventory as a starting point for air quality and impact modelling, it is possible to analyse the contribution of a specific source (or source group) to the exceedance of thresholds (e.g., critical loads for ecosystems) by taking into account predefined background conditions.

Application of water quality and aquatic fate modelling

Similar to air quality models discussed in the previous section, water quality or aquatic fate models such as the Geography-Referenced Regional Exposure Assessment Tool for European Rivers (GREAT-ER 1.0; Feijtel et al. 1997; Schulze et al. 1999) can be used to evaluate the impact from surfacewater releases. GREAT-ER calculates the concentrations of 'down-the-drain' chemicals from point-source emissions in receiving riverine water by coupling a geographical information system (GIS) with fate models. The river fate model is based on the CemoS/Water model, which calculates a spatial concentration profile for a point emission in a river stretch (Trapp and Matthies 1998). So far, incorporated regions include catchment areas from parts of England, Italy, Belgium, and Germany, but more regions from different European countries are in preparation. A similar model, Better Assessment Science Integrating Point and Nonpoint Sources (BASINS), is available for the U.S. (Lahlou et al. 1998).

GREAT-ER 1.0, which was developed in the context of risk assessment of chemicals, can be applied in LCA in different ways. The most straightforward way is to calculate fate factors for dispersive use of down-the-drain chemicals, based on reference emissions taking place in reference regions. The concentration increase averaged over all river stretches can then be used as a fate factor that is specific to both substance and catchment. The required parameters are the pollutant elimination rate in wastewater treatment and in the surface water, the method of wastewater treatment, as well as the hydrological data, which determines the dilution ratios in the different stretches.

Alternatively, in cases in which the functional unit can be assigned to a time span, an approach proposed by Schulze (2001) can be used and has been applied to detergents. All these approaches are suited only for cases in which a spatially explicit analysis is desired.

Use of the Models in LCIA

Multimedia models have been used in LCA to derive generic, location-independent TEPs, based on average landscape parameters and assuming a closed system (Guinée and Heijungs 1993). This reflects the traditional use of multimedia models for gaining qualitative insights into the fate of pollutants, such as pollutant partitioning, persistence, and travel distance. Single-medium models, on the other hand, are traditionally used in site-specific risk assessments to assess the increased dose in specific populations. They have been used in LCA in several countries to derive marginal dose increases resulting from an incremental increase in the average emissions of a single country (Krewitt et al. 2001). Spadaro and Rabl (1999) have suggested an approach that uses the results of air quality models to derive generic impact factors. Multimedia models usually consider the dose increment in a single, average individual, whereas medium-specific models calculate the incremental population dose.

Despite this difference in traditional use, the multimedia approach can also be used to derive country-average population dose increases or even to characterise the location-dependent distribution of the dose in the population. Single-medium models, on the other hand, can be used to derive large-scale averages, which can be used for generic characterisation factors (Table 4-3). Both models can be used to calculate individual or population dose. The use of the models in LCA, and the model setup and parameterisation, hence need to be distinguished from the general modelling capabilities.

Subsequently, we will consider the use of both modelling approaches for generic as well as location-dependent characterisation.

Evaluation of Fate and Exposure Methods

In this section, the fate and exposure modelling approaches discussed above are assessed against a set of suggested criteria. We address multimedia models and single-medium models as general approaches. Particular models are not addressed here because a detailed investigation is beyond the scope of this chapter. Our evaluation, however, is based on close familiarity with a number of the most prominent models.

Environmental relevance and comprehensiveness

1) What is the depth of analysis, that is, how large a part of the cause–impact chain is included? Is the indicator that is modelled at midpoint or endpoint level? (See Figure 4-2.)
2) What is the sophistication of the approach, understood as the degree of detail and accuracy in the modelling of the individual elements of the impact chain, for example, comprising spatial and temporal differentiation? (See Figure 4-2.)
3) What is the comprehensiveness of the model? Does it address all the relevant impacts associated with the impact category?

Table 4-3 Two different modelling approaches in comparison

Model characteristics	Multimedia models	Single-medium models
Strengths	Account for intermedia transfer and large number of possibly relevant exposure pathways	Account explicitly for chemical reactions and formation of secondary pollutants (e.g., photooxidants, particulates)
Traditional use	Generic evaluation using simple, uniformly mixed representations of environmental compartments	Risk assessment based on characterisation of local or regional distribution of pollutants
Emerging use	Global distribution, region-specific dose increments, and location-specific assessments	Calculation of generic characterisation factors based on averages across countries or continents

All the models discussed in 'Medium-Specific Approaches' (p 109) and 'Use of the Models in LCIA' (p 112) have the same depth of analysis: They are able to calculate ambient concentrations and presented dose to humans. They differ, however, in their sophistication and comprehensiveness.

Multimedia models consider many pathways, including those mediated through other compartments, such as the ingestion of pollutants that were originally deposited from air onto plants, or deposited onto soil and then absorbed by plants. These pathways are the most important for many pollutants, for example, cadmium and dioxin (Hertwich et al. 2001). Air quality models such as EcoSense explicitly include chemical reactions and thus cover the generation of secondary pollutants, which sometimes are of greatest concern. Secondary aerosols formed from sulphates, nitrates, and ammonia, but also from organic and inorganic carbon, may represent some of the most important human health impacts of all but are not considered in many current multimedia fate models. Photooxidant formation is currently the highest concern in the attainment of air quality standards. While SO_2 and NO_2 are included in the TEPs derived by Huijbregts, Thissen, Guinée et al. (2000) and Hertwich et al. (2001), several studies show that the consideration of secondary aerosol formation significantly increases the HTP for these substances (Hofstetter 1998). All approaches are able to consider spatial detail and timing, but not all current models do so. The consideration of spatial differentiation is currently more common in single-medium models.

Multimedia models currently address many more impacts and are appropriate for a larger number of pollutants. Single-medium models, however, address some important impacts more accurately. Modular multimedia approaches may prove valuable by linking single-medium models to provide multimedia solutions.

Scientific validity and reliability

1) To what extent have the model modules or overall model been externally peer reviewed and accepted within the relevant international scientific community?
2) To what extent have the model modules or the overall model been validated against empirical data?
3) To what extent have the uncertainties underlying the model, the scenarios represented, and the parameters it is based on been characterised?
4) To what extent are value choices present in the model, and what is the sensitivity to these?

The approaches described in this report are widely used to screen chemicals, assess risks, and derive LCA characterisation factors. Models such as EcoSense, GREAT-ER, EQC, CalTOX, and the European Union System for the Evaluation of Substances (EUSES) have received significant peer review, and some have even weathered regulatory review processes. Many of the underlying mechanisms are empirically based or have been confirmed in experiments. In spite of their wide acceptance, validation — or more appropriately, evaluation — has been conducted only to a limited extent. The modelling approaches are usually evaluated for only a few chemicals released under specific conditions. Air quality modelling has been subjected to more intense empirical tests, thanks to better monitoring data and a smaller number of relevant pollutants. Classic organic pollutant concentrations (polychlorinated biphenyl [PCB], hexachlorocyclohexane [HCH]) have been compared to multimedia model results, yielding significant insights into pollutant origin and behaviour. It is unclear, however, to what extent multimedia models also appropriately represent all relevant factors for a wider range of chemicals and release conditions. A stronger effort is now required to compare model results with empirical data.

Generic multimedia models are commonly adopted in LCA, reflecting the often unknown location of the emissions and lack of spatial differentiation within the model. Associated uncertainty has been estimated by looking at Monte Carlo propagation of parameter uncertainties for different regions (McKone 1994; Huijbregts, Thissen, Jager et al. 2000), although only a limited number of comparisons between spatial and nonspatial models have been conducted (Pennington 2001). An exploratory assessment of scenario and model uncertainty has been conducted by Hertwich, McKone, and Pease (2000), for example, and it reiterates the importance of comparison between modelling alternatives and with empirical data.

Accuracy is often highly dependent on the quality of input data. This area requires much greater attention, for example, in improving prediction techniques for degradation rate constants. There is a similar need for creating a peer-reviewed database, using defined sources, establishing hierarchical selection criteria, and quantifying parameter uncertainty.

All the methods presented to quantify fate and exposure in LCA involve some assumptions, value choices, and expert judgements. Model-based approaches are presently dependent on assumptions made by experts, for example, the identification of which environmental media need to be taken into account and which elements of a food web are important. These assumptions are of a technical nature and could be supported empirically if sufficient knowledge were available. Other choices, such as the significance of the timing of exposure, involve value choices. Infinite time without discounting is often used, reflecting time-integrated exposure, but other studies have integrated over shorter time periods of 20 or 100 years (Huijbregts et al. 2001) or used a discounting of 1% per year (Hertwich et al. 2001). The lesser consideration of later exposure, reflecting an uncertainty about whether it will actually occur, is especially important for metals and other highly persistent pollutants.

Transparency and reproducibility

1) Is there a well-defined and transparent procedure of calculation in the model and for collecting the necessary underlying data?
2) Will different practitioners arrive at the same results for a given substance? (Practitioners will often not be experts in environmental chemistry or toxicology.)

Many of the reviewed models are well documented with clear instructions. Some of the models are openly available, and the underlying equations can be reviewed by anyone who is interested. Other models are proprietary, or at least unpublished.

Because of model complexity, significant efforts are often required to understand and back-calculate model results.

Data collection is typically conducted by experts but, to date, often has resulted in discrepancies amongst practitioners. Differences in input data values, and the underlying quality of measurements, continue to plague the field and result in different calculated characterisation factors.

Relevance to the decision context

1) Is it possible to interpret the result in terms of a communicable environmental impact or damage?
2) Is it possible to aggregate the impact score with scores for other impact categories?

Fate and exposure methods provide insights into the relative behaviour of chemicals and emissions in the environment, which can be used directly in decision-making. For example, it may be desirable to reduce the emission of persistent chemicals even in the absence of toxicological information. Such midpoint insights are not, however, commonly addressed in isolation in current LCA practice. These measures provide input only to the subsequent methodological stages of toxicological characterisation in LCA.

Feasibility

1) Is the approach feasible for all the potentially important substances in the product life cycle? Many of these substances will not be identified as priority pollutants at a national level and hence the information may be limited.
2) What are the data requirements? To a large extent, the feasibility is governed by the requirements for underlying parameter data: the more specific the data, the fewer the substances or settings that the approach can be used for.

Sufficient data are available for evaluating a number of well-studied pollutants in single-medium and/or multimedia models. For most chemicals, however, the availability of data for evaluating fate and exposure remains limited in terms of quality. While methods are improving to predict partitioning coefficients used in multimedia models, measured transformation rates are often not available or not representative, and predicted values are based on simplistic models. Transformation products and environmental chemistry are still poorly understood, hence commonly ignored. Few chemicals (maybe a dozen) have been extensively studied. Measured and model insights exist for about 300 to 400 chemicals, primarily those of interest to different government environmental agencies and commonly found in LCAs.

Despite the limitations, the fate and exposure characteristics of any chemical can be, and are, quantified with existing data and data prediction capabilities. Even if significant uncertainties remain, modelling offers substantial benefits to decision-makers as long as uncertainties are reported. Research is required to continue improving the current practice. The development of a continually updated, peer-reviewed database on physical–chemical properties, degradation rates, and chemical reactions is considered essential.

Recommendations

If we accept that integrated modelling approaches are used to calculate characterisation factors, we must address the following issues:
- the degree to which spatial variations are considered,
- the timing of exposure or predicted effect,
- the standard of evidence required for information to be included in the assessment,
- the degree to which uncertainty should be quantified and how it should be treated,
- whether one or more models for the fate and exposure assessment should be used, and
- which models should be applied.

An approach to the location issue

Today, most LCAs use site-generic characterisation factors. Methods that include spatial variation range from simply ignoring emissions believed not to lead to exposure that exceeds a threshold (Hogan et al. 1996) to assessing of the number of people exposed (Spadaro and Rabl 1999). In principle, the following factors may be important:
- the background concentration and mixture composition,
- the sensitivity of the receptor,
- variations in population density,
- variations in landscape characteristics,
- variations in meteorological conditions, and
- the presence of mitigating factors that can neutralise or counteract the action of a pollutant.

Some recent methods ignore small-scale variations in spatial factors, which are relevant only for pollutants with a small spatial range, but include variations at large regional, continental, and global scales (Hofstetter 1998; Goedkoop and Spriensma 1999; Huijbregts 1999; Huijbregts, Thissen,

Guinée et al. 2000). Others have conducted preliminary investigations into local versus regional differences (Crettaz 2000; Nigge 2000).

A number of factors must be taken into account when we consider whether, and how, to best account for variations in the toxic effects associated with the location of a release. These factors concern the availability of information on the point of release in life-cycle inventories (LCIs), the availability of information characterising local differences in exposure and effect, as well as the significance of these differences and the desirability of including local differences.

The risk posed by specific emissions depends on, among other factors, the release site. This dependency is especially significant for pollutants that are not very mobile or persistent and for emissions to stationary compartments, for example, for soil injection. The characteristics of the release site (e.g., whether land is used for agriculture, the dilution volume of a receiving water) may therefore be relevant for the impact assessment.

A site-specific characterisation of toxic emissions in LCA requires knowledge of the release site or region and its characteristics. Most LCAs today lack information on the release site, at least for some of the processes in the life cycle, either because they assess generic products, because the products require generic inputs that are globally traded commodities, or because suppliers may change frequently. The characterisation of all the release sites may hence be infeasible. Even if it is possible, it may be too expensive and the variations associated with releases at different regions may negate each other's effects, to yield predictions similar to those of the generic models.

To justify the additional effort of a site-specific approach, significant variation in exposure and/or effect must exist and information on these variations must be available. If the differences from site to site are small or difficult to characterise, the gains of information from considering the release site will be too small to justify the effort.

A consideration of site-specific variations must be desirable within the context of a decision. For the assessment of human toxicity, variations in population densities can play an important role and affect exposure for criteria air pollutants by a factor of 3 to 30 between densely populated and rural areas (Spadaro and Rabl 1999; Crettaz 2000; Nigge 2000). However, outside of the quantification of uncertainty, some individuals may find the consideration of the local population density undesirable because it may lead to the relocation of polluting industries to less populated areas, and acceptable risk to individuals may be higher in low-density areas compared to high-density areas.

In line with the charge to the task force by WIA-2 (Udo de Haes et al. 1999, section 3.4), we propose the following approach to the issue of location dependency.[4]

Step 1
Generic indicators for the comparison of toxic compounds are used for an initial assessment. These generic indicators account for the general properties of the chemical, such as its toxicity, its persistence, and its availability to human and ecological receptors under generic conditions. These indicators do not take into account local differences in population density, landscape conditions, receptor sensitivity, or variations in exposure factors such as diet and activity patterns. We expect that these indicators will be sufficient for many purposes. If a more detailed analysis is warranted to achieve finer distinctions among alternatives and resources are available, proceed to Steps 2 and 3.

Step 2
Uncertainty factors are used to assess whether a site-specific assessment of toxic compounds of high concern would significantly change the conclusions

---

[4] Chapter 3 suggests a different procedure for the categories of acidification, eutrophication, and photooxidation.

of the LCA. If this is the case, proceed to Step 3. One of the techniques available to systematically assess the potential impact of site-specific assessment on the conclusions of an LCA is value-of-information analysis (Raiffa and Schlaifer 1961; Morgan and Henrion 1990). In many cases, a simpler sensitivity analysis may be sufficient.

Step 3

A site-specific assessment (e.g., USEPA 1999; Crettaz 2000; Hauschild and Potting 2000; Huijbregts, Thissen, Guinée et al. 2000; Nigge 2000; Krewitt et al. 2001; Schulze 2001) is used to evaluate releases considered most important. The spatial resolution necessary for this assessment will depend on the spatial range of the pollutant and the spatial scale at which important regional parameters vary.

It remains to be decided whether regional and continental approaches, such as those suggested by Hofstetter (1998) and Huijbregts, Thissen, Guinée et al. (2000), should be taken as generic indicators. The situation in Europe assessed by these approaches may be seen as representative enough for the whole world. Medium-specific models such as EcoSense and GREAT-ER are inherently more site-specific and can serve to identify the degree of spatial and temporal variation in exposure.

Temporal considerations

The accumulation of persistent chemicals, especially metals, in the environment has very long time constants. Steady-state models, designed to model cumulative exposure, treat immediate exposure the same way as exposure that may occur in thousands of years.[5] Udo de Haes et al. (1999) proposed that, as baselines, characterisation factors be calculated for an infinite time (possibly approximated by, e.g., 500 y for global warming) without discounting or a cutoff period. This reflects the time-integrated exposure scenario. In addition, shorter cutoff periods that yield considerably different results could be calculated in a sensitivity analysis (e.g., 100-y period). Assessments could hence introduce a preference for the timing of exposure. Huijbregts et al. (2001), for example, calculated toxicity potentials based on the integrated exposure over 20, 100, and 500 years. Hertwich et al. (2001) discount future exposure at 1% per year.

We recommend a further investigation and discussion of the timing of exposure. Our task force noted the importance of this choice for the assessment of metals, but we could not agree on what is essentially a value issue: how to address the rights of future generations (Hertwich, Hammitt, Pease 2000).

Uncertainty and the standard of evidence

How well we need to understand a pollutant before we consider it in decision-making depends on, among other things, the treatment of uncertainty. From a decision-making perspective, it is desirable to include all potential effects and hence a large number of pollutants. Scientists, however, are often reluctant to accept the evaluation of a pollutant if they judge the quality of the evidence as low and think that further research would yield significantly different results (Shrader-Frechette 1996). Treating poorly characterised pollutants as zero risk, however, is likely to result in larger errors than is accepting a low standard of evidence to assess releases for which higher-quality evidence is not available. In this case, it would be desirable for both scientists and decision-makers to indicate the uncertainty associated with an assessment.

---

5 For some metals, the concentration in the ores mined today is not too far from the concentration in the earth's crust. Steady-state models may calculate environmental concentrations that cannot be achieved with the abundance of these substances on the earth, because they ignore the 'recycling' of metals that will eventually be dominant.

We recommend that efforts include a large number of pollutants in the assessment and that these efforts also consider data sources thought to be of lower quality, such as property estimation procedures and unverified literature data. At the same time, the underlying data and data sources should be published, and third parties should be encouraged to provide better data. We recommend conducting further efforts in uncertainty analysis to evaluate the uncertainty in the characterisation of toxic chemicals and how it affects actual LCA case studies. We do not think that a routine description of the uncertainties associated with all releases will become feasible soon, but we suggest that the uncertainties in the most important impacts in an LCA should be characterised.

Model selection

No single model, or modelling approach, is superior in all situations or for all pollutants. We recommend combining results from single-medium and multimedia models to increase the relevance and comprehensiveness of the assessment. These models could eventually be integrated into a model library, similar to EcoSense, which already combines 3 different air pollution models. For the generic assessments, the combination of an improved multimedia model with the approach to air pollutant modelling proposed by Spadaro and Rabl (1999) appears promising. It will be important to select similar model parameters (modelled area, wind speed, population density, etc.) so that results can indeed be compared. For site-specific assessments, present single-medium results offer a viable approach for those pollutants that are not multimedia in nature. They can be combined with assessments of regional multimedia models or even detailed site-specific models such as Total Risk Integrated Methodology (TRIM).

For the further method development of the UNEP–SETAC Life-Cycle Initiative, we recommend developing both generic and site-specific models to address chemical reactions as well as multimedia exposure pathways. This model development should be sufficiently coordinated so that results can be compared across model groups. For human health, the individual exposure and population intake fraction — expressing the fraction of an emitted pollutant that is eventually absorbed by the entire population — can serve as metrics for comparison. These concepts also offer a promising basis for comparing model results with measurements.

Conclusions

This chapter outlines 2 approaches to chemical fate and exposure assessment in the development of characterisation factors for toxicological impacts in LCA. It describes choices that must be made in the toxicity categories in general, as well as in fate and exposure assessment specifically. These choices relate to the indicator form (comprehensiveness, sophistication, transparency, and accuracy); the question of local, regional, global, or generic assessments; and the selection of the models used for the assessment.

At this point, multimedia fate and exposure models are considered the most developed approach to fate and exposure assessment in LCA. They are the method of choice for generic fate and exposure assessment if all releases are supposed to be assessed by a single model. Medium-specific models, however, are more comprehensive in explicitly accounting for chemical reactions such as oxidation and ozone generation, which are important for some pollutants. Single-medium models lend themselves more naturally to a site-specific assessment, although site-specific multimedia models exist. We recommend using results of multiple models for the assessment of chemical fate and exposure in LCA. This will improve the comprehensiveness and quality of LCA characterisation methods. Careful model calibration will be needed to ensure that the results of complementary models are indeed comparable.

References

Bare JC, Hofstetter P, Pennington DW, Udo de Haes HA. 2000. Life cycle impact assessment workshop summary; Midpoints versus endpoints: The sacrifices and benefits. *Int J LCA* 5(6): 319–326.

Bare JC, Udo de Haes HA, Pennington DW. 2000. An International Workshop on Life-Cycle Impact Assessment Sophistication. Cincinnati OH, USA: United States Environmental Protection Agency (USEPA) and United Nations Environment Programme (UNEP). EPA/600/R-00/023. http://www.epa.gov/clariton/clhtml/pubtitle.html. Accessed 26 Apr 2002.

Brandes LJ, den Hollander H, van de Meent D. 1996. SimpleBOX 2.0: A nested multimedia fate model for evaluating the environmental fate of chemicals. Bilthoven, NL: National Institute of Public Health and the Environment (RIVM). Report Nr. 719101029.

Brode R, Wang J. 1992. Users' guide for the Industrial Source Complex (ISC2) dispersion models. Volumes I–III. Research Triangle Park NC, USA: U.S. Environmental Protection Agency (USEPA). EPA-450/4-92-008a. EPA-450/4-92-008b. EPA-450/4-92-008c.

Campfens J, Mackay D. 1997. Fugacity-based model of PCB bioaccumulation in complex aquatic food webs. *Environ Sci Technol* 31(2):577–582.

Cowan CE, Mackay D, Feijtel TCJ, van de Meent D, Di Guardo A, Davies J, Mackay N. 1995. The multimedia fate model: A vital tool for predicting the fate of chemicals. Pensacola FL, USA: Society of Environmental Toxicology and Chemistry (SETAC).

Crettaz P. 2000. From toxic releases to damage on human health: A method for life cycle impact assessment, with a case study on domestic rainwater use [PhD thesis N_2242]. CH-1015 Lausanne, CH: Federal Institute of Technology (EPFL). http://dgrwww.epfl.ch/GECOS/DD. Accessed 10 Apr 2002.

Crommentuijn T, Polder MD, Van de Plassche EJ. 1997. Maximum permissible concentrations and negligible concentrations for metals, taking background concentrations into account. Bilthoven, NL: National Institute of Public Health and the Environment (RIVM). Report Nr. 601501001.

Derwent RG, Nodop K. 1986. Long-range transport and deposition of acidic nitrogen species in north-west Europe. *Nature* 324:356–358.

European Commission. 1999. Externalities of energy : Methodology 1998 update. Brussels, Belgium: EC DG XII. ExternE Report No. 7. http://ExternE.jrc.es/. Accessed 4 May 2002.

Feijtel T, Boeije G, Matthies M, Young A, Morris G, Gandolfi C, Hansen B, Fox K, Holt M, et al. 1997. Development of a geography-referenced regional exposure assessment tool for European rivers: GREAT-ER contribution to GREAT-ER #1. *Chemosphere* 34(11):2351–2373.

Goedkoop M, Spriensma R. 1999. The Eco-Indicator 99. Amersfoort, NL: PRe Consultants. www.pre.nl. Accessed 10 April 2002.

Guinée J, Heijungs R. 1993. A proposal for the classification of toxic substances within the framework of life cycle assessment of products. *Chemosphere* 26(10):1925–1944.

Hauschild M, Potting J. 2000. Guideline on spatial differentiation in life cycle impact assessment: The EDIP 2000 methodology. Copenhagen, DK: Danish Environmental Protection Agency. Forthcoming.

Heijungs R. 1995. Harmonization of methods for impact assessment. *Environ Sci Pollut Res* 2(4):217–224.

Heijungs R, Wegener Sleeswijk A. 1999. The structure of impact assessment: Mutually independent dimensions as a function of modifiers. *Int J LCA* 4(1):2–3.

Hertwich EG, Hammitt JK. 2001. A decision-analytic framework for impact assessment. Part I: LCA and decision analysis. *Int J LCA* 6(1):5–12.

Hertwich EG, Hammitt JK, Pease WS. 2000. A theoretical foundation for life-cycle assessment: Recognizing the role of values in environmental decision making. *J Ind Ecol* 4(1):13–28.

Hertwich EG, Mateles SF, Pease WS, McKone TE. 2001. Human toxicity potentials for life-cycle assessment and toxics release inventory risk screening. *Environ Toxicol Chem* 20(4):928–939.

Hertwich EG, McKone TE, Pease WS. 2000. A systematic uncertainty analysis of an evaluative fate and exposure model. *Risk Anal* 20(4):437–452.

Hertwich EG, Pease WS, Koshland CP. 1997. Evaluating the environmental impact of products and production processes: A comparison of six methods. *Sci Tot Environ* 196:13–29.

Hertwich EG, Pease WS, McKone TE. 1998. Evaluating toxic impact assessment methods: What works best? *Environ Sci Technol* 32(5):A138–A144.

Hofstetter P. 1998. Perspectives in life cycle impact assessment: A structured approach to combine models of the technosphere, ecosphere and valuesphere. Boston MA, USA: Kluwer.

Hogan LM, Beal RT, Hunt RG. 1996. The threshold inventory interpretation methodology: A case study of three juice container systems. *Int J LCA* 1(3):159–167.

Huijbregts MAJ. 1999. Ecotoxicological effect factors for the terrestrial environment in the frame of LCA. Amsterdam, NL: Univ Amsterdam. http://www.leidenuniv.nl/interfac/cml/lca2/index.html. Accessed 10 April 2002.

Huijbregts MAJ, Guinée JB, Reijnders L. 2001. Priority assessment of toxic substances in life cycle assessment; III: Export of potential impact over time and space. *Chemosphere* 44(1):59–65.

Huijbregts MAJ, Thissen U, Guinée JB, Jager T, Kalf D, van de Meent D, Ragas AMJ, Wegener Sleeswijk A, Reijnders L. 2000. Priority assessment of toxic substances in LCA I: Calculation of toxicity potentials for 181 substances. *Chemosphere* 41(4):575–588.

Huijbregts MAJ, Thissen U, Jager T, van de Meent D, Ragas AMJ. 2000. Priority Assessment of Toxic Substances in LCA II: Assessing parameter uncertainty and variability. *Chemosphere* 41(4):541–573.

Jia CE, Di Guardo A, Mackay D. 1996. Toxic release inventories: Opportunities for improved presentation and interpretation. *Environ Sci Technol* 30(2):86A–91A.

Jolliet O, editor. Assies J, Bovy M, Finnveden G, Guinée J, Hauschild M, Heijungs R, Hofstetter P, Potting J, Udo de Haes HA, Wrisberg N. 1996. Impact assessment of human and ecotoxicity in life-cycle assessment. In: Udo de Haes HA, editor. Towards a methodology for life-cycle impact assessment. Brussels, B: Society of Environmental and Toxicology and Chemistry (SETAC) Europe. p 49–61.

Jolliet O. 2000. Human toxicity and ecotoxicity : Modelling versus scoring. In: Bare JC, Udo de Haes HA, Pennington DW, editors. UNEP/USEPA LCA sophistication workshop. Washington DC, USA: United Nations Environment Programme/U.S. Environmental Protection Agency (UNEP/USEPA). EPA/600/R-00/023, July 2000.

Jolliet O, Crettaz P. 1997. Fate coefficients for the toxicity assessment of air pollutants. *Int J LCA* 2(2):104–110.

Jolliet O, Crettaz P. 2000. Human toxicity and ecotoxicity: Modelling versus scoring. In: Bare J, Pennington D, Udo De Haes H, editors. An International Workshop on Life Cycle Impact Assessment Sophistication. Washington DC, USA: United States Environmental Protection Agency (USEPA).

Krewitt W, Friedrich R, Heck T, Mayerhofer P. 1998. Assessment of environmental and health benefits from the implementation of the UN-ECE protocols on long range transboundary air pollution. *J Hazard Mater* 61(1–3):239–247.

Krewitt W, Heck T, Trukenmüller A, Friedrich R. 1999. Environmental damage costs from fossil electricity generation in Germany and Europe. *Energy Policy* 27(3):173–183.

Krewitt W, Hurley F, Trukenmüller A, Friedrich R. 1998. Health risks of energy systems. *Risk Anal* 18(4):377–383.

Krewitt W, Mayerhofer P, Trukenmüller A, Friedrich R. 1998. Application of the impact pathway analysis in the context of LCA: The long way from burden to impact. *Int J LCA* 3(2):86–94.

Krewitt W, Trukenmüller A, Bachmann TM, Heck T. 2001. Country specific damage factors for air pollutants: A step towards site dependent life cycle impact assessment. *Int J LCA* 6(4):199–210.

Lahlou M, Shoemaker L, Choudhury S. 1998. The BASINS version 2.0 users manual. Washington DC, USA: U.S. Environmental Protection Agency (USEPA) Office of Water. EPA 823-B-92-006. http://www.epa.gov/OST/BASINS/. Accessed 10 April 2002.

Mackay D. 1991. Multimedia environmental models: The fugacity approach. Chelsea MI, USA: Lewis.

Mackay D, Boethling RS. 2000. Handbook of property estimation methods for chemicals : Environmental and health sciences. Boca Raton FL, USA: Lewis.

Mackay D, Diamond M. 1989. Application of the QWASI (Quantitative Water Air Sediment Interaction) fugacity model to the dynamics of organic and inorganic chemicals in lakes. *Chemosphere* 18:1343–1365.

Mackay D, Di Guardo A, Paterson S, Cowan CE. 1996. Evaluating the environmental fate of a variety of types of chemicals using the EQC model. *Environ Toxicol Chem* 15(9):1627–1637.

Mackay D, Di Guardo A, Paterson S, Kicsi G, Cowan CE, Kane M. 1996. Assessment of chemical fate in the environment using evaluative, regional and local-scale models: Illustrative application to chlorobenzene and linear alkylbenzene sulfonates. *Environ Toxicol Chem* 15(9):1638–1648.

Mackay D, Paterson S. 1981. Calculating fugacity. *Environ Sci Technol* 15(9):1006–1014.

Margni M, Jolliet O, Pennington DW. 2001. Degree of coupling between environmental compartments in multimedia models. 11th Annual Meeting of SETAC Europe; 2001 May 6–10; Madrid, Spain.

McKone TE. 1993. CalTOX, a multimedia total exposure model for hazardous-waste sites. Livermore CA, USA: Lawrence Livermore National Laboratory. UCRL-CR-111456PtI-IV. http://www.cwo.com/~herd1/caltox.htm. Accessed 10 April 2002.

McKone TE. 1994. Uncertainty and variability in human exposures to soil contaminants through home-grown food: A Monte-Carlo assessment. *Risk Anal* 14(4):449–463.

Morgan MG, Henrion M. 1990. Uncertainty: A guide to dealing with uncertainty in quantitative risk and policy analysis. Cambridge, GB: Cambridge Univ Pr.

Nigge K-M. 2000. Life cycle assessment of natural gas vehicles: Development and application of site-dependent impact indicators. Berlin, D: Springer.

Pennington DW, Bare JC. 2001. Comparison of human health screening and ranking approaches: The Waste Minimization Prioritization Tool (WMPT) and Toxic Equivalency Potentials (TEPs). *Risk Anal* 21(5):897–913.

Pennington DW, Yue PL. 2000. Options for comparison of process design alternatives in terms of regional environmental impacts. *J Clean Prod* 8(1):1–9.

Pennington DW. 2001. Multi-region multimedia chemical fate and exposure model for use in life-cycle assessment in Japan. 11th Annual Meeting of SETAC Europe; 2001 May 6–10; Madrid, Spain.

Potting J, Hauschild M. 1997. Predicted environmental impact and expected occurrence of actual environmental impact. *Int J LCA* 2(4):209–216.

Pratt GC, Gerbec PE, Livingston SK, Oliaei F, Bollweg GL, Patterson S, Mackay D. 1993. An indexing system for comparing toxic air pollutants based upon their potential environmental impacts. *Chemosphere* 27(8):1359–1379.

Raiffa H, Schlaifer R. 1961. Applied statistical decision theory. Boston MA, USA: Harvard Univ.

[RIVM, VROM, VWS] National Institute of Public Health and the Environment; Ministry of Housing, Spatial Planning and the Environment; Ministry of Health, Welfare, Sport. 1998. Uniform System for the Evaluation of Substances 2.0 (USES 2.0). The Hague, NL: RIVM, VROM, WVC.

[RIVM, VROM, WVC] National Institute of Public Health and the Environment; Ministry of Housing, Spatial Planning and the Environment; Ministry of Welfare, Health, and Cultural Affairs. 1994. Uniform System for the Evaluation of Substances 1.0 (USES 1.0). The Hague, NL: RIVM, VROM, WVC. VROM Distr. Nr. 11144/150.

Schulze C. 2001. Modelling and evaluating the aquatic fate of detergents. Osnabrueck, D: Univ Osnabrueck.

Schulze C, Matthies M, Trapp S, Schroder FR. 1999. Georeferenced fate modelling of LAS in the Itter stream. *Chemosphere* 39(11):1833–1852.

Shrader-Frechette K. 1996. Methodological rules for four classes of scientific uncertainty. In: Lemons J, editor. Scientific uncertainty and environmental problem solving. Cambridge MA, USA: Blackwell Science. p 12–39.

Simpson D, Olendrzynski K, Semb A, Storen E, Unger S. 1997. Photochemical oxidant modelling in Europe: Multi-annual modelling and source-receptor relationships. Oslo, N: Norwegian Meteorological Office. EMEP/MSC-W Report 3/97.

Spadaro JV, Rabl A. 1999. Estimates of real damage from air pollution: Site dependence and simple impact indices in LCA. *Int J LCA* 4(4):229–243.

Steen B. 1999. A systematic approach to environmental priority strategies in product development (EPS). Version 2000: General system characteristics. Göteborg, S: Chalmers Univ Centre for Environmental Assessment of Product and Material Systems (CPM). CPM Report 1999:4. www.cpm.chalmers.se. Accessed 10 April 2002.

Swanson MB, Socha AC, editors. 1997. Chemical ranking and scoring: Guidelines for relative Assessments of chemicals. Pensacola FL, USA: Society of Environmental Toxicology and Chemistry (SETAC).

Trapp S, Matthies M. 1998. Chemodynamics and environmental modelling: An introduction. Berlin, D: Springer Verlag.

Trukenmüller A, Friedrich R. 1995. Die Abbildung der großräumigen Verteilung, chemischen Umwandlung und Deposition von Luftschadstoffen mit dem Trajektorienmodell WTM. In: Jahresbericht 1995. Stuttgart, D: ALS Universität. p 93–108.

Udo de Haes HA. 1996. Discussion of general principles and guidelines for practical use. In: Udo de Haes HA, editor. Towards a methodology for life-cycle impact assessment. Brussels, B: Society of Environmental Toxicology and Chemistry (SETAC). p 7–30.

Udo de Haes HA, Jolliet O, Finnveden G, Hauschild M, Krewitt W, Müller-Wenk R. 1999. Best available practice regarding impact categories and category indicators in life cycle impact assessment. *Int J LCA* 4(2):67–74.

[USEPA] United States Environmental Protection Agency. 1997. Waste minimization and prioritization tool: Beta test version 1.0. Washington DC, USA: USEPA Office of Solid Waste and Office of Pollution Prevention and Toxics (OPPT). EPA 530-R-97-019.

[USEPA] United States Environmental Protection Agency. 1999. Total risk integrated methodology. Status report. Research Triangle Park NC, USA: Office of Air Quality, Planning and Standards. EPA 453/R-99-010.

Wania F, Mackay D. 1995. A global distribution model for persistent organic-chemicals. *Sci Tot Environ* 161:211–232.

Wania F, Mackay D. 1999. The evolution of mass balance models of persistent organic pollutant fate in the environment. *Environ Pollut* 100(1–3):223–240.

Woodfine DG, Seth R, Mackay D, Havas M. 2000. Simulating the response of metal contaminated lakes to reductions in atmospheric loading using a modified QWASI model. *Chemosphere* 41(9):1377–1388.

Indicators for Human Toxicity in Life-Cycle Impact Assessment

Wolfram Krewitt, David W. Pennington, Stig I. Olsen, Pierre Crettaz, Olivier Jolliet

Acknowledgements — The following members of this working group and reviewers have directly contributed verbal and written comments to this document. Most, though not necessarily all, are reflected in the text by the authors: Suzanne Efting (PRé Consultants), Patrick Hofstetter (USEPA-ORISE), Willie Owens (P&G), Edgar Hertwich (NTNU), Michael Hauschild (DTU), Helias A. Udo de Haes (CML), Ruedi Müller-Wenk (IWÖ), Stephan Volkwein (CAU).

Abstract — The main objectives of this task group under the Society of Environmental Toxicology and Chemistry (SETAC) Europe Second Working Group on Life-Cycle Impact Assessment (WIA-2) were to identify and discuss the suita-bility of toxicological impact measures for human health for use in characterisation in life-cycle impact assessment (LCIA). The current state of the art of defining health indicators in LCIA is summarised in this document, promising approaches are addressed in further detail under the headings 'Potency' and 'Severity', and then the suitability of the approaches is discussed with the aid of selected criteria.

Toxicological potency factors are based on test data such as no observable effect levels (NOELs). NOELs and similar data are determined in laboratory studies of rodents and are then extrapolated to more relevant human measures. Many examples also exist of measures and methods beyond potency-based indicators that attempt to account for differences in expected severity, as well as potency. Quantitative severity-based indicators yield measures in terms of years of life lost (YOLLs), disability-adjusted life years (DALYs), quality-adjusted life years (QALYs), and other similar measures. DALYs and QALYs are examples of approaches that attempt to account for both YOLLs (mortality) and years of impaired life (morbidity). Qualitative severity approaches tend to arrange potency-based indicators in categories, avoiding the need to express differences in severity quantitatively. Based on the proposed criteria and current state of the knowledge, toxicological potency indicators are pre-selected as a minimum default. Addressing accuracy and ensuring consistency, particularly when data are being extrapolated, are seen as some of the key issues that are beginning to be addressed in LCIA. While associated approaches are still in their infancy, we encourage taking into account relative severity whenever possible, using qualitative and/or quantitative approaches.

Introduction

This position paper was prepared by the Task Group on Human Toxicity, which was established under the Society of Environmental Toxicology and Chemistry (SETAC) Europe Second Working Group on Life-Cycle Impact Assessment (WIA-2). The objective of WIA-2 is 'to contribute to the establishment of best available practice regarding impact categories, together with category indicators [CIs], and lists of concomitant characterisation factors to be used in life-cycle impact assessment [LCIA]' (see WIA-2 background document, Udo de Haes et al. 1999). The main objective of our task group was to identify and discuss the suitability of toxicological impact measures for human health for use in characterisation in LCIA.

Toxicological characterisation factors for human health are calculated by taking into account time-integrated fate, exposure of a unit mass of chemical released into the environment (including, in many cases, the size of the exposed population), toxicological potency (a quantitative measure related to the dose–response of a chemical, such as the lowest observable effect level [LOEL]), and toxicological severity (a measure or description, qualitative or quantitative, of the effect incurred, such as bladder cancer or skin irritation). These stages are illustrated in Figure 5-1. The Task Group on Human Toxicity, hence this chapter, addressed current and developing practice in the areas of toxicological potency and severity for LCIA. With minor exceptions such as the consideration of exposed population characteristics and size, fate and exposure are addressed separately in Chapter 4.

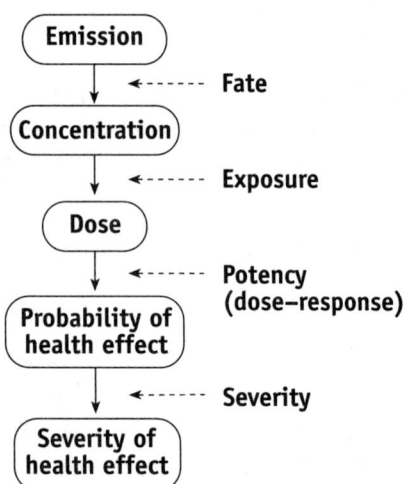

Figure 5-1 Outline of stages for calculating characterisation factors for human health

Within LCIA, the assessment of effects related to the impact category 'human toxicity' focuses on effects that result from direct exposure to chemicals. Health effects caused by other agents, or by other mechanisms of action, either are not clearly allocated to one of the impact categories suggested, for example, in the SETAC WIA-2 background document (Udo de Haes et al. 1999; e.g., impacts from radiation, fine particles, from noise), or are covered by other impact categories (e.g., health effects from increased tropospheric ozone concentrations). While the mechanism of exposure in these cases might differ from that of toxic chemicals, the resultant effects on human health are best described using comparable indicators. Therefore, we attempt in this chapter to extend the human toxicity impact category to a human health category by taking into account effects from fine particles, tropospheric ozone, and radiation. In the case of tropospheric ozone, the photochemical ozone creation potential (POCP), which primarily reflects fate and exposure, is discussed in detail in the related SETAC position paper by Potting et al. (2002). Health effects from stratospheric ozone depletion and global warming are not addressed here, but the indicators discussed in principle can be applied equally for these impact categories.

Following the recommendations of the WIA-2 background document, one of the major challenges for the assessment of human health effects in the context of LCIA is to extend current indicators that are based on a substance's potency by providing information on the severity of the expected effect in the environment. Human health indicators are classified and presented here as toxicological potency-based indicators (reflecting the likelihood or probability of an effect, which is sometimes termed *hazard*) and severity-based indicators (reflecting both the likelihood and the consequences or resultant damage[1]). Taking severity into account is expected theoretically to increase the amount of information available to decision-makers, hence to improve the basis of their decisions. This additional information is also

---

[1] The term *damage* is sometimes used to describe an economic damage resulting from an environmental impact, while here we refer to a physical damage. We adopt *severity* for precision and its current acceptance in the toxicological literature. For example, Murray and Lopez (1996) refer to the severity of disease.

expected to help improve subsequent weighting and valuation steps in life-cycle assessment (LCA). We therefore discuss both qualitative categorisation and quantitative scaling approaches.

Severity indicators provide information that is sometimes considered more relevant environmentally (i.e., more directly linked to society's concerns), but the relationship to the environmental interventions can be more uncertain. However, both potency-based and severity-based indicators are considered as so-called *endpoint-related indicators*.[2] For a given population, potency-based indicators provide a measure of the likelihood that people potentially will be affected by an emission. Severity-based indicators provide a measure that takes into account resultant hardships that may be experienced in terms of, for example, years of quality life lost due to death or injury. Severity-based factors therefore go a step further in specifying the effect that is of societal concern. Noting the conclusion of the SETAC–U.S. Environmental Protection Agency (USEPA)–Centre of Environmental Science (CML) Brighton workshop (Bare et al. 2000), that indicators should be presented at different points in the environmental mechanism to provide different insights with differing types of uncertainty, it is expected that potency- and severity-based approaches will be used in a complementary way. The relative merits of each indicator basis are discussed in this chapter, with the aid of selected criteria.

In the second section, we outline current practice in LCA. In the third and fourth sections, we discuss the various approaches and related issues under the headings 'Potency' and 'Severity', respectively. We present an evaluation in the fifth section, and concluding remarks in the sixth.

Current Practice in LCIA

A range of indicators is in current use. Eco-indicator 99 (Goedkoop et al. 1999) is one example of a tool in which both potency and severity are taken into account. The Environmental Priority Strategies (EPS) methodology uses a severity-based indicator, where morbidity and nuisance[3] are then weighted by a willingness-to-pay (WtP) approach (Steen et al. 1999). Although not an LCA methodology, the ExternE project (European Commission 1999) also adopted severity-based indicators weighted by using data as an indication of individuals' preferences. Most other methodologies currently use potency-based indicators. Several methodologies for assessing toxicological effects to humans using potency-based indicators have been proposed by Guinée et al. (1996; updated by Huijbregts et al. 2000), Jolliet and Crettaz (1996), Hauschild et al. (1997), and Hertwich (1999).

Available methodologies follow the framework presented by the first SETAC working group (Jolliet 1996), that is, that the impact score for each substance is presented as the product of an effect factor, a fate and exposure factor, and the total mass loading of the emissions (see also Figure 5-1). This framework is adapted from the principles of risk assessment.

The fate and exposure measure in LCIA is a predicted daily intake, that is, a daily dose (ingested or inhaled). More recently, this has been expressed as an exposure efficiency or dose-fraction (e.g., fraction of mass released that is either inhaled or ingested). In line with the doctrines of LCA, the fate and exposure measures account for the time-integrated concentration of the substances in each environmental compartment and associated exposure related to factors such as inhalation and con-

---

[2] Factors can reflect measures at midpoints or endpoints, depending on whether they reflect differences at midpoints or endpoints in the environmental mechanism or cause–effect chain (Bare et al. 2000).

[3] Here, *nuisance* is understood as a mild form of nuisance that does not constantly irritate people. Visibility reduction, dirty surfaces, or a moderate noise level is regarded as a nuisance. (Steen et al. 1999)

sumption rates (of vegetables, beef, milk, and fish, etc., associated with complex food webs). Dermal exposure is not addressed in most current approaches because of a lack of associated toxicological data for this route and the common opinion that it is often a relatively negligible pathway (which may not be true in all cases).

The toxicological potency measure in LCIA approaches is usually a slope factor based on risk per unit dose of a given effect for carcinogens and, either implicitly or explicitly, on the dose–response gradient between 0 and a given measure for non-carcinogens. No threshold is generally assumed, or taken into account, and both gradient measures are usually on a linear scale. These *no-threshold* and *linear* assumptions eliminate the need to account for background concentrations at specific sites when the marginal change in effect associated with a given emission is estimated. Crettaz et al. (2002) and Pennington et al. (2002) summarise arguments in support of these assumptions. LCA is expected to continue to adopt such no-threshold, linear approaches, while complementary methodologies such as risk assessment may address whether acceptable adverse risk levels are exceeded for specific emissions at specific sites.

The toxicological potency measure selected for a chemical may be a value derived from an extensive review of the available toxicological literature by an expert panel, as reflected in the derivation of regulatory measures such as

- the allowable daily intake (ADI) by the Joint Food and Agriculture Organization of the United Nations and the World Health Organization Expert Committee on Food Additives (JECFA/WHO),
- the oral reference doses (RfDs) by USEPA, or
- a survey of databases such as the Registry of Toxic Effects of Chemical Substances (RTECS) or the Hazardous Substances Data Base (HSDB).

In both cases, the potency is usually based on laboratory studies conducted on experimental animals and then extrapolated using factors to arrive at a relevant measure (e.g., in some methodologies, acute data from tests on rodents are extrapolated to chronic measures for humans). Problems for LCA practitioners include the inconsistency in the degree of conservatism adopted for extrapolations in and across regulatory applications and the availability of multiple measures for some compounds.

Life-cycle assessment usually does not permit the quantification of actual effects, primarily due to practices adopted in the inventory phase, but does enable the characterisation of the relative impact of emissions associated with a product's life cycle (Udo de Haes 1996). The toxicological indicator of each substance emitted must be additive, in order to aggregate the indicator results of the many different emissions that occur throughout a life cycle. If population is taken into account, when added, current potency indicators provide a count of the number of people affected by a given emission (if not, the risk of an effect to an individual). In current practice for potency-based indicators, these are then summed across the different emissions, noting that an individual in the population may be affected by more than one emission. Relative severity is not considered. Reflecting some extreme stances, the relevance of such potency-based indicators is therefore questioned (e.g., Owens 1996, 1998).

The damage to human health by a chemical may differ substantially, depending on whether the effect causes, for example, a gastrointestinal (GI) inflammation that lasts for a few days or weeks or a severe foetal malformation. This is sometimes regarded as a limitation of current practice (Burke et al. 1996). Some practitioners have therefore adopted, or proposed, measures to take into account the relative severity of a chemical's effects (European Commission 1999; Goedkoop and Spriensma 1999). The categorisation of potency-based indicators and the so-called *severity-based indicators* account for both potency and severity, sometimes using epidemiological insights (Hofstetter 1998).

Different formats have been adopted to convey results to decision-makers. In some methodologies, the result is presented in the form of an equivalency factor. In the equivalency factor approach, the combined fate–exposure–potency factor of a substance is divided by that of a reference substance (e.g., 1,4-dichlorobenzene used by Guinée et al. [1996]). The final indicator value then reflects the result in terms of equivalents of the reference substance (e.g., the relative toxicological impacts associated with one life cycle compared to another in terms of 1,4-dichlorobenzene equivalents). Other methods, such as Environmental Design of Industrial Products (EDIP; Hauschild and Wenzel 1997), avoid the use of a reference substance but reflect the results instead in terms of a critical volume approach. The *critical volume* is the volume of water, air, or soil in which the pollutant must be diluted in order not to exceed a given potency (or severity) level. The choice between these approaches may have consequences for interpretation in the valuation and weighting phases.

In the next 2 sections, 'Toxicological potency' and 'Severity' (p 133), we discuss the merits and limitations associated with different potency and severity measures. For additional discussions that provide greater depth on some of the summarised issues, the reader is advised to consult, for example, Burke et al. (1996), Jolliet (1996), Owens (1996, 1998), Udo de Haes (1996), Barnthouse et al. (1997), Olsen and Hauschild (1998), Crettaz et al. (2002), and Pennington et al. (2002).

Toxicological Potency: Dose–Response

We commonly use toxicological potency as an indicator for toxic effects, as a first step before considering severity, or as an endpoint indicator in

its own right. In principle, all chemical substances can cause adverse effects in humans. In contrast to other impact categories, human toxicity therefore theoretically includes all substances (and emissions), and it includes many different toxicological effect mechanisms. The detailed mechanism of action is not known, however, for most substances.

There is a long tradition of using toxicological information, especially for regulatory purposes (e.g., authoritative approval of drugs, pesticides, and other chemicals). For the detailed assessment of the toxicological effects of a substance to humans, knowledge of the following issues is desirable or necessary:

- Uptake, distribution, metabolism, and excretion of the substance in the human organism — Metabolism rates and pathways of the same substance can vary significantly between individuals, which is one parameter that gives rise to differences in sensitivity.
- Effects of the substance — Effects vary on the basis of characteristics such as acute toxicity by inhalation, oral, or dermal exposure; irritation and sensitisation properties; systemic toxicity; carcinogenicity, genotoxicity, reproductive toxicity, neurotoxicity, and immunotoxicity.
- Dose–response relationships — The toxicological response normally follows a sigmoid curve as a function of the intake dose. Both the dose levels resulting in response and the steepness of the curve are important.
- Biological mechanisms by which the substance exerts its effect — A substance may have several effects, and although it has been tested to describe one effect, it is not possible to draw parallels to other types of effects because different mechanisms may be involved.

Although a well-established scientific background for retrieving toxicological information and data exists, experimental toxicological information is available for only a small percentage of the marketed chemicals (see Table 5-1). In the absence of such data and for ethical reasons, tools are becoming more widely available to predict toxicological potency, but their scope and reliability often remain limited.

Most toxicological studies aim at deriving information that may be used to establish a virtually safe dose for regulatory purposes, such as an ADI, below which no unacceptable risk of adverse effects is expected. The toxicological information often is established through experiments, which are not without fundamental problems (National Research Council [NRC] 1994). The experiments use small groups of test animals and usually are conducted for chemicals in isolation. The purpose is to provide a qualitative understanding of the effect mechanism and a quantitative determination of the dose–response relationship for known effects that are considered hazardous or critical. Given the conditions of the experiments, often short-term mortality studies of rodents, it is commonly necessary to extrapolate to a more relevant basis, namely humans. Given such insights, the relevance of the results in the context of human populations, with various levels of resistance, exposed to such chemicals as part of complex, often-interacting mixtures, is still widely debated.

In the next 2 sections ('Noncarcinogens' and 'Carcinogens', p 131), we provide a more detailed discussion of the merits and limitations of currently available potency measures, with the objective of identifying a path towards best available practice in LCIA. Note that this distinction between noncarcinogens and carcinogens is rather historic in nature, and it may not reflect the relative severities of the chemicals or the differences in how potency (dose–response) measures should be addressed, as we will discuss later.

Table 5-1 Availability of toxicity data for high production volume (HPV) substances

Data type	Availability (%)	
	Estimated by European Chemical Bureau[a]	Estimated by USEPA[b]
Acute toxicity	90	49
Subacute toxicity	53	—
Carcinogenicity	10	—
Mutagenicity	62	34
Fertility	20	23
Teratogenicity	30	
Chronic toxicity	—	14
Acute ecotoxicity (fish or daphnia)	55	—
Short-term toxicity (algae)	20 to 30	—
Toxicity to terrestrial organisms	5	—
Environmental fate	—	31

[a] >1000 t in the European Union (EU). For all marketed chemicals, it is estimated that only 5% to 10% are studied for acute toxicity and 1% for longer-term toxicity such as cancer, reproductive toxicity, etc. (Bro-Rasmussen et al. 1996).

[b] A study by USEPA (1998) reached similar conclusions regarding data availability for HPV chemicals.

Noncarcinogens

In traditional toxicological safety assessments, the ultimate aim is to estimate a virtually safe dose, for example, ADI or RfD. This is achieved by determining the critical effect and estimating the no observable adverse effect level (NOAEL) or the lowest observable adverse effect level (LOAEL) for this effect. The NOAEL is the highest dose that does not cause a statistically differentiable effect of interest in the test population, which is then often extrapolated to humans. The toxicologist's interpretation of the experimental data is crucial for the determination of the adverse effect. The designation of a given effect as adverse becomes increasingly complex as increasingly subtle effects are identified by more sophisticated techniques and assays (DeRosa et al. 1989).

The NOAEL (or LOAEL) can be divided by extrapolation factors to account for differences in sensitivity between humans and animals, between humans (i.e., specifically sensitive individuals), and between short-term test periods and long-term exposure periods of human populations. Furthermore, a modifying factor can be employed on the basis of a judgement about the study's quality and relevance (Barnes and Dourson 1988). In essence, the derivation of virtually safe doses requires professional judgement, and each individual toxicologist or panel of toxicologists develops their own judgement of principal studies, critical effects, extrapolation factors, etc. Virtually safe doses developed by different institutions and levels of conservatism across chemicals are therefore not always similar, which poses a problem in LCIA.

In 1992, the U.S. Society of Toxicology held a symposium that addressed such issues in 'Improvements in quantitative noncancer risk assessment' (Beck et al. 1993). The basic thrusts of the symposium were these:

- How can mechanistic and other data be used on a case-by-case basis to avoid the use of default extrapolation (uncertainty or safety) factors?
  - The safety factors are arbitrarily set at 10, meaning that the level of protection may differ from chemical to chemical and that risk assessment and risk management are inappropriately combined (Baird et al. 1996).
- How can the full set of experimental data be incorporated into the determination of toxic potency?
  - The ADI is based on only one value and therefore is dependent on the conditions of one particular experiment.
  - Experiments involving fewer animals tend to produce larger NOAELs and consequently may produce larger virtually safe doses.
  - The slope of the dose–response curve plays little or no role in determining the NOAEL.

In the next 3 subsections, we discuss the problems associated with extrapolation, the measures adopted to calculate the potency measure in LCIA, and the existence of thresholds.

Extrapolation

Toxicologists' traditional use of factors to extrapolate between species, extrapolate to sensitive individuals, and extrapolate from less-than-chronic studies is the subject of considerable discussion (Baird et al. 1996; Dourson et al. 1996; Pennington 2002). For commonly used safety factors, data are often divided by a safety factor of 10 for each required extrapolation. However, the real differences have been found to vary considerably between substances. For example, in many cases, a factor of 3 may be adequate for the extrapolation from subchronic to chronic exposure, whereas interindividual variability in susceptibility has been shown to be as high as 10 000 (Beck et al. 1993).

In addition to the difficulties of extrapolation, ADIs and RfDs are intended to provide a virtually safe dose. Test results are commonly extrapolated to 'safe doses'. This practice can result in bias for chemicals with few data, which is sometimes considered desirable in regulatory screening applications. Because the need in LCA is a realistic effect potential to be comparable with the other impact categories, Burke et al. (1996) argued that using an NOAEL directly without safety factors would be better for LCA purposes. However, the very limited data availability emerges as a significant drawback because a reliable chronic NOAEL is available only for relatively few substances, as illustrated in Table 5-1.

Extrapolation is still a widely accepted practice amongst toxicologists and is considered to remain necessary in LCA. To support this practice, best-estimate extrapolation factors can be calculated from empirical insights with associated uncertainty distributions (Jager et al. 1997; Pennington 2002).

Potency measures: NOAEL or NOEL versus benchmark dose versus effective dose

The NOAEL is the highest dose that does not cause a statistically differentiable effect of interest in a tested population. In the U.S., the benchmark dose (BMD) is a progressing alternative measure, although it is not yet widely applied. The BMD is an estimate of the lower confidence limit of the dose that affects a small percentage of the population (e.g., 1%, 5%, or 10%) compared to the control group. The percentage chosen can depend on the severity of the critical effect (e.g., GI inflammation 10% and teratogenic effects 1%). One of the goals in selecting a benchmark risk (BMR, the level of increased response that the BMD represents) is to make it as small as practical without the BMD becoming too model dependent. The estimation is based on statistical modelling of the dose–response curve, which is continuous, whereas an NOAEL will always be one of the experimental doses and, thus, in part is chosen by the study investigators. The USEPA (1995) referred to 3 studies in which the NOAEL and BMD were compared for an array of datasets. In almost all cases, the BMD turned out to be smaller or similar to the NOAEL.

Crettaz (2000), Crettaz et al. (2002), and Pennington et al. (2002) adapted the risk assessment concept of BMD for LCIA, proposing to adopt the effective dose to 10% of test organisms (ED10) measure (subscripted with h to denote humans or a to denote animals). ED10 is the best estimate of the dose that induces a 10% added risk over background for humans (see Figure 5-2). This proposed ED10 approach differs from the USEPA's application of the BMD10 at 2 levels:

1) The ED10 is considered, instead of the BMD10, to obtain the best estimate of the risk rather than an upper bound. The lower confidence limit provides one estimate of the associated uncertainty range.

2) While the USEPA proposes to use the BMD to derive an RfD (i.e., an estimate of daily oral exposure that is likely to be without an appreciable risk of deleterious effects during a lifetime), Crettaz et al. (2002) and Pennington et al. (2002) instead use the ED10 to quantify the risk of toxic effects, assuming a linear dose–response curve without threshold (β_{ED10} = 0.1/ED10, risk per unit dose; Figure 5-2).

The nonthreshold assumption is considered justified by the growing recognition that 'no evidence' does not necessarily mean 'no effect' and that bioassays cannot give real insights on linearity or nonlinearity at low doses, which only depend on the extrapolation model adopted. In agreement, recent epidemiological studies have suggested that there are no safe levels for some compounds. This issue is discussed further in the next section, 'Thresholds'.

The ED10 is highly correlated with the BMD and to the more widely available NOAELs for animals (Crettaz 2000; Crettaz et al. 2002; Pennington et al. 2002). Pennington et al. (2002) present ED10s and corresponding β_{ED10} slope factors for more than 600 chemicals. Given the high uncertainty, extrapolation of ED10 from the lethal dose data (e.g., LD50) unfortunately cannot be considered to be reliable beyond preliminary screening (Crettaz 2000; Crettaz et al. 2002; Pennington et al. 2002; Pennington 2002).

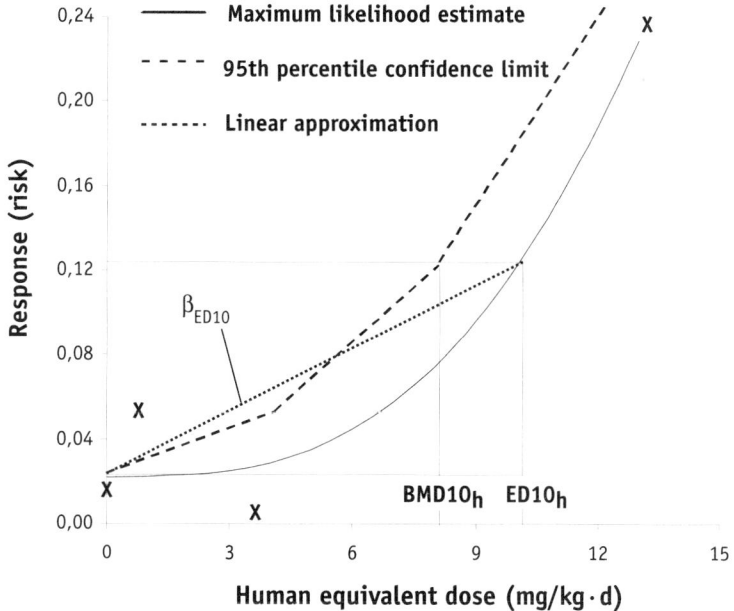

Figure 5-2 Dose–response measure β_{ED10} for acephate insecticide predicted by fitting the Crouch (1985) multistage model to data (X) observed in a mice bioassay, reported in the Integrated Risk Information System (IRIS) database (USEPA 1998; Crettaz et al. 2002; Pennington et al. 2002).

Thresholds

Toxicologists argue that mechanistic threshold concentrations or doses may exist for human health effects for many substances, noting that statistically differentiable thresholds are observed for individual substances tested on small populations of carefully maintained test animals. Bioassays cannot provide insights into low dose responses (below, say, a risk of 10^{-2}), and below acceptable risk thresholds (usually a risk of 10^{-4} to 10^{-6} or a measure based on the NOAEL for noncarcinogens), the extent of additive, synergistic, or antagonistic effects remains unknown. In epidemiological studies, it usually has not been possible to establish the existence of mechanistic thresholds (e.g., European Commission 1999).

In LCA, accounting for thresholds will require site-dependent background information and the consideration of interactions within complex mixtures. The derivation of such thresholds and accounting for them will long remain beyond the scope of LCA. A precautionary stance of no-thresholds is appropriate, following the doctrines of the precautionary principle. LCA will continue to provide a powerful tool to identify areas for improvement and, possibly, to focus resources for further site-dependent investigations.

Carcinogens

At least for genotoxic carcinogens, it is believed that there is no virtually safe dose because there is no mechanistic threshold for the carcinogenic effect; hence, up to now, associated substances have been addressed separately in traditional risk assessments. Known nongenotoxic carcinogenic effects, as well as necessary extrapolations for genotoxic effects, are treated analogously to noncarcinogens, as described in the previous section, 'Noncarcinogens'.

The risk that a substance causes cancer at different doses is estimated by models using bioassay data, for example, based on the low-dose slope factor of a dose–response curve, extrapolated from data in the observable range. The term *slope factor*, used by the USEPA, is based on a similar concept and describes the increased cancer risk from a lifetime exposure to an agent per milligram substance intake per kilogram body weight (risk per unit dose per kg body weight). The WHO (1987) adopts a similar measure known as the *unit lifetime risk* or *unit risk*. The unit risk factor is an estimate of the probability that an average individual will develop cancer when continuously exposed to an agent at a concentration of 1 µg/L in water, or 1 µg/m³ in air over the individual's life (70 y).

Various international and national organisations, including the WHO and the USEPA, suggest unit risk factors and slope factors for a wide range of different substances. The Integrated Risk Information System (IRIS; http://www.epa.gov/ngispgm3/iris/) database of the USEPA provides a quite comprehensive compilation of unit risk and slope factors. Hofstetter (1998), Crettaz et al. (2002), and Pennington et al. (2002) provide a compilation of unit risk factors and slope factors from various sources.

As mentioned in 'Current practice in LCIA' (p 125), current LCIA methods for characterising carcinogenic effects are based upon the principles developed in risk assessment. These principles and their application in LCA have been reviewed by, for example, Crettaz (2000), Crettaz et al. (2002), Pennington et al. (2002), and Pennington (2002). Most methodologies, for example, the Eco-indicator 99 approach (Hofstetter 1998; Goedkoop et al. 1999), adopt the low-dose slope factor q_1^* measure to quantify the risk of cancer. However, there are some drawbacks to their application in LCIA. The dose–response assessment is based upon the extrapolation towards low dose using a mathematical model, most commonly the linear multistage model. Using such models is open to criticism, since associated extrapolations can lead to large differences in the projected risk at low doses, depending on the model selected (Crettaz 2000; Crettaz et al. 2002; Pennington et al. 2002). The high associated uncertainties of extrapolating the dose–response curve towards low doses are usually hidden.

The application of the BMD10 has recently been proposed (USEPA 1996). As they did for noncancer effects, Crettaz (2000), Crettaz et al. (2002), and Pennington et al. (2002) proposed the related ED10 measure for use in LCIA to provide a consistent basis for comparison. The ED10 is defined as the best estimate of the dose corresponding with a 10% risk over background. A linear dose–response curve without threshold is retained for low-dose extrapolation ($\beta_{ED10}=0.1/ED10$) in LCA.

Using the IRIS database, Crettaz et al. (2002) and Pennington et al. (2002) demonstrated that β_{ED10} and q_1^* are strongly correlated. ED10s are also strongly related to the USEPA's BMD10. A correlation was similarly found between ED10 and the toxic dose 50 (TD50, or ED50) values reported in the Carcinogenic Potency Database (Gold and Zeiger 1997). As a result, values of β_{ED10} were calculated for 600 substances (Crettaz 2000; Crettaz et al. 2002; Pennington et al. 2002). No robust correlations with LD50 data (from short-term studies resulting in 50% mortality within the study period) have been suggested in the literature.

Radiation

Although radiation effects in general are not considered toxic effects, as a response to Udo de Haes et al. (1999), we suggest including radiation effects under the human toxicity impact category in order to provide a harmonised conceptual framework for assessing different impacts on human health. Ionising radiation leads to an increased probability of cancer, so that indicators similar to those for carcinogenic substances can be used.

The manifestation of health effects after exposure to ionising radiation is governed by different biological mechanisms and has been classified into 2 categories: 1) deterministic effects and 2) stochastic effects. Deterministic effects occur above a threshold level of radiation exposure (approximately 1 Sievert), and the severity of the impact increases with increasing exposure. Below the threshold level of deterministic effects, we become concerned with stochastic effects. In this range of doses, an increase in dose increases the probability of an effect but not the severity of the effect. In the field of LCA, we generally deal with low doses that result in stochastic effects.

Calculating the expected occurrence of cancer or fatalities following radiation exposure is facilitated by mathematical models with age-specific baseline rates and a small number of regression parameters estimated from epidemiological data. Most information available for estimating the effects of exposure to ionising radiation comes from the follow-up study of survivors of the atomic bomb explosions at Hiroshima and Nagasaki. The International Commission on Radiological Protection (ICRP 1990) has recommended risk factors that establish a relationship between the exposure and the occurrence of fatal cancer, nonfatal cancer, and hereditary effects. An operational model for including human health damages caused by radiation in LCIA has been developed by Frischknecht et al. (2000) for use in Eco-indicator 99.

Severity

The International Life Sciences Institute (ILSI) Health and Environmental Sciences Institute (HESI) convened an expert panel to give recommendations on a methodology for addressing toxicological impacts in LCA (Burke et al. 1996). One of their main reservations about existing methodologies was the use of toxicity values that incorporate safety factors (e.g., the RfD used by the USEPA or the ADI used by JECFA/WHO).

The other main reservation was the disregard for the severity of the effects (Burke et al. 1996). The methodologies for assessing human toxicological impacts in LCA are sometimes criticised for not reflecting the severity of an effect because they aggregate toxicological measures based only on the NOAEL (Burke et al. 1996; Jolliet 1996; Owens 1996), or similar measures. The procedure can provide a relative measure of the number of individuals affected, for example, but irritative substances are implicitly valued as equal to substances that have irreversible effects such as foetal malformations. In an effort to overcome such limitations, both qualitative and quantitative methodologies have been proposed.

Qualitative approaches to account for severity

To better account for the severity of different health effects, the ILSI panel proposed a classification of substances in 3 subcategories according to the severity of the related effects (Table 5-2). Jolliet (1996) proposed the use of similar subcategories, or if this is not possible, a classification into known mechanisms (e.g., acute toxicity, irritation, carcinogenicity, teratogenicity). More recently, Owens (2002) has expanded on the ILSI proposal for subcategorisation, testing the applicability of this proposal with USEPA-IRIS datasets. The ILSI panel approach is summarised here to exemplify some of the benefits and problems associated with categorisation.

The ILSI panel procedure has both scientific and subjective (i.e., value-laden) components and requires input from experts because 'informed valuation' is a critical part of the process, for example, when substances are classified into subcategories. As mentioned earlier, the scientific part also includes professional judgement. The subcategories can be weighted in the characterisation or valuation step; Subcategory 1, for instance, by a factor of 100; Subcategory 2 by a factor of 10; and Subcategory 3 by a factor of 1 (Burke et al. 1996). The

Table 5-2 Proposal for the definition of human toxicity subcategories according to the severity of effect[a]

Defining characteristics	Impact category: Human toxicity		
	Subcategory 1	Subcategory 2	Subcategory 3
Severity of effect	Irreversible or life-shortening	May be reversible or life-shortening	Generally reversible or generally not life-shortening
Example endpoints	Cancer Reproductive effects Teratogenic effects Acute fatal or acute severe and irreversible effects (e.g., fatal poisoning) Mutagenicity	Immunotoxicity Neurotoxicity Kidney damage Liver damage Heart disease Pulmonary disease (e.g., asthma)	Irritation Sensitisation Reversible acute organ effects (e.g., GI inflammation)

[a] Burke et al. 1996.

procedure involves several other steps before toxicity and persistency equivalence is reached. The main point of interest here is the classification step as a possible way of reflecting the difference in severity of toxic effects.

Each of the 3 subcategories reflects a more homogeneous endpoint whose relevance is easier to interpret in terms of consequences than is the relevance of the consequences of a single, aggregated, potency-based endpoint. The subcategories therefore provide better input for decision-making. However, the classification involves value judgement, as does the definition of which effect types should be included in each subcategory. Even the ILSI panel had to note that, by some, neurotoxicity is ranked equally or more severely than cancer (Burke et al. 1996). Furthermore, the severity of effect may not be equal even within the same subcategory of effects. An example of this is cancer, as different cancer types have different survival probabilities and times.

The introduction of subcategories can make us question whether a substance should be included only in the subcategory of its critical effect or in all subcategories to which it contributes. It is highly uncertain that a substance will be studied for all effects. Furthermore, there is a risk of double counting if different effects related to the same substance occur in several subcategories. We recommend including only the most severe effect.

As illustrated in Table 5-1 and described in 'Severity' (p 133), the availability of toxicological data is very restricted, and for many of the inventory items, it may not be possible to find suitable data for the classification into effect-type subcategories. Classification into the subcategories proposed by the ILSI panel is feasible for those substances reviewed by USEPA and listed in IRIS because a thorough evaluation of effects has been performed (Owens 2002). Based on a newly proposed classification, Owens (2002) tested the applicability of his proposal with the IRIS datasets for 2200 high production volume (HPV) chemicals in an attempt to explore whether a relatively limited number of toxicity categories broken into cancer and noncancer was at all feasible with regard to 1) classification and 2) compatibility and feasibility with existing regulatory toxicity databases (and implicitly with the large body of data in the literature, European Union (EU) and Organisation for Economic Co-operation and Development (OECD) databases, industry files, etc., because the same or similar bioassays are used). He found that all noncancer endpoints are rather easily classifiable and that there is a good starting set for classifying into genotoxic and nongenotoxic carcinogens. But even in the IRIS database, there is wide variability among the data. However, those substances for which less valid and incomplete information has been found in RTECS (and partly HSDB) cannot be classified into one or the other subcategory because of lack of knowledge. For these substances, it would be necessary to perform a more thorough literature search and evaluation of the toxicological properties of the substance if such data exist. It would therefore be of great value to generate a list of accepted NOAELs or ED10s (or other measures of toxicological potency) and associated endpoints, evaluated by toxicologists, for those substances most frequently encountered in LCA inventories, as also mentioned by the ILSI panel.

Quantitative approaches to account for severity

The approach of quantifying human health effects on the endpoint level of the environmental mechanism, that is, in terms of physical impacts such as loss of life expectancy or cough days, is an attempt to increase the relevance of the indicators and thus make them more useful for the interpretation and valuation of results. The assessment of health endpoints requires indicators that are appropriate to measuring change in health status.

In the fields of health management and environmental economics, there is a long tradition of using severity-oriented health indicators to help measure the health status of individuals or a society. Rosser's often-cited index of QALYs (Rosser 1987) was an early attempt to measure an individual's well-being on a single score, which was used for decision support in the UK health system. Many similar indicator schemes have been developed; the most recent and well-accepted one is the DALY concept by Murray and Lopez (1996), which is also supported by the WHO. Similar to the QALY approach, the DALY concept translates nonfatal adverse health effects that can be classified according to a multidimensional scheme into a single score; it also establishes a tradeoff between premature death (expressed as YOLLs) and years lived disabled, that is, the time spent suffering from a negative health effect. The driving force behind the development of QALY- or DALY-type indicators was the increasing need for such indicators in health management to measure the health status of a given population.

There is also a growing interest in valuing health and environmental impacts in monetary units for policy-oriented decision support, which is based on the theory of neoclassical welfare economics. In the U.S., a formal cost–benefit analysis is mandatory for the evaluation of various environmental policy measures, and in Europe, there is an increasing demand for cost–benefit analysis to justify new environmental regulation. The consideration of health and environmental impacts within a cost–benefit analysis requires the quantification of health and environmental impacts as far as possible on the endpoint level to facilitate a subsequent valuation. Fortunately, developments in the fields of health management and environmental economics both led to comparable health indicators; in particular, the YOLL (or YLL) is a key indicator used in both fields. Hofstetter (1998) introduced the use of the DALY concept in the LCA community, and it is adopted in the Eco-indicator 99 methodology (Goedkoop and Spriensma 1999) and by Crettaz et al. (2002) and Pennington et al. (2002).

Fatal effects: Years of life lost indicator

The YOLL seems an appropriate indicator to measure increased mortality risk on the endpoint level. The YOLL indicator measures the reduction in life expectancy resulting from an increased level of exposure to pollutants in the environment. There is some discussion of the fact that the YOLL concept, because it measures the loss of life expectancy resulting from a fatal event rather than 'death' per se, puts a higher weight on the premature death of, for example, a 40-year-old person than on the premature death of a 70-year-old person (because the lost life expectancy is higher). Some people argue that the death as such is what matters, and that we should not a priori give a lower weight to the premature death of an old person. However, we know that the physical effect from an exposure to a chemical, for example, is the reduction of life expectancy because the probability of death is equal to 1 for any individual. Therefore, we consider the YOLL an appropriate indicator, on the damage level, that can be quantified on the basis of natural science.

At least in the economics literature, the importance of the loss of life expectancy for valuation is controversial. Although putting a value on a life year is increasingly used in environmental economics, a study by Rowlatt et al. (1998) for the UK Department of Environment, Transport and the Regions and the Department of Trade and Industry strongly suggests using a context-specific Value of Statistical Life (i.e., different values of statistical life for, e.g., road safety and in the air pollution context) rather than a Value of Life Year for the valuation of mortality risks from air pollution. Rowlatt et al. (1998) argue that people's WtP to reduce mortality risks depends upon a great deal more than life span. The length of expected future life span is undoubtedly one factor for determining people's WtP to reduce the risk of premature death. It appears, however, to be only one of many factors, and one that is far from the dominant factor in determining how people's WtP to reduce this risk changes with age. For the same reasons, Rowlatt et al. (1998) suggest that the QALY indicator is better suited to the comparison of morbidity impacts than it is to the handling of risks of mortality (the same applies to the DALY indicator).

From our point of view, this criticism is valid, and important for our discussion, because the health indicators to be used in LCIA should reflect society's concern towards the effect at stake. However, with regard to the open and unresolved discussion on the valuation of mortality risk, the loss of life expectancy from our point of view still is the most sensible natural science–based indicator for the quantification of increased mortality risk. Both the YOLL and the DALY indicator require the quantification of the loss of life expectancy.

Nonfatal effects

We face a very large number of different, nonfatal, adverse health effects, and there is no natural science–based indicator that allows the aggregation of different endpoints to a single score. A set of morbidity endpoints linked to air pollutants, which can be quantified by using dose–response functions, was derived from a review of the recent epidemiological literature in the ExternE project (European Commission 1999). This list of endpoints is not comprehensive, and it might be criticised for reflecting what is quantifiable rather than what is relevant for decision-making. Although this is true in principle, we strongly believe that the underlying epidemiological studies have addressed health endpoints that are of direct social concern. Therefore, taking a pragmatic approach, we suggest assuming that the list of morbidity endpoints is incomplete but provides a reasonable approximation of the most important known effects that are of direct social concern.

From a scientific perspective, we are required to quantify and report results for every individual health endpoint. However, this inflation of subcategory indicators is not manageable in an LCA study, so there is a strong need to represent nonfatal health effects with a single indicator. The large scientific literature on the valuation of different health effects is used in national and international policy-making, and thus also might be used as a basis for aggregation in LCIA. Because it is supported by the WHO, the DALY concept would be the most appropriate one to be recommended as a best available practice. One potential problem is that the DALY scheme does not provide weighting factors for many health effects that result from increased exposure to chemicals and that are of interest in LCA.

A weighting between some of the relevant health endpoints can also be derived from the economic valuation literature. The ExternE study recommends monetary values that are based mainly on recent contingent valuation studies in Europe and the U.S. for a wide range of morbidity endpoints (European Commission 1999). However, although ExternE results are currently used in policy-oriented decision support, no international body has authorised these values. From a comparison of weights allocated to different health endpoints, it seems that monetary valuation studies put a lower weight on nonfatal effects, compared to the risk of death, than does the DALY approach.

Availability of data to address expected severity

Carcinogenic severity data

The assessment of the carcinogenic potency of substances was discussed in 'Carcinogens' (p 131). The operationalisation of the YOLL indicators requires quantitative information on the actual increase in cancer risk resulting from an increased exposure to a given substance, and on the expected loss of life expectancy per case of fatal cancer. Dose–response assessment involves describing the quantitative relationship between the amount of exposure to a substance and the extent of toxic injury or disease. Data are derived from animal studies or, less frequently, from studies in exposed human populations.

For many recognised carcinogens, the target tissue and thus the type of cancer that is expected to develop is known. Hofstetter's review (1998) of studies quantifying the survival rate and the loss of life expectancy suggests that different cancer types result in about 15 to 20 YOLLs per fatal cancer, which is consistent with findings from the ExternE study or from the field of radiation protection (e.g., Ehrhardt et al. 1995). Together with a slope factor or unit risk factor, which gives the probability of effect, the information on the loss of life expectancy per case of fatal cancer can be used to estimate the YOLLs per unit change in the concentration of a carcinogenic substance.

The ICRP (1990) has recommended risk factors that establish a relationship between the exposure (collective dose) and the occurrence of fatal cancers, nonfatal cancers, and hereditary effects. Because there are estimates on the loss of life expectancy for different fatal types of cancer, these risk factors can be used to calculate the YOLLs, the number of nonfatal cancers, and the number of hereditary effects per unit increases in collective dose.

Noncarcinogenic severity data

The assessment of the noncarcinogenic potency of substances was discussed in 'Noncarcinogens' (p 129). There are only relatively few noncarcinogenic substances for which it is currently possible to link the biological mechanism to a specific effect in humans, although the type of effect related to the potency measurements is reported in animal tests. Crettaz (2000) therefore proposed a preliminary approach in which DALYs are assigned to qualitative categories such as those described in 'Qualitative approaches to account for severity' (p 133). However, a large number of epidemiological studies have analysed the correlation between various health endpoints and the concentration of 'classical' air pollutants such as particles, SO_2, and ozone.

Results from epidemiological studies were used to derive dose–response functions, which allow the quantification of a wide range of health effects, including both mortality and morbidity impacts (Hofstetter 1998; European Commission 1999). Taking into account the age-specific death rate within a given population, the increase in mortality risk observed in the epidemiological studies can be translated into YOLLs per unit change in concentration levels.

There is substantial epidemiological evidence of adverse acute health effects from particulate air pollution and strong, but much less widespread, epidemiological evidence of chronic health effects (Hurley et al. 1999). The particles of main interest come from 2 principal sources:

1) direct emissions from combustion processes, and
2) the formation of secondary particles (sulphate aerosols and nitrate aerosols, from the emission of gaseous SO_2, NO_X, and NH_3).

Based on a thorough literature review, the ExternE study (European Commission 1999) provided a set of dose–response functions for the quantification of fatal and nonfatal health endpoints related to exposure to fine particles, SO_2, NO_X, and ozone.

Although causality of acute health effects is somewhat accepted, and that of chronic health effects quite widely accepted, there is no well-established mechanism of action for particulate air pollution. Epidemiological studies so far have analysed the relation between the mass of fine particles and various health effects. Questions remain open about the extent to which the chemical composition of particulate matter influences the magnitude of its effect and about the influence of particle size on the ability of particles to induce effects (see, e.g., Harrison and Yin 2000). Correspondingly, there is little strong evidence about the relative effect of various kinds of inhalable particles. However, there is some evidence — and strong conjecture — that, per unit mass of ambient concentration, the relatively fine fractions (fine particles with an aerodynamic diameter < 2.5 µm [PM2.5] and sulphates) generally are associated with greater risks than are PM10s (fine particles with an aerodynamic diameter < 10 µm). It may also be true that the toxicity of particles is greater according to their acidity and less according to their solubility. Others regard particulates as indicators of overall pollution, rather than the sole cause of the associated impacts. The latter point implies that existing potency estimates for particulate matter may significantly exaggerate their true importance.

Because the ambient air concentration of SO_2 and fine particles (which include sulphates and nitrates, among other substances) are often strongly correlated, it is difficult to separate the effects of individual pollutants in epidemiological studies. Results from epidemiological studies in the early 1990s were interpreted in a way that the role of particles was more fundamental than that of SO_2. However, a reanalysis of data from some U.S. cities, sponsored by the U.S. Health Effects Institute (HEI), together with new findings from the European Air Pollution and Health: A European Approach (APHEA) studies strengthened the case for an association between daily ambient SO_2 and acute health effects.

The principal epidemiological studies linking ambient concentrations of ozone with acute health effects have been carried out on the west coast of the U.S. and in the northeastern U.S. and southeastern Canada, and there are some recent data available from the APHEA study cities. Results of these studies provide substantial evidence of the acute health effects of ambient ozone.

Relatively few epidemiological studies report exposure–response relationships linking ambient NO_X with mortality or morbidity. In those that do, particles are generally also implicated, and there is some evidence that the apparent NO_2 effect is best understood not as causal, but as NO_2 being a surrogate for some mixture of traffic-related pollution. However, NO_X is a precursor to the formation of ozone and nitrate aerosols, so that an effect on health by secondary pollutants can also be considered.

Evaluation of Human Toxicity Indicators

In this section, the health indicators discussed in 'Current Practice in LCIA' (p 125), 'Toxicological Potency' (p 127), and 'Severity' (p 133) are assessed against a set of criteria to help identify the most appropriate indicators. The criteria were specified by the WIA-2. The main issues arising from the use of the different indicators are summarised in Tables 5-3a and 5-3b.

Scientific validity and reliability

Procedures for establishing ADI values on the basis of the NOAEL, for example, are scientifically accepted, although new approaches such as the BMD are emerging with adaptations for LCA. Uncertainty can be addressed quantitatively. Dose–effect curves from animal studies and epidemiological studies in general provide confidence intervals. Similarly, probabilistic extrapolation factors can be adopted. The overall uncertainty of the indicator value also depends on the uncertainty linked to the fate and exposure modelling. There is not, however, a common effect mechanism or mode of action between different effect types and even within effect types. The degree of additivity therefore is sometimes questioned, but at a minimum, it provides an indicator or a score (somewhat related to a count of the number of cases in a population).

There is unresolved debate on the existence of or ability to measure thresholds and on the extrapolation of the dose–effect curve towards low doses, particularly for noncarcinogens. Several scientific bodies have concluded that there is no scientific basis for assuming a threshold for genotoxic carcinogens, and in general, a linear extrapolation of the dose-response curve (which might overestimate the effect) towards 0 is recommended.

Dose–effect models that assess the physical health effects (often in the form of slope and unit risk factors) of a number of organic and inorganic substances and of ionising radiation have been adopted or recommended by international or national authorities (e.g., WHO, USEPA), so that

Table 5-3a Summary review of different human health indicators with respect to their use in calculating characterisation factors in LCIA: Potency-based indicators

Type of indicator	Key advantages	Key issues	LCIA application
Regulatory-based dose–response potency measures, such as ADIs, RfDs, RfCs	Widely adopted basis for site-dependent risk assessment to ensure regulatory compliance	Inconsistent levels of conservatism, reflection of politically acceptable adverse effect risk levels rather than low-dose risk response measures	Hertwich 1999; Huijbregts et al. 2000; Goedkopp and Spriensma 1999; partly used by Hauschild et al. 1997
Slope factors based on benchmark doses, such as β_{ED10}	Introduced to provide a consistent basis for the derivation of low-dose risk response measures for carcinogens and non-carcinogens from a measure in the observable range	Not currently widely adopted by regulatory agencies for risk assessment. While implicit in most measures for noncarcinogenic effects in LCA, adopting low-dose response curves for noncarcinogens remains somewhat debated.	Crettaz et al. 2002; Pennington et al. 2002, suggesting values for approximately 600 carcinogens and 400 noncarcinogens.
Acute toxicity data, such as LD50s and LC50s	Widely available data	Relevance of acute data when calculating time-integrated exposures to populations is poor, and relative acute to chronic importance is unlikely to be consistent across chemical emissions.	Partly used in Hauschild et al. 1997. Extrapolations from acute to chronic data are widely adopted.

Table 5-3b Summary review of different human health indicators with respect to their use in calculating characterisation factors in LCIA: Severity-based indicators

Type of indicator	Key advantages	Key issues	LCIA application
Qualitative indicators			
ILSI classification: Health endpoints allocated to 3 categories according to reversibility and life-shortening of effect. Classification based on panel procedure (Burke et al. 1996)	The 3 categories represent a somewhat homogeneous group of health effects with different levels of severity. Provides additional information for decision support.	Allocation to categories might be controversial. Use of 3 categories allows rough severity ranking only. Weighting between categories currently requires value judgement.	Demonstrated by Owens (2002) and adopted by Pennington et al. (2002), who proposed an initial hybrid use of the categories with assigned DALYs.
Quantitative indicators			
DALYs, based on Murray and Lopez (1996), supported by WHO, World Bank	Allows aggregation of any health effects (mortality and morbidity) on a single cardinal scale.	No final consensus on weighting factors for different health effects. Different approaches (e.g., panels, willingness-to-pay [WtP]) partly lead to different weights. Quantification of DALYs may not be possible for all relevant substances, particularly where effects are unknown when exposure is to environmental mixtures.	Hofstetter 1998; Goedkopp and Spriensma 1999 (Eco-indicator 99); Crettaz et al. 2002; Pennington et al. 2002, who suggest preliminary defaults.
QALYs (e.g., Rosser 1987)	Similar to DALY.	Similar to DALY.	Not currently used in LCIA but adopted in some comparative risk assessments.
YOLLs (also included in DALY and QALY calculations)	Allows aggregation of different mortality effects (different reduction of life expectancy) on a single cardinal scale based on a physical measure.	Giving the same value to any life year is a value choice that might not be commonly shared. Does not cover nonfatal effects. Quantification of YOLLs may not be possible for all relevant substances, as with DALYs.	Key indicator in ExternE-type applications (European Commission 1999).

we can assume a sufficient general acceptance of the approach.

Although the valuation of increased mortality risk based on the loss of life expectancy is still controversial in some of the literature, it is gaining growing acceptance, and the YOLL indicator is commonly used to measure increased mortality risk. There exist different approaches for the aggregation of nonfatal health effects to a single indicator. The selection of a specific approach is a value choice, rather than an issue of scientific validity, and up to now, there is no consensus on which is the most appropriate one.

Using indicators that account for severity requires quantifying health effects that result from an increase of exposure to a given substance. Most dose–effect models that link a change in ambient concentration level of a pollutant to a health effect are based on epidemiological studies and on animal studies. In most cases, the actual mechanism that leads to the negative effect is not fully understood, so that the dose–effect model is based on a statistical association, which in the case of biological plausibility is interpreted as causal. In spite of remaining uncertainties, this procedure can be considered as science-based in the sense of the International Organization for Standardization (ISO).

The use of the YOLL indicator is well established in some scientific and policy-oriented areas. Reliable and widely accepted science-based models exist (although uncertainties might be significant) for the quantification of YOLLs from different substances and for radiation. Although the reduction in life expectancy expressed as YOLLs is a 'physical' measure, the use of the YOLL indicator for aggregation includes a strong value choice, namely the assumption of equal value for any life year, irrespective of the affected person. While this view is not without controversy, we consider it sufficiently accepted by society so that it is justified within LCA. However, we note that YOLLs still can be quantified for only a limited number of chemicals, using epidemiological data.

In contrast to mortality, the treatment of nonfatal effects is much more problematic. It is obvious that an aggregation of nonfatal health endpoints is mandatory to achieve operational indicators, but this aggregation requires value choices. An aggregation scheme authorised by an international body (like the DALY concept supported by the WHO) is desirable. Different approaches for weighting and aggregation currently discussed in the literature lead to different results. The future LCIA activities planned under the United Nations Environment Programme (UNEP)–SETAC Life-Cycle Initiative might lead to a consensus on weighting factors for nonfatal health effects or a decision not to use them.

Based on the concept of risk assessment, in some LCA studies, assessing health effects focuses on the potential risk to a hypothetical individual, while estimating cumulated impacts expected to occur within an exposed population may be more appropriate to LCA. Small individual risks summed over a large population might result in an unacceptably large total impact, while a small collective risk might include an unacceptably high risk to the most exposed individuals. The choice between a measure of either individual risk or collective (population) risk is certainly a value choice that can affect the outcome of an LCA study. Both individual and/or collective risk might be relevant in a specific decision context. However, assuming that current legislation helps to prevent unacceptable risks to the most exposed individual at individual sites and from specific emissions, environmental policy is increasingly concerned with the reduction of collective risks. We therefore conclude that the consideration of collective risk in LCIA will be of increasing importance.

Transparency and reproducibility

The interpretation of results from animal studies might differ between professionals. Furthermore, the use of safety factors requires professional judgement. Consequently, virtually safe doses from various institutions may differ. We therefore recommend establishing a single set of virtually safe doses, or similar potency measures such as ED10s, for use in LCA. These calculations can be made transparent, as demonstrated by the derivation of RfDs and inhalation reference concentrations (RfCs) by USEPA expert panels (all relevant information is published in IRIS).

The estimation of YOLLs and nonfatal effects depends on dose–effect models published in the literature or in relevant toxicity database systems. In particular, risk factors for carcinogens are recommended by various organisations, and they differ between sources. As with potency, available data must be reviewed, and a single set of factors should be recommended.

The consideration of severity can require some modelling linked to fate and exposure modelling — in particular, the estimation of site-dependent actual impacts — but this does not necessarily mean that the process cannot be presented in a transparent way.

Comprehensiveness and sophistication

An indicator based on a virtually safe dose implies a perception that adverse effects from chemicals are unwanted. It provides a highly relevant measure of each individual chemical's potential to cause an effect in humans. Although based on risk assessment principles, it does not provide a measure of risk.[4] The actual morbidity and mortality effects of environmental chemicals on humans are quite difficult to interpret, however, but for the most part, this is due to insufficient exposure information.

Severity-oriented indicators aim to describe the physical impacts that are expected to occur within an exposed group of persons, a measure often assumed to be the endpoint of the cause–effect chain (or environmental mechanism) for toxicological impacts. Both qualitative and quantitative approaches to accounting for severity are assumed to facilitate improved interpretation of results. To help avoid misleading conclusions associated with high uncertainties, however, it is currently recommended that results be presented with and without severity insights. Practitioners and decision-makers should also be aware of the implications of the choice between individual and population-based effect measures.

Dose–effect models generally give information on the change of the incidence rate of a specific effect as a function of the concentration level. The resulting effect therefore depends on the background incidence rate (e.g., mortality rate), which is influenced by many other parameters (e.g., lifestyle) and might differ between countries or regions. Providing the relevant country-specific data generally is very resource intensive. It is often assumed, in the absence of alternative insights, that the error introduced by using a constant risk factor is relatively small compared to other uncertainties.

Fate and exposure might be strongly influenced by local conditions (e.g., meteorology, population distribution). Most models that quantify health impacts at the endpoint level operate in a site-dependent way. Results can be generalised stepwise (e.g., damage factors on the country level, continental level, or global level; see discussion in Chapter 4).

There might be a significant time between the release of a substance and the negative health effect because of latency time (e.g., cancer; several years) or long-living radioactive decay products (several thousand years). For the impact assessment phase, we strongly recommend presenting physical impacts without discounting. If discounting is required in the interpretation phase of LCA, the impact assessment phase needs to provide information on the time distribution of effects.

Although there is still some controversy about the shape of dose–effect curves at low doses and about the existence of, or ability to measure, thresholds for most of the effects discussed above, several scientific bodies have concluded that there is no scientific basis for assuming such thresholds or no-effect levels for genotoxic carcinogens, ionising radiation, fine particulates, and ozone. Arguments have similarly been proposed for non-carcinogens, citing that LCA should take into account residual risks below acceptable adverse effect thresholds. The linear extrapolation of the dose–response curve towards 0 provides the most straightforward approach to the estimation of low doses, if we note that neither the shape of the low dose–response curve nor the existence of mechanistic thresholds can be determined in most bioassay studies.

---

4 Some practitioners consider that it provides an estimate of the time-integrated risk associated with a given functional unit (basis: time-integrated exposure combined with a linear dose–response gradient yields time-integrated risk of an effect).

In the case of linear dose–effect functions without threshold, there basically is no difference between an average and a marginal analysis for the effect assessment. However, the formation of secondary pollutants (e.g., ozone) might strongly depend on background conditions, so that the difference between marginal and average analysis mainly affects the exposure modelling.

Relevance to the decision context

The degree of additivity and the relevance of potency-based indicators are sometimes questioned. We note, however, that potency-based indicators can provide insights into the time-integrated risk of an emission associated with a functional unit. Following the principles of risk analysis, we can add such measures. These measures will not differentiate between severities of the associated risks. Lacking such relevance, added potency-based measures should be interpreted with caution and may be misleading.

The distinction amongst chemicals in terms of severity may improve the information presented to decision-makers, although practitioners have noted the need to provide potency- and severity-based factors in parallel. The YOLL is a physical measure and allows the aggregation of different types of increased mortality risk from different substances, which, however, implies a value choice. Quantitative severity-oriented indicators such as DALYs or QALYs provide one approach that compares morbidity across impact categories. Qualitative approaches, such as categorising potency-based indicators, add information to the decision-making process. The future LCIA activities planned under the UNEP–SETAC Life-Cycle Initiative might lead to a consensus on weighting factors for nonfatal health effects or a decision to use one of the alternatives.

We propose a 3-step procedure:
1) Use a potency-based toxicity indicator, as is done in current practice. This indicator could, for example, be the dose–response slope derived from ED10, to bring together current practice for noncarcinogens and for carcinogens. These indicators do not provide information on the severity of effects.
2) If information on the severity of effects is desirable, divide indicators into subcategories on the basis of the critical effect of the substance.
3) Calculate severity-based indicators such as YOLLs and DALYs.

Feasibility

There is a severe lack of toxicological data. It is therefore not possible to estimate a potency measure for all substances in a product life cycle, particularly for chemicals that are not yet of interest to an environmental agency. However, for most chemicals recognised as being severely hazardous, we are well informed about their toxicological properties.

Researchers such as Crettaz et al. (2002) and Pennington et al. (2002) have compiled slope factors suitable for LCA, for more than 600 carcinogens and more than 400 noncarcinogens. Nevertheless, we now require methodologies to expand beyond such datasets. Estimation tools such as quantitative structure-activity relationships (QSARs) and extrapolations from more readily available acute toxicity data may prove useful, although they may be suitable only for initial screening LCA studies because of their high uncertainties.

The estimation of actual impacts (e.g., YOLLs in the actually exposed population) requires information on the spatial distribution of both the change in concentration and the population; such information generally is not available. Specific models that provide this information can be used to produce damage factors, but such models currently do not cover all the relevant substances, and they are available only for selected regions.

In general, the effect models that quantify severity-oriented health indicators require ambient concentration data as inputs from fate and exposure modelling. Severity-oriented indicators can be used in 2 different ways:

1) to calculate potential impacts within a predefined standard population (expressed, e.g., as YOLLs or DALYs per person per unit change in concentration level) or
2) to calculate actual impacts in absolute terms (e.g., YOLLs or DALYs per unit change in concentration level).

In the first case, severity-oriented indicators are used in a way similar to potency-based indicators. To assess actual impacts in absolute terms, we need to link the concentration data to data on the exposed population, which is a significant extension of current LCIA practice. The models currently used for fate and exposure modelling in LCIA generally do not provide the level of spatial resolution required to consider site-dependent exposure. However, models that have been developed for other purposes can be used to derive site-dependent impact factors on different levels of spatial resolution (see, e.g., Potting et al. 2000; Krewitt et al. 2001). In the field of ionising radiation, complex models have been used to derive site-dependent impact factors for different source types (e.g., uranium mining, power plant, reprocessing plant) for representative sites (United Nations Scientific Committee on the Effects of Atomic Radiation [UNSCEAR] 1993). However, the models that are designed to quantify actual health impacts by accounting for site-dependent conditions up to now do not cover the wide range of substances commonly addressed in LCA. (See Chapter 4 for more details).

Conclusions and Recommendations

A main driving force for the further development of human toxicity indicators for LCIA is the desire to improve the environmental relevance of indicators, to decrease uncertainty, and to increase the number of substances that can be covered. Such indicators are then more useful for valuation and weighting and for addressing impacts from substances that currently are not well covered (e.g., radiation, fine particles). Using a toxicological potency-based indicator to group health effects into subcategories can be seen as a step forward because doing so may differentiate levels of severity. We suggest delving further into the feasibility and relevance of using such qualitative approaches to account for severity. Allocating the different effects to categories leaves some practitioners with the question of how to derive weighting factors across the categories. There is currently no consensus on such weighting in the literature. DALYs, QALYs, or monetary measures could offer possibilities for quantifying severity.

The use of severity-based indicators that describe the expected health effect in physical units is considered a sensible way of further increasing the environmental relevance of the indicator value. In the case of mortality effects, the YOLL indicator is a natural science–based indicator that allows the aggregation of different types of mechanisms from different substances into a single physical indicator. Because the YOLL indicator is used in the fields of health management and environmental economics, we feel that there is a good basis for its use in LCIA. It seems that the YOLL indicator can be quantified for a reasonable, but not yet sufficient, number of substances.

The DALY- or QALY-type indicators that are currently used by some national and international organisations provide a theoretically consistent framework for the aggregation of 'all' health effects (including mortality, measured as YOLL, and morbidity effects). However, the weighting of different health endpoints on a cardinal scale is a prerequisite for using DALY- or QALY-type indicators, and this explicitly includes value choices. Current weighting schemes for mortality and morbidity endpoints do not lead to similar results, so that an internationally accepted and 'authorised' weighting scheme is desirable before we suggest DALY- or QALY-type indicators as best practice for LCIA. Further discussion might explore whether the YOLL indicator (which is already operational in many cases) can be used as part of a DALY- or QALY-type indicator that combines fatal and nonfatal effects, or whether it can be used as an independent indicator, which some might interpret as 'omitting' nonfatal effects.

It was beyond the scope of the WIA-2 human toxicity subgroup to analyse in detail the tradeoffs between the use of a more sophisticated method on a perhaps limited subset of chemicals in an LCA versus the use of a less sophisticated approach that includes the majority of chemicals in an LCA. The decision towards a more or less sophisticated method will certainly depend on the context of the LCA study. Specific case studies like those carried out under the ongoing Operational Models and Information Tools for Industrial Applications of Eco/toxicological Impact Assessments (OMNIITOX) project funded by the European Commission (http://www.omniitox.net) will provide insights about the usefulness of either approach for a range of different applications.

As a conclusion on the methodology, we recommend the following stepwise approach, which might be applied according to the objectives of a particular study and the resources available. Step 1 is mandatory, while Steps 2, 3, and 4 are considered complementary.

1) Use a potency-based toxicity indicator for relative weighting between substances, similar to what is done in current practice. If desirable for decision-making, differentiate between subcategories to reflect different levels of severity. Potency indicators for substances like fine particles or radiation can be derived from published risk factors. Existing datasets for potency measures such as the ED10, which already cover up to 1000 chemicals, should be reviewed by experts, taking into account best-estimate extrapolation factors with confidence intervals and associated critical effect information suitable for subsequent categorisation. The use of a sole peer-reviewed database will help to provide consistency within LCA and will be a valuable resource for other types of comparative assessment applications.
2) Calculate the YOLL indicator as far as possible for relative weighting between substances. This is sensible only if it is possible to quantify YOLLs for the key substances of the analysed processes in the LCA study. Check for changes in the resulting score compared to Step 1, and if applicable, discuss the implications. If there is insufficient information to quantify YOLLs for all the relevant key substances, then we suggest using only potency-based indicators.
3) Calculate the DALY or QALYs indicators for relative weighting between substances, as far as possible. As in the case of YOLLs, this is sensible only if quantification is possible for the key substances. Check for changes in the resulting score compared to Step 1, and if applicable, discuss the implications.
4) If the objectives of the study make it desirable, and if information from the fate and exposure modelling is available, use endpoint indicators (YOLL and DALYS or QALYs) to estimate health effects that take into account site-dependent characteristics (e.g., total exposed population).

References

Baird SJS, Cohen JT, Graham JD, Shlyakhter AI, Evans JS. 1996. Noncancer risk assessment: A probabilistic alternative to current practice. *Human Ecol Risk Assess* 2(1):79–102.

Bare JC, Hofstetter P, Pennington DW, Udo de Haes HA. 2000. Life cycle impact assessment workshop summary; Midpoints versus endpoints: The sacrifices and benefits. *Int J LCA* 5(6):319–326.

Barnes DG, Dourson M. 1988. Reference dose (RfD): Description and use in health risk assessment. *Regul Toxicol Pharmacol* 8:471–486.

Barnthouse L, Fava J, Humphreys K, Hunt R, Laibson L, Noesen S, Norris G, Owens J, Todd J, Vigon B, Weitz K, Young J. 1997. Life-cycle impact assessment: The state-of-the-art. Pensacola FL, USA: Society of Environmental Toxicology and Chemistry (SETAC). Report of the SETAC Life-Cycle Assessment Impact Assessment Workgroup, SETAC LCA Advisory Group.

Beck BD, Conolly RB, Dourson ML, Guth D, Hattis D, Kimmel C, Lewis SC. 1993. Symposium overview: Improvements in quantitative noncancer risk assessment. *Fund Appl Toxicol* 20:1–14.

Bro-Rasmussen F, Boyd HB, Jørgensen CE, Kristensen P, Laursen E, Løkke H, Nielsen KM, Grundahl J. 1996. The non-assessed chemicals in EU. Report and recommendations from an interdisciplinary group of Danish experts. Copenhagen, DK: Report from the Danish Board of Technology 1996/5.

Burke TA, Doull J, McKone T, Paustenbach D, Scheuplein R, Udo de Haes H, Young JS. 1996. Human health impact assessment in life cycle assessment: Analysis by an expert panel. Washington DC, USA: International Life Science Institute (ILSI).

Crettaz P. 2000. From toxic emissions to human health impact: A generic model for life cycle impact assessment [Dissertation]. Lausanne, CH: Swiss Federal Institute of Technology (EPFL).

Crettaz P, Pennington D, Brand K, Rhomberg L, Jolliet O. 2002. Assessing human health response in life cycle assessment using ED10s and DALYs. Part 1: Cancer effects. *Risk Anal*. Forthcoming.

Crouch E. 1985. MSTAGE, version 1.1 [software]. 44 Radcliffe Road, Somerville MA 02145, USA.

DeRosa CT, Dourson ML, Osborne R. 1989. Risk assessment initiatives for non cancer endpoints: implications for risk characterization of chemical mixtures. *Toxicol Ind Health* 5(5):805–824.

Dourson ML, Velazquez SF, Robinson D. 1996. Evolution of science-based uncertainty factors in noncancer risk assessment. Cincinnati OH, USA: Toxicology Excellence for Risk Assessment (TERA). http://www.tera.org. Accessed 15 Mar 2002.

Ehrhardt J, Hasemann I, Matzerath-Boccaccini C, Steinhauer C, Raicevic J. 1995. COSYMA: Health effects models. Karlsruhe, D: Forschungszentrum Karlsruhe. Wissenschaftliche Berichte FZKA 5567.

European Commission. 1999. Externalities of fuel cycles. European Commission, DG XII, Science, Research and Development, JOULE. ExternE: Externalities of Energy. Volume 7, Methodology 1998 update. Luxembourg, L: European Commission. EUR 19083.

Frischknecht R, Braunschweig A, Hofstetter P, Suter P. 2000. Human health damage due to ionising radiation in life cycle impact assessment. *Environ Impact Assess Rev* 20(2000):159–189.

Goedkoop M, Spriensma R. 1999. The Eco-indicator 99: A damage oriented method for life cycle impact assessment. The Hague, NL: Ministry of Housing, Spatial Planning and the Environment (VROM).

Goedkoop M, Müller-Wenk R, Hofstetter P, Spriensma R. 1999. The Eco-indicator 99 explained. *Int J LCA* 3(6):352–360.

Gold LS, Zeiger E. 1997. Handbook of carcinogenic potency and genotoxicity databases. Boca Raton FL, USA: CRC Pr. ISBN 0-8493-2684-2. 754 p.

Guinée J, Heijungs R, van Oers L, van de Meent D, Vermeire T, Rikken M. 1996. LCA impact assessment of toxic releases: Generic modelling of fate, exposure, and effect for ecosystems and human beings with data for about 100 chemicals. The Hague, NL: Ministry of Housing, Spatial Planning and the Environment (VROM). Report Nr. 1996/21.

Harrison R, Yin J. 2000. Particulate matter in the atmosphere: Which particle properties are important for its effects on health? *Sci Tot Environ* 249(2000):85–101.

Hauschild M, Olsen SI, Wenzel H. 1997. Human toxicity as a criterion in the environmental assessment of products. Chapter 7. In: Hauschild M, Wenzel H, editors: Environmental assessment of products. Volume 2, Scientific background. London, GB: Chapman and Hall.

Hauschild M, Wenzel H. 1997. Environmental assessment of products. Volume 2, Scientific background. London, GB: Chapman and Hall.

Hertwich EG. 1999. Toxic equivalency: Addressing human health effects in life cycle impact assessment [thesis]. Berkeley CA, USA: Univ California-Berkeley.

Hofstetter P. 1998. Perspectives in life cycle impact assessment: A structure approach to combine models of the technosphere, ecosphere and valuesphere. Boston MA, USA: Kluwer.

Huijbregts MAJ, Thissen U, Guinée JB, Jager T, Kalf D, van de Meent D, Ragas AMJ, Wegener Sleeswijk A, Reijnders L. 2000. Priority assessment of toxic substances in life cycle assessment. Part 1: Calculation of toxicity potential for 181 substances with the nested multi-media fate, exposure and effect model USES-LCA. *Chemosphere* 41(2000):541–573.

Hurley F, Donnan P, Miller B, Pilkington A. 1999. Health effects of PM10, SO_2, NO_X, O_3, and CO. In: Externalities of fuel cycles. Luxembourg, L: European Commission, DG XII, Science, Research and Development, JOULE. ExternE Project, Report No 7 Methodology 1998 Update.

[ICRP] International Commission on Radiological Protection. 1990. 1990 Recommendations of the International Commission on Radiological Protection. Oxford, GB: Pergamon Pr. ICRP Publication 60.

Jager T, Rikken MGJ, van der Poel P. 1997. Uncertainty analysis of EUSES: Improving risk management by probabilistic risk assessment. Bilthoven, NL: National Institute of Public Health and the Environment (RIVM).

Jolliet O, editor. Assies J, Bovy M, Finnveden G, Guinée J, Hauschild M, Heijungs R, Hofstetter P, Potting J, Udo de Haes HA, Wrisberg N. 1996. Impact assessment of human and ecotoxicity in life cycle assessment. Part IV. In: Udo de Haes HA, editor. Towards a methodology for life cycle impact assessment. Brussels, B: Society of Environmental Toxicology and Chemistry (SETAC) Europe.

Jolliet O, Crettaz P. 1996. Critical surface-time 95 (CST 95), a life cycle impacts assessment methodology including exposure and fate. Paper 1/96, prepared for the Workshop on Impact Assessment of the Concerted Action on LCA in Agriculture. Revised version 2, November 1, 1996. Lausanne, CH: Swiss Federal Institute of Technology (EPFL), Institute of Soil and Water Management.

Krewitt W, Trukenmüller A, Bachmann TM, Heck T. 2001. Country specific damage factors for air pollutants: A step towards site dependent life cycle impact assessment. *Int J LCA* 6(4):199–210.

Murray C, Lopez A, editors. 1996. The global burden of disease. Published by the Harvard School of Public Health on behalf of the World Health Organization (WHO) and the World Bank. Cambridge MA, USA: Harvard Univ Pr.

[NRC] National Research Center. 1994. Science and judgment in risk assessment. Washington DC, USA: NRC: National Academy Pr.

Olsen SI, Hauschild M. 1998. Assessing toxicological impacts in life-cycle assessment. *Arch Toxicol Suppl* 20:331–345.

Owens JW. 1996. LCA impact assessment categories, technical feasibility and accuracy. *Int J LCA* 1(3):151–158.

Owens JW. 1998. Life cycle impact assessment: The use of subjective judgements in classification and characterisation. *Int J LCA* 3(1):43–46.

Owens JW. 2002. Chemical toxicity indicators for human health: Case study for classification of chronic non-cancer chemical hazards in life cycle assessment. *Environ Toxicol Chem* 21(1):207–225.

Pennington DW. 2002. Uncertainty associated with the human health toxicological component in relative comparisons. Submitted to *Risk Anal*.

Pennington D, Crettaz P, Brand K, Rhomberg A, Tauxe A, Jolliet O. 2002. Assessing human health response in life cycle assessment using ED10s and DALYs. Part 2: Non-cancer effects. *Risk Anal*. Forthcoming.

Potting J, Klöpffer W, editors. Taylor A, Seppälä J, Risbey J, Meilinger S, Norris G, Lindfors LG, Hudson L, Henshaw C, Goedkoop M. 2002. Best available practice in life cycle assessment of climate change, stratospheric ozone depletion, acidification, eutrophication and tropospheric ozone formation. Backgrounds on general issues. Report by SETAC Europe Scientific Task Group on Global and Regional Impact Categories (STG-GARLIC). Forthcoming.

Potting J, Trukenmüller A, Jaarsveld H, Hauschild M. 2000. Human exposure from air emissions. In: Potting J, editor. Spatial differentiation in life cycle impact assessment. Utrecht, NL: Utrecht Univ, Dept of Science, Technology and Society. ISBN 90-393-2326-7.

Rosser R. 1987. A health index and output measure. In: Walker S, Rosser R, editors Quality of life: Assessment and application. Lancaster, England: MTP Pr.

Rowlatt P, Spackman M, Jones S, Jones-Lee M, Loomes G. 1998. Valuation of deaths from air pollution. A report for the Department of Environment, Transport and the Regions and the Department of Trade and Industry. London, GB: National Economic Research Associates.

Steen B. 1999. A systematic approach to environmental priority strategies in product development (EPS). Version 2000: Models and data of the default method. Göteborg, S: Chalmers Univ of Technology Centre for Environmental Assessment of Products and Material Systems (CPM). CPM report 1999:5.

Udo de Haes HA. 1996. Discussion of general principles and guidelines for practical use. In: Udo de Haes HA, editor. Towards a methodology for life cycle impact assessment. Brussels, B: Society of Environmental Toxicology and Chemistry (SETAC) Europe.

Udo de Haes HA, Jolliet O, Finnveden G, Hauschild M, Krewitt W, Müller-Wenk R. 1999. Best available practice regarding impact categories and category indicators in life cycle impact assessment. Background document for the SETAC Europe Second Working Group on Life Cycle Impact Assessment (WIA-2). Part 1, *Int J LCA* 4(2); Part 2, *Int J LCA* 4(3).

[UNSCEAR] United Nations Scientific Committee on the Effects of Atomic Radiation. 1993. Sources and effects of ionizing radiation. New York NY, USA: UNSCEAR. UNSCEAR 1993 report to the General Assembly, with scientific annexes.

[USEPA] United States Environmental Protection Agency. 1995. The use of the benchmark dose approach in health risk assessment. Washington DC, USA: USEPA Risk Assessment Forum. EPA/630/R-94/007.

[USEPA] United States Environmental Protection Agency. 1996. Proposed guidelines for carcinogen risk assessment. Washington DC, USA: USEPA. EPA 600/P-92/003C.

[USEPA] United States Environmental Protection Agency. 1998. Chemical hazard data availability study. What do we really know about the safety of high production volume chemicals? EPA's 1998 baseline of hazard information that is readily available to the public. Washington DC, USA: Office of Pollution Prevention and Toxics (OPPT).

[WHO] World Health Organization. 1987. Air quality guidelines for Europe. Geneva, CH: WHO Regional Publications. European Series Nr. 23.

Indicators for Ecotoxicity in Life-Cycle Impact Assessment

Michael Hauschild, David W. Pennington

Acknowledgements — The following members of the working group have contributed verbal and written comments to this document. Most, though not necessarily all, are reflected in the text by the authors: Jerome Payet (EPFL), Beate Escher (EAWAG), Kristin Becker van Slooten (EPFL), Mark Huijbregts (IVAM), Almut Beck (ETHZ), Patrick Hofstetter (USEPA/ORISE).

Abstract — This chapter reflects the views and opinions of the Task Group on Ecotoxicity under the Society of Environmental Toxicology and Chemistry (SETAC) Europe Second Working Group on Life-Cycle Impact Assessment (WIA-2). We present the current state-of-the-art in the assessment of ecotoxicological effects for chemicals emitted to the environment, as considered in the context of life-cycle assessment (LCA). Being the charge of another working group, fate and exposure measures are not addressed in this chapter. Starting with a discussion of the use of acute ecotoxicity measures as a relative comparison basis, the main emphasis of the chapter is on the use of the predicted no-effect concentration (PNEC) or the hazardous concentration affecting 5% of species (HC5) to provide a measure of expected change in the potentially affected fraction (PAF) of species that is associated with a small change in exposure concentration. In addition, we consider a provisional approach, which is based on the use of median data (HC50s and effective concentrations affecting 50% of test organisms [EC50s]), in an attempt to better reflect the change in the potentially disappeared fraction (PDF) of species for a given change in exposure concentration. We discuss the use of an average gradient versus a marginal gradient on the concentration effect curve, how mixtures are taken into account, and the choice of endpoint: the affected fraction of species versus the disappeared fraction of species. To help evaluate the options, we perform a preliminary evaluation of the approaches using criteria to reflect their suitability for life-cycle impact assessment (LCIA). We conclude that all approaches currently proposed are encumbered with significant problems within one or more of the criteria, and it remains unclear which is the most scientifically justifiable. The marginal measures from combi-PAF curves representing the effect of an emission on a background of chemicals already present in the environment and the marginal measures of the PNEC style of approaches may yield similar results in practice, and they seem to have similar parameter and model uncertainties. We therefore acknowledge that the potential of a combi-PAF approach for representing mixture toxicity is interesting, particularly from the perspective of improved environmental relevance. However, the fundamental principles require further investigation before it can be considered for best practice in LCIA.

Introduction

The Society of Environmental Toxicology and Chemistry (SETAC) Europe First Working Group on Life-Cycle Impact Assessment (WIA-1) proposed the following framework for characterising ecotoxic or human toxic impacts in life-cycle impact assessment (LCIA; Jolliet et al. 1996):

$$S_i^{mn} = E_i^m F_i^{nm} M_i^n \qquad (6\text{-}1).$$

The impact score S is presented as the product of an effect factor, E, a fate and exposure factor, F, and the total mass loading of the emissions, M. The subscript i denotes the chemical, n the environmental compartment to which the emission is released, and m the route of exposure of the ecosystem, species, or human. In order to obtain the total impact score within an impact category for all emissions in the life cycle or life-cycle step, the individual impact scores are summarised across chemicals, compartments of release, and routes of exposure:

$$S = \sum_{i,m,n} S_i^{nm} \qquad (6\text{-}2).$$

Under the SETAC Europe Second Working Group on Life-Cycle Impact Assessment (WIA-2), a separate task group on fate and exposure modelling explicitly covered the fate and exposure assessment (see Chapter 4). Ecotoxicity effect assessment is thus considered to cover only those aspects that relate to the inherent ecotoxic properties of a substance. The fate and exposure modelling gives the restriction to the ecotoxicity effect indicator that it must match a possibly time-integrated concentration increase in the relevant environmental compartment. In other words, the ecotoxicity effect indicator must be a concentration in the media exposing the organisms of the ecosystems (and not, e.g., a dose).

Reflecting current practice, this chapter predominantly focuses on aquatic water column species. The effect measures are based on either bulk or dissolved exposure concentrations in the water column. The exposure measures reflect the time-integrated mass in a medium of the environment (in the form of a change in concentration) to which a species or ecosystem will be exposed. At any given time, the change in concentration exposure associated with an emission in the inventory will usually be unknown in life-cycle assessment (LCA). The resultant measure is therefore a time-integrated effect per unit mass of chemical released into the environment. This is consistent with the basis of emissions inventory data in LCA, which is reported in the form of mass released per functional unit (e.g., the mass of benzene released associated with 10 m^2 of wall painted for 10 y). It is therefore necessary that the effect indicator similarly reflects long-term, or time-integrated, effects associated with a change in mass.

Terrestrial species and aquatic sediment-dwelling species, for example, may be exposed to concentrations in more than one environmental medium or compartment. Concentrations in these surrounding exposure media can vary by orders of magnitude, which makes it important to consider exposure pathways. Such multimedia exposures are not addressed in Chapter 6 but are in Chapter 4, 'Fate and Exposure Assessment in the Life-Cycle Impact Assessment of Toxic Chemicals'. Similarly, biomagnification (the build-up of a chemical in a food web associated with consumption) in food webs and, hence, the secondary poisoning of predators at higher trophic levels are issues of importance to both ecosystems and human health that are not addressed here.

According to the preparatory background document of the WIA-2 (Udo de Haes et al. 1999), the task groups were to explore the possibilities for modelling the causality chain between emission and ultimate damage to an endpoint (see example in Figure 6-1). The environmental relevance of the result increases as the model includes larger parts of the causality chain, but so can the uncertainty. If the current state of science permits, characterisation should be based on modelling that covers the whole causality chain. Often, however, uncertainties in the chain are so large that the additional information obtained by endpoint modelling does not justify the uncertainty that is introduced concomitantly to the model. For several impact categories, the best current practice is thus anticipated to be midpoint modelling, that is, modelling to a midpoint somewhere along the causality chain. However, the task groups were, in all cases, expected to explore the possibilities for endpoint modelling (Udo de Haes et al. 1999).

Impact modelling in LCA may be based on a marginal perspective (i.e., what is the additional

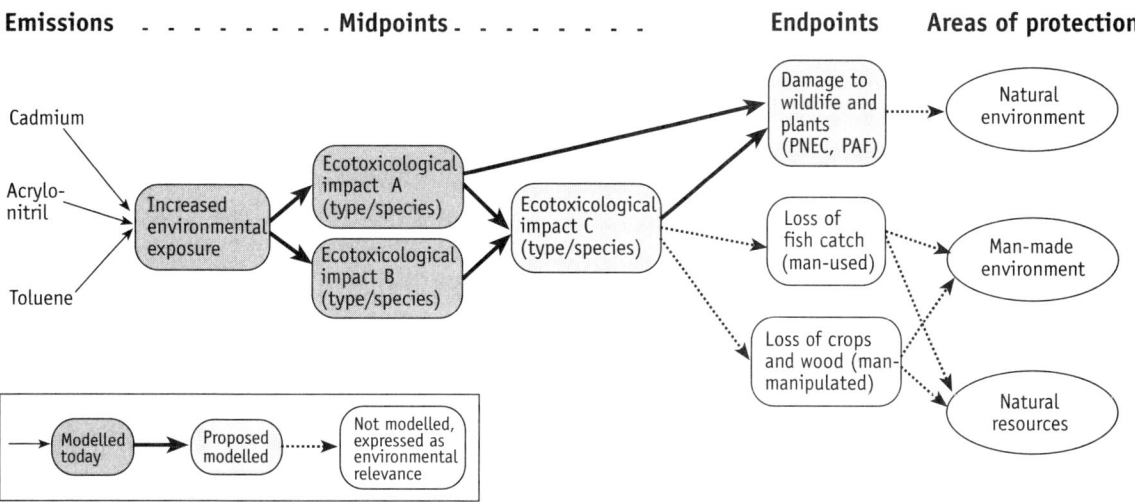

Figure 6-1 Example of causal relationships in the environmental mechanism between environmental emissions, midpoints, and (category) endpoints for ecotoxicity (Based on Udo de Haes et al. 1999. Reprinted by courtesy of *Int J LCA* 4(2):1–15.)

impact caused by the emission to an actual situation) or an average perspective (what is the relative share of the emission in the overall impact). The background document of WIA-2 states that the choice for LCIA depends on the type of application of LCA. In change-oriented assessments (Should we choose alternative A or alternative B?), the marginal approach is preferable, while the average approach is considered preferable for more descriptive LCA studies. In practice, however, the marginal approach to impact assessment may require such information as background chemical compositions, which are not usually available, and an average approach may have to be applied for pragmatic reasons (Udo de Haes et al. 1999). This chapter discusses these concepts and reviews ecotoxicity indicators that belong to both average and marginal impact assessments.

Impact assessment for ecotoxic chemicals in LCA seeks its inspiration from environmental risk assessment, but there are important differences. Rather than a site-dependent assessment, the object of an LCA is to calculate the impacts attributable to a system of processes (the product system) that provide a required product or service (the functional unit of a study). The modelling of environmental impacts is typically integrated over time and space, addressing multiple emissions from multiple sites with impacts over multiple generations. In risk assessment, the scope is usually narrower in time and space, the focus being on an individual chemical or a group of chemicals, and the modelling often being more time dependent. Risk assessment is generally performed in a legislative context, where the purpose can be to help ensure that there is no unacceptable risk to the environment from a given emission at a given site. A conservative approach is often followed in such risk assessments, and the indication of a possible concern triggers more detailed assessments of the fate and effect of the substance. LCIA, on the other hand, attempts to address all relevant environmental impacts associated with a product, not just site-dependent, localised impacts from specific emissions. To avoid an unintentional bias in the comparison of the different impacts, LCIA thus aims at providing the best estimation for each type of impact. Therefore, a conservative estimate of the ecotoxic effect of a substance is unwanted in this relative comparison context. We must aim for the best estimate of the actual effect properties of the substance, based on current knowledge and data.

Similarities with many risk assessments lie in the fact that LCIA of ecotoxicity rarely includes
- the consideration of site or temporal variation of ecosystems,
- the potential for biomagnification (because only diffusive uptake [bioconcentration] is addressed),
- the importance of species co-dependence in an ecosystem, or
- information on the background concentration.

In the following 3 sections, we provide an overview of available data ('Laboratory Test Data for Toxicity to Key Species'), outline the derivation of measures such as 'Predicted No-Effect Concentrations' (PNECs), and in 'Marginal Measures, Mixtures, and Endpoint Relevance' (p 159), discuss proposed methods to account for complex mixtures, the shape of dose–response curves (average versus marginal modelling), and the selection of endpoints (affected versus disappeared fraction of species). The key differences associated with these approaches are then compared against 'Criteria to Evaluate Ecotoxicity Indicators' (p 166).

Laboratory Test Data for Toxicity to Key Species

The simplest effect indicator is the proper laboratory test data for acute or chronic effects measured on selected individual species. An overview of data sources and prediction tools is presented in the Annex to this chapter (p 174).

Examples of environmental assessment schemes that use this kind of information include these:
- Chemical screening, where it is used for selection, on the basis of effect properties, of those substances for which a more detailed risk assessment will be performed
- Risk assessment of pesticides, where the effect assessment for nontarget organisms can be based on the lethal concentration to 50% of test organisms (LC50) for earthworms as a representative of important soil-dwelling nontarget organisms and the lethal dose to 50% of test organisms (LD50) for bees as a representative of nontarget insects (European and Mediterranean Plant Protection Organisation [EPPO] 1993)
- Labelling of chemicals, where the criteria for the risk phrases R50 to R52 concerning toxic effects in the aquatic environment are based on the acute toxicity expressed as the LC50 value to fish, crustaceans, or algae (Organisation for Economic Co-operation and Development [OECD] 1998).

Examples of the use of this indicator in the LCIA of ecotoxicity are found for a semiquantitative assessment method in Finnveden et al. (1992), and as a tool for identifying the most prominent contributions to ecotoxicity from a product system in Wenzel et al. (1997) and Walz et al. (1996).

Predicted No-Effect Concentration

The desire to protect the functioning of the ecosystem as a whole, rather than particular species, can lead to the use of a different effect indicator: the PNEC. The PNEC is the ecotoxicity indicator most frequently used in LCIA and risk assessment today (Guinée et al. 1996; Jolliet and Crettaz 1997; Wenzel et al. 1997; Huijbregts 1999). The PNEC is the highest environmental concentration expected to cause no effects — acute or chronic — on the structure or functioning of ecosystems (European Commission 1996). Estimating the PNEC requires data on long-term toxicity to several functionally different species, preferably belonging to 3 or more different phyla at different trophic levels of the ecosystem.

In current practice, characterising the risk to an ecosystem from a chemical's use is determined by the risk quotient as the ratio of the predicted environmental concentration to the PNEC, as described, for example, in the European Union (EU) technical guidance document on risk assessment (European Commission 1996). The 5th percentile of the so-called *potentially affected fraction* (PAF) of species is usually adopted as a basis for the PNEC (also termed *hazardous concentration* [HC5] in the U.S.; the implications of adopting PAF as an endpoint measure are discussed in 'Ecosystem endpoint measure relevance', p 164). The U.S. Environmental Protection Agency (USEPA) generally favours the median estimate of the 5th percentile (Stephan et al. 1985), a basis considered appropriate in the context of LCA. Others have required the 95th percentile confidence limit that the 5th percentile will not be exceeded (Suter 1998).

Differences will exist in sensitivity between different types of ecosystems (type, real species composition, state of pollution, adaptation), and the relevance of one common PNEC based on standard laboratory species is therefore questionable. Nevertheless, in the absence of a practical alternative, this remains common practice in risk assessment and LCA. Several different methods for estimation of PNEC have, however, been developed. They are reviewed and compared in the following sections.

Measuring PNEC in field meso- and microcosm tests

The PNEC will depend both on the properties of the substance in question and on the exposed ecosystem, and in principle, it can be measured only through field tests with the substance in this ecosystem (the target ecosystem). The best approximation to such in situ determination of the no-effect concentration consists of field tests with model ecosystems that have as many features in common with the target ecosystem as possible (mesocosm test or multiple-species test). These tests study the target ecosystem's functioning expressed by parameters such as the quantity of biomass at various trophic levels, fluxes of nutrients, numbers of individuals of various species, or species diversity. Field tests with model ecosystems are generally very expensive to perform. Their results are often specific to local conditions, difficult to interpret, and difficult to transfer to ecosystems other than the one under study, hence limiting their suitability in the context of most LCA studies.

Estimating PNEC on the basis of laboratory tests performed on individual species

Ecotoxicity data determined in laboratory tests on selected species are much more available than are data from field studies. Various methods have therefore been developed and standardised for estimating PNECs through extrapolation from the results of laboratory tests performed on individual species. Extrapolation approaches and the associated uncertainties are addressed in subsequent sections.

The extrapolation from laboratory test concentrations to field concentrations is supposed to cover several differences:
- From observed effects (lethality, reproduction) for a limited number of individual species to the possible effect on the functioning and structure of the ecosystem. The latter can depend on the interaction between species (competition, predation) and biomagnification.
- From laboratory test conditions with optimal physicochemical conditions for the experimental animals and controlled concentrations of the test substance to field conditions where the organisms on one hand are often exposed to other stressors at the same time, but where the bioavailability of the test substance on the other hand may be reduced because of adsorption and degradation or other types of loss.
- From short-term exposure (acute effects) to long-term exposure (chronic effects), as well as

from high lethality (50% from laboratory LC50) to low lethality (0% in the test population for no observable effect concentration [NOEC]) or other types of nonlethal effects (behavioural changes, etc.).

In addition, the extrapolation must cover the variability in the determined ecotoxicity levels within the same laboratory as well as between laboratories.

Extrapolation from statistical distribution of single-species sensitivities

For a given substance, an ecosystem species sensitivity distribution (SSD) can be determined when the sensitivity is known for a sufficient number of species. From the distribution, it is possible to estimate the HCp that protects $(1 - p)$% of all species present in the ecosystem (Figure 6-2). The extrapolation requires that the individual species for which ecotoxicity data are available can be regarded as mutually independent random samples of the total quantity of species in the ecosystem.

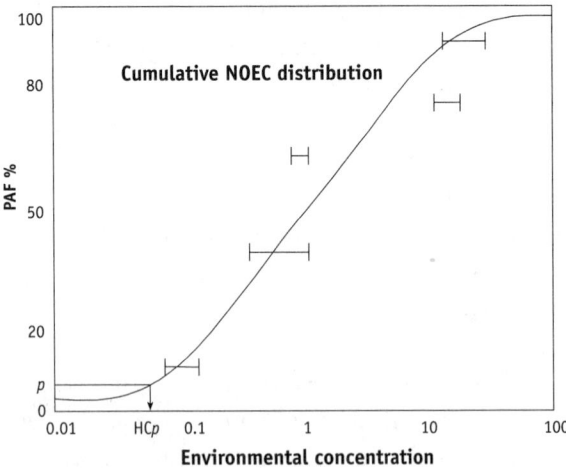

Figure 6-2 Determination of the environmental concentration HCp that affects p% of all species in the ecosystem through extrapolation from a statistical distribution estimated from measured species sensitivities to a substance (adapted from Klepper and Van de Meent 1997)

Several distribution-based extrapolation methods have been proposed for estimation of PNECs (Hauschild et al. 1998; Pennington and Payet 2001). They differ primarily in the choice of the type of statistical distribution. Van Straalen and Dennemann (1989) assume that species sensitivity follows a logarithmic logistic distribution; Wagner and Løkke (1991) assume a logarithmic normal distribution (as illustrated in Figure 6-2), while Stephan et al. (1985) assume a logarithmic triangular distribution. These distributions do, however, resemble each other between the 5th and 95th percentiles. Hence, although debated (Newman et al. 2000), the influence of the distribution selection on the PNEC is generally insignificant. All 3 approaches determine the PNEC as the 5th percentile protection level, also referred to as *HC5*.

Because generally our intention in risk assessment, as well as in LCA, is to model the ecosystem effects from a long-term exposure, the statistical extrapolation methods are based on results of tests on chronic exposure of individual species or predictions of chronic effects. Although sometimes questioned (Newman et al. 2000), results from chronic tests or predictions for at least 4 to 8 different species are often believed to give reasonable predictions in the determination of the distribution.

The USEPA 5th percentile estimate (HC5) is based on at least 8, mostly randomly selected genus mean values (GMVs; Stephan et al. 1985; USEPA 1999). The 4 most sensitive are then modelled to calculate the HC5, reducing the error associated with fitting a curve to the entire distribution. Species within a genus are suggested to be toxicologically much more similar than species in different genera, and so the use of GMVs may help prevent datasets from being biased by an overabundance of data for species in one genus or consideration of a limited number of genera. Stephan and Erickson (1988) noted, however, that the final acute values (FAVs, the median acute HC5s) calculated from species means are similar to those from genus means with a geometric mean ratio of 1.04 and a maximum ratio of 2.8 (genus to species).

Slooff (1992), RIZA (1999), Huijbregts (1999), and others have proposed estimation of the 5th percentile using NOECs from as few as 4 different taxa. However, for the majority of substances, this data requirement poses a problem, even for priority chemicals on regulatory inventory lists such as the U.S. Toxic Release Inventory (TRI). An environmental hazard or risk assessment is therefore often forced to alternative methods that enable estimation of the PNEC on the basis of a weaker database using general empirical insights.

In an attempt to reduce the required input of chronic toxicity data in statistical extrapolation methods, De Zwart and Sterkenburg (1999) and Van de Meent and Van Straalen (1999) proposed an approach to estimate generic dose–response curves for ecosystems. The approach is based on the analysis of a large number of organic and inorganic chemicals (about 40 000 acute toxicity data for 642 chemicals and about 2000 chronic data for 277 chemicals). A common underlying assumption in the derivation of the dose–response curve is that the distribution is unimodal, with nonspecific mechanisms of toxicity having narrow distributions. For more specific mechanisms, the distribution can be much wider because of either pharmacokinetic effects or pharmacodynamic effects. De Zwart and Sterkenburg (1999) and Van de Meent and Van Straalen (1999) suggested approximately 20 values for the distribution spread, or standard deviation (SD), each based on the results for 10 or more species with a potential use for PNEC estimation. Substances with the same mode of action are assumed to have the same distribution form (i.e., SD is equal, only HC50s differ). From the results of a linear regression for 89 chemicals, De Zwart and Sterkenburg (1999) and Van de Meent and Van Straalen (1999) further suggested that a correlation may exist between the spread and the medians (HC50) for acute and chronic SSDs. Hence, given only limited acute data and insights into the mode of action, the SSD for a chronic ecosystem can be predicted.

As in most other methods, the associated uncertainties are dependent on the sample size, the degree of acute-to-chronic extrapolation, and the extent of data prediction. With a sample size of 3 acute, measured data, the median estimate of the PNEC or HC5 can frequently be predicted with statistical techniques that have a 95th percentile confidence interval of approximately a factor of 100 (Pennington and Payet 2001).

Extrapolation with assessment factors

Fish, crustaceans, and algae (e.g., European Commission 1996) are standard taxonomic groups commonly considered in toxicity tests for aquatic species that predominantly live in the water column. Acute data (e.g., short term, high mortality), often predicted, are more readily available than chronic (e.g., long-term, low observed morbidity effects). These short-term data require extrapolation to yield a measure that is relevant in the context of long-term exposure.

Assessment factors (extrapolation factors that sometimes are termed *uncertainty factors* or *application factors*) can be used to adjust for differences such as exposure duration between available (e.g., acute) and desired (e.g., chronic) effects measurements in the estimation of measures like PNECs (Wagner and Løkke 1991; Toet and Van de Meent 1992; Suter 1993; Bro-Rasmussen et al. 1994; USEPA 1994; European Centre for Ecotoxicology and Toxicology of Chemicals [ECETOC] 1995; Zeeman 1995; European Commission 1996; Van Leeuwen and Hermens 1996; Jager et al. 1997; Molak 1997; Smerchek and Zeeman 1997; Chapman et al. 1998; Vermeire et al. 1998). Because the availability and relevance of toxicological data are commonly limited, the use of such factors reflects typical practice, and indeed, the predominant practice to estimate PNECs is the use of assessment factors. The magnitude of the assessment factors is determined on the basis of the quality and relevance of the available ecotoxicity test data. The PNEC value is usually obtained by dividing the lowest of the available ecotoxicity data by the relevant assessment factor. As with statistical approaches, confidence intervals on the best estimate can be predicted to reflect uncertainty.

A number of different assessment factor methods have been developed for chemical screening, but they show only minor differences. Table 6-1 shows the assessment factors proposed by the European Commission, as an example.

Table 6-1 Assessment factors for determination of $PNEC_{water}$ from ecotoxicological test data (European Commission 1996)

Assessment factor	Criterion
1000	At least 1 short-term LC50 or EC50 from each of 3 trophic levels (fish, *Daphnia*, algae)
100	One long-term NOEC (either fish or *Daphnia*)
50	Two long-term NOECs from species representing 2 trophic levels (fish and/or *Daphnia* and/or algae)
10	Long-term NOECs from at least 3 species (normally fish, *Daphnia*, and algae) representing 3 trophic levels
case-by-case	Field data or data of model ecosystems

Direct use of test data to estimate a PNEC (i.e., application of an assessment factor of 1) is warranted only when the relevant field data are available for the chemical, or when the chemical has been tested in a model ecosystem considered sufficient to represent real ecosystems likely to be exposed to the chemical. This is done on a case-by-case basis and will be relevant for only a very limited number of substances. The lowest assessment factor of 10 is used to extrapolate from chronic laboratory data for an adequate number of individual species to cover the different conditions in the recipient and the interaction among all species in the ecosystem as expressed in the ecosystem's functioning. The intermediate assessment factor of 100 also covers extrapolation from acute laboratory data for an adequate and representative number of species to chronic laboratory data for the same substance. In addition to what was covered by the other assessment factors, the highest assessment factor of 1000 covers extrapolation from sparse laboratory data for acute toxicity to an adequate database for assessment of acute ecotoxicity.

Other factor methods have been developed by the OECD (1992) and the USEPA (1994). The methods all follow the same pattern and have strong similarities. The most significant differences from the EU method are these:

- The USEPA method permits the use of quantitative structure-activity relationships (QSARs) for estimating acute ecotoxicity data in the first group with an assessment factor of 1000. QSAR covers tools that, on the basis of a substance's structural molecular properties, attempt to predict its behaviour, in this case, its acute toxicity to selected aquatic species, on the basis of a knowledge of the ecotoxicity of structurally related substances. The USEPA method also permits use of an assessment factor of 1 if results of toxicity tests on ecosystems are available (USEPA 1984). The method requires data only for Crustacea and fish.
- The OECD method permits use of QSARs at all 3 levels.
- Neither the USEPA nor the OECD method operates with an intermediate assessment factor of 50.

These systems of assessment factors parallel the factor systems used in the determination of human toxicology-based threshold values such as acceptable daily intake (ADI). The most significant difference is that, within human toxicology, additional factors are applied to cover the extrapolation to particularly sensitive or exposed individuals. It could be argued that additional factors should also be introduced in the estimation of the ecotoxicologically tolerable concentration, to cover extrapolation to

- the interaction among species in the ecosystem, including effects on functions other than those observed in the single-species toxicity tests used; and
- ecosystems that are already stressed because of other physical or chemical impacts, including the possibility of increased ecotoxicity of the substance on simultaneous occurrence with other substances (*synergism*).

The historical application of assessment factors, usually order of magnitude *safety factors*, is rooted in chemical screening and helps to provide extrapolated results that are likely to be conservative. The degree of conservatism and the uncertainty are not specified. Jager et al. (1997), for example, demonstrated the use of probabilistic extrapolation factors as an alternative. Such probabilistic extrapolation factors are based on existing data (e.g., ratios of LC50s to NOECs). Median factors can be adopted to provide best estimates of, say, the HC5 (or PNEC or HC50). The 5th and 95th percentiles of these mean values, for example, provide insights into the associated uncertainty. To initiate consensus building for applications such as LCA and comparative risk assessment, Pennington and Payet (2001) proposed probabilistic extrapolation factors on the basis of an extensive literature survey.

Evaluation and comparison of PNEC estimation methods

The ECETOC (1997) considered the relationship between the more similar NOECs for single species (fish, invertebrates, *Daphnia*, and algae), model ecosystems and the 'real world'. Although again based on a limited number of samples, the results of ecosystem studies were considered to be of sufficient complexity to be representative of the real world. Because there may be significant variation between sites and models, the EU recommends judging the applicability of experimental ecosystem data on a case-by-case basis (European Commission 1996).

The ECETOC (1997) compared the results of 248 static freshwater, flowing freshwater, and marine ecosystem models covering 34 compounds with single-species data from the ECETOC Aquatic Toxicity (EAT) database. It was demonstrated that the median ratio of the lowest single-species NOEC to the ecosystem NOEC was 1.5, with a 90th percentile of 8.1. Data were generally available for the same most-sensitive species in both the ecosystem and single-species tests. The range of ratios was 0.02 to 77.5 and approximated a log-normal distribution. The distribution and median values for nonpesticides and pesticides were similar (nonpesticides: median = 1.8, 90th percentile = 6.1; pesticides: median = 0.8, 90th percentile = 9.3). The majority (26 of 34) of the results were for flow-through freshwater ecosystem models. The median lowest single-species test result to the ecosystem NOEC for these flow-through systems alone is 1.6, with a 90th percentile of 10.

Among more than 3000 references to tests in multiple-species (MS) freshwater mesocosm systems, Emans et al. (1993) found reliable NOEC values (MS-NOECs) for only 29 different substances, 10 of which were metals. Based on 17 data pairs, they suggested that species tested in multi-species experiments appear to be statistically as sensitive as similar or related species in single-species tests. For most comparisons, the NOECs from the multiple-species and the single-species experiments for similar species varied between 0.2 and 5.0. Exceptions were typically associated with differences in test conditions or unreliability of the tests. Emans et al. (1993) presented a comparison of NOECs from multiple-species tests, based on statistical significance from a control, with PNEC predictions using the Aldenberg and Slob (1993) and the Wagner and Løkke (1991) SSD methods and the USEPA assessment factors approach. The SSD methods were adopted only if at least 4 single-species NOECs were available. The 95th percentile confidence estimates of the PNEC using the SSD methods were similar to those of the assessment factor approach, and all applied methods gave an estimate of the PNEC for all 29 substances, which was of the same order of magnitude as the experimentally determined MS-NOEC value, typically on the lower side. Although data for all 19 organic compounds and 10 metals could not be considered because of reliability issues, based on the geometric mean ratio, the 50th percentile statistical estimates of the PNEC were thus typically a factor of 3 to 4 lower than the NOECs from multiple-species tests. The authors concluded that the levels for the experimentally determined MS-NOEC values and PNEC values determined by the distribution methods showed the greatest agreement at a confidence level of 50%. Based on the reported data and considering only reliable or moderately reliable MS-NOECs, the ratio with the Aldenberg and Slob (1993) 50th percentile confidence limit estimate of the PNEC varied between approximately 0.07 and 50. The 95th percentile confidence limits of the PNECs calculated using the SSD methods were lower than the 50th percentile by a factor of approximately 10.

Even if this kind of comparison explores only a small selection of the possible combinations of emitted chemicals and exposed ecosystems, they nevertheless give an indication that it may be reasonable to use one of the extrapolation methods described in the preceding sections for estimating a PNEC in situations where the available data do not support SSD-based extrapolation.

The statistical methods can be considered to use the available data better than the assessment factor methods when the mass of data influences only the order of magnitude of the assessment factor and strong weight is put on the lowest toxicity data in determining the PNEC value. On the other hand, even if the factor methods may seem rather pragmatic and less rigorous than the statistical extrapolation methods, the distribution methods are still encumbered with a number of problems:

- It is difficult to fulfil the condition that the species whose sensitivity to the substance is the basis for determining the sensitivity distribution are mutually independent, random samples of the total number of species in the ecosystem.
- There are only empirical, but not particularly theoretical, arguments for selecting precisely the proposed statistical distribution types for the sensitivity of species in natural ecosystems. In addition, the sensitivity distribution may change if the ecosystem is simultaneously exposed to other forms of stress.

Furthermore, statistical methods tend to yield a PNEC estimate that is close to the lowest value of the NOECs involved. The estimated PNEC is usually within a factor of 3 of the lowest species or GMV (Stephan and Erickson 1988; Emans et al. 1993). Hence, given the high level of remaining uncertainty and the additional complexities, the added benefits of statistical approaches can be questioned.

Estimation of PNEC for terrestrial ecosystems and sediments

For terrestrial ecosystems and sediments, data are generally much more scarce than for aquatic systems. Within the EU, there is no requirement for new notified chemicals to provide data on toxicity to sediment-dwelling organisms, while there is an option to require toxicity data for higher plants and earthworms. This reflects the current situation in which there is no internationally accepted set of standardised ecotoxicological tests (European Commission 1996). In the general absence of readily available alternatives, both terrestrial and sedimentary PNECs are commonly derived from aquatic data for water column organisms, applying an equilibrium partitioning approach based on the assumptions that

- sediment-dwelling organisms and terrestrial organisms show a sensitivity distribution similar to water column organisms,

- exposure of the sediment- or soil-dwelling organisms occurs exclusively through contact with the water phase in the sediment or soil, and
- the distribution of the substance between the phases in the soil or sediment is in thermodynamic equilibrium.

None of these assumptions has been shown to have general validity. Particularly for highly lipophilic substances and for organisms primarily exposed through food, the exposure will not be dominated by water contact. To reflect this aspect, in the EU risk assessment scheme, for substances with log K_{OW} higher than 5, a correction factor of 10 is suggested to be applied on the corresponding predicted environmental concentration (European Commission 1996).

The $PNEC_{soil}$ for terrestrial ecosystems or sediments is estimated from the $PNEC_{water}$ for the water column by using the adsorption coefficient $K_{soil\text{-}water}$ or $K_{sed\text{-}water}$ and the density of the soil or sediment as shown in Equations 6-3 or 6-4 (European Commission 1996):

$$PNEC_{soil} = \frac{K_{soil-water}}{Density_{soil}} \times PNEC_{water} \quad (6\text{-}3),$$

or

$$PNEC_{sed} = \frac{K_{sed-water}}{Density_{sed}} \times PNEC_{water} \quad (6\text{-}4).$$

The equilibrium partitioning method has been found valid for estimating $PNEC_{soil}$ for some organic chemicals (Di Toro et al. 1991) and metals (Ankley et al. 1996) and, more specifically, for short-term toxicity of several chlorinated aromatics to earthworms (Van Gestel and Ma 1993).

In the European scheme for risk assessment of chemicals, a proposed extrapolation system uses assessment factors to estimate a PNEC from test data on terrestrial organisms. It is stressed that the factors should be seen as indicative because they are based on limited experience, and we need a deeper understanding of the difference between short-term and long-term toxicity for several taxonomic groups as well as the difference between laboratory and field tests. The choice of taxonomic groups on which to base the PNEC estimation is also still a point of discussion. The assessment factors are presented in Table 6-2.

Table 6-2 Assessment factors for determination of $PNEC_{soil}$ from ecotoxicological test data (European Commission 1996)

Assessment factor	Information available
1000	LC50 or EC50 short-term toxicity tests (e.g., plants, earthworms or microorganisms)
100	NOEC for long-term toxicity tests (e.g., plants)
50	NOEC for additional long-term toxicity tests of 2 trophic levels
10	NOEC for additional long-term toxicity tests for 3 species of 3 trophic levels
case-by-case	Field data or data of model ecosystems

If only one piece of terrestrial toxicity data is available (assessment factor 1000), the PNEC should be determined using the equilibrium partitioning approach as well, and the lowest of the 2 resulting values should be chosen for further use.

Marginal Measures, Mixtures, and Endpoint Relevance

At least 4 methods and related indicators that account for multiple species are currently proposed for use in LCIA:

1) The PNEC method (described above and adopted in Guinée et al. 1996; Jolliet and Crettaz 1997; Wenzel et al. 1997; Huijbregts 1999),
2) the Assessment of the Median Impact (AMI) method (Payet et al. 2001),

3) the HC5 approach (adapted from USEPA and described by Pennington 2000), and
4) the Eco-indicator 99 approach (Goedkoop and Spriensma 1999).

These approaches differ in terms of the gradient selected (marginal versus average), how mixtures are taken into account, and the choice of endpoint (affected versus disappeared fraction of species). A discussion of these fundamental differences between the methodologies is outlined in the following 3 subsections.

Existing marginal ecosystem effect measures

From the dose–response curve in Figure 6-2, comparisons of chemicals can be performed using at least 3 approaches:
1) Division by a concentration at a given PAF or potentially disappeared fraction (PDF) of species in the ecosystem, such as the PNEC, HC5, or HC50 (as described in the preceding sections)
2) The secant gradient, a linear scale between the origin and a defined working point on the dose–response curve; analogous to the slope factor in human health carcinogenic evaluations and termed the *average gradient* in Udo de Haes et al. (1999)
3) The tangential gradient, based on tangent at a defined working point on the curve; termed the *marginal gradient* in Udo de Haes et al. (1999).

Division by a given concentration, such as the PNEC or the HC5, and the corresponding secant gradient are often referred to as *average approaches* (Udo de Haes et al. 1999). There is no difference in practice, however, between the secant gradient approach and comparison by division using the concentration at a given PAF (such as the PNEC, the HC5, or the HC50). In Figure 6-3, the secant gradient at a concentration X_1 is Y/X_1, and at X_2, it is Y/X_2. The secant gradient at PAF = 5% is 0.05/HC5, for example. Based on the secant gradients, the relative ratio is therefore X_2/X_1, which is equal to the concentration ratio. In the remainder of the discussion, we therefore consider only the 2 gradient measures.

The tangential gradient provides a marginal measure of the change in PAF associated with a small change in exposure concentration (changes in other measures such as PDF can be considered analogously; see 'Ecosystem endpoint measure

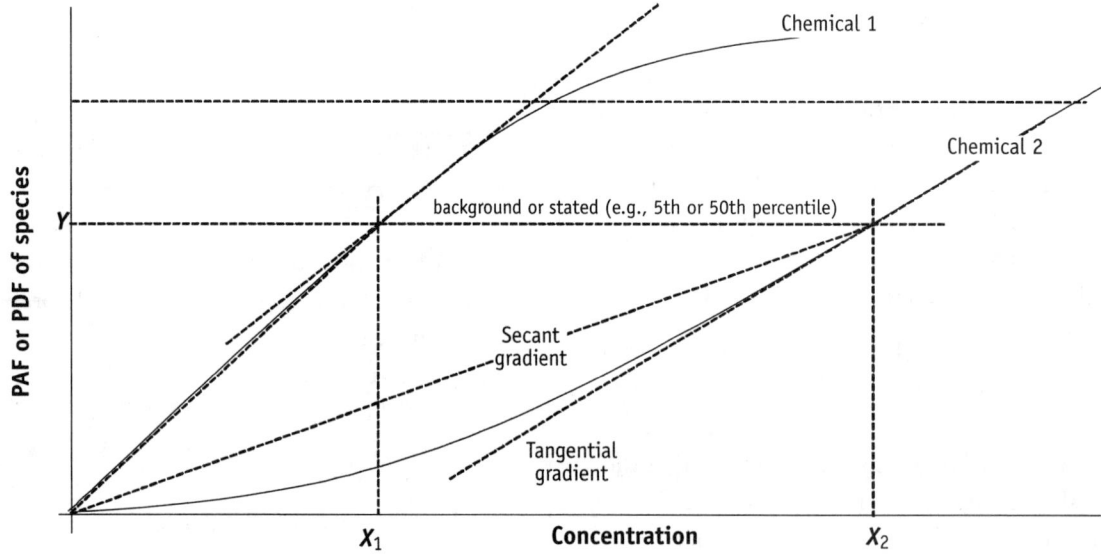

Figure 6-3 Tangential (marginal) versus secant (average) gradient measures

relevance', p 164). It is often argued that using a marginal gradient approach is preferable to the average measures from a relevance perspective for change-oriented LCA studies (e.g., Udo de Haes et al. 1999). However, as discussed below, the tangential gradient can approximate the secant gradient at low PAFs, and then a preference between the marginal and the average measures (i.e., between the tangential and the secant gradients) is no longer so straightforward.

It can be demonstrated that the uncertainty associated with the tangential gradient and the preference over the secant gradient is dependent on the location of the working point on the SSD curve (denoted by X_1 or X_2 in Figure 6-3) and the expected magnitude of the change in exposure concentration. The range of validity of a given gradient should be confirmed on a case-by-case basis. This will, however, rarely be feasible in LCA.

If the background PAF is less than 0.05, then the gradient prediction is a strong function of the distribution model adopted (see Figure 6-4). In the typical absence of insights at such low effect levels to justify any particular distribution model and to warrant complexity, it can be considered preferable to perform comparisons in terms of the secant gradient. Reflecting common risk assessment policy, chemicals or emissions could be compared using an effect measure of 0.05/HC5, for example (Pennington 2000). In this case, uncertainty depends on the accuracy of the secant gradient model (*model uncertainty*), the accuracy of the HC5 estimate (*parameter uncertainty*; described in 'Predicted No-Effect Concentration', p 152), and whether the PAF is in actuality below 0.05 (*scenario uncertainty*). Similarly, Payet et al. (2001) proposed the measure 0.5/HC50 in the AMI method. Given parameter, scenario, and model uncertainty, the most appropriate HCx basis for the secant gradient calculation is unknown, but Payet et al. (2001) note that the median estimate (HC50) has the minimum parameter uncertainty. Critics have raised concerns related to the model uncertainty of a 0.5/HC50 gradient, however, if environmental exposures are expected to be low.

Above PAF 0.05 (and below 0.95), the uncertainty associated with the tangential gradient may

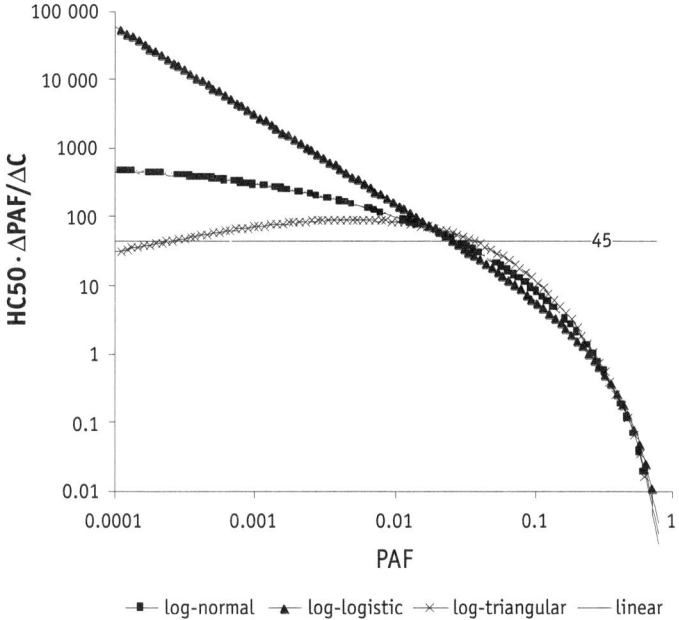

Figure 6-4 Tangential gradient to the SSD curve (ΔPAF/ΔC) as a function of the PAF of species for 4 types of distribution model

not be a strong function of the distribution model selected, as illustrated for one situation in Figure 6-4 (this may hold true only for unimodal models, which are commonly adopted to describe SSDs at this time). For a given range of background PAFs such as 0.1 to 0.5 (Goedkoop and Spriensma 1999) and a typical range of distribution spreads (log SD = 0.36 to 1.8; De Zwart and Sterkenburg 1999; Van de Meent and Van Straalen 1999), the tangential gradient is not a strong function of the SSD spread nor of the actual site-specific PAF. Model uncertainty may be smaller if this PAF range is more applicable than assuming the PAF were less than 0.05.

The tangential gradient is dependent on 2 parameters:
1) the spread or distribution of the dose–response curve and
2) a toxicity measure such as the median, the HC50.

The curves are shown here for a spread of log SD = 1.8. The secant gradient is calculated from 0.05/HC5. To avoid the need to plot a separate set of curves for each chemical's median (HC50), the median-independent gradient (HC50·ΔPAF/ΔC) is plotted on the y axis. We also note that the results and differences are very different for different spreads (log SD values).

The background PAF range of 0.1 to 0.5 considered in Goedkoop and Spriensma (1999) reflects a range of PAFs associated with chemical mixtures and calculated for the Netherlands (termed *combi-PAFs*, as discussed in the next section). Any one chemical is not expected to result in such a PAF, in general, except in extreme site-specific cases. If the combined effect of chemicals in mixtures can be described using a single distribution, such as those in Figure 6-4, then the marginal change associated with the addition or removal of such a single chemical into or from the environment can also be readily estimated by the above approach. Based on such logic, Goedkoop and Spriensma (1999) adopt a value of 0.6/HC50

in Eco-indicator 99 to describe the marginal effect gradient (PAF/C) for all chemicals. The model uncertainty of this gradient is assumed to be low. The parameter uncertainty depends on the accuracy of the HC50 estimate. Given the obvious attractions of such a method, we need to consider the validity of the underlying assumptions.

Based on these discussions alone, 3 gradient measures are currently proposed:
1) 0.6/HC50,
2) 0.5/HC50, and
3) 0.05/HC5 (or 0.05/PNEC).

The measures are based on different fundamental arguments but, in practice, will have very similar parameter uncertainty and may yield similar results (Pennington 2000). The applicability of the fundamental arguments, hence the model uncertainties, are discussed in the next section for 2 of the approaches (the basis of the 0.5/HC50 and the 0.05/HC5 secant gradient measures are generally similar). In 'Ecosystem endpoint measure relevance' (p 164), we discuss the use of the effective concentration affecting 50% of test organisms (EC50) rather than NOEC data to calculate the HC50, providing a measure which Payet et al. (2001) argue is more representative of the disappeared PDF than of the PAF.

Potentially affected fraction of species in mixtures: Combi-PAF

Exposures to chemicals in the environment occur in the presence of complex mixtures. These mixtures must be considered when we determine the contribution of a chemical emission to the actual stress on an ecosystem, in order to quantify the marginal change in the stress level caused by that emission. (This is also one of the many arguments against assuming the existence of thresholds for toxicological effects on ecosystems.)

Consider 2 chemicals that for a given concentration in isolation result in PAFs of 3% and 5% respectively. The combi-PAF, or the total number of species stressed in an ecosystem by introducing them together, can range from less than 5% to more than 8%. The exact effect will depend partly on whether the chemicals interact, or not, and whether the chemicals affect the same or different groups of species in the ecosystem.

If the 2 chemicals considered in this simple example interact toxicokinetically and/or toxicodynamically, then the interactions are commonly described as either *antagonistic* (less than additive) or *synergistic* (greater than additive). However, interactions are usually assumed to be negligible at low doses or ignored in common practice.

Assuming no interactions, we may want to consider 3 additive scenarios:

- Scenario 1 — If the 2 chemicals behave similarly, then the tangential change associated with the addition of the second chemical can be described using a single SSD curve (termed *dose-additive* or, as exposure concentrations are often addressed, *concentration-additive*).
- Scenario 2 — If the 2 chemicals affect completely independent groups of species, then the combi-PAF can be as high as 8% (3% + 5%). Whether the additive effects on an ecosystem are treated as probabilities (combi-PAF = 1 − (1 − PAF_1)·(1 − PAF_2)) or are added (combi-PAF = PAF_1 + PAF_2) will be of little theoretical consequence because these mathematical representations are essentially identical if each PAF is small (termed *response additive*, and sometimes referred to as *independent action*).
- Scenario 3 — In the case of response-additivity, if one of the chemicals affects species already affected by the other, then the combined effect will be a combi-PAF 5%. Removal of the chemical that exerts stress on 3% of the species will have no effect on the combi-PAF (i.e., the change in the number of stressed species is 0, but note that the level of stress on 3% of the species may be reduced).

Scenarios 1 and 2 are illustrated in Figure 6-5. If all chemicals behave as described by Scenario 1, then a single SSD curve can be adopted (PAF/HC50 versus concentration for a given SD).

Assuming more than, say, 5% of species are already affected by chemical impacts, the marginal change associated with any one chemical can be described using straightforward measures such as 0.6/HC50, as demonstrated by Goedkoop and Spriensma (1999) and described in 'Existing marginal ecosystem effect measures' (p 160). If all chemicals in a mixture are response additive (Scenario 2), then it is unlikely that the marginal effect associated with a chemical is described by such an approach. If the concentration of each chemical is low and the chemicals are generally response-additive, resulting in a combi-PAF less than 0.05, then the associated marginal effect may be best described by the secant gradient (e.g., 0.05/HC5), as described in the previous section.[1] Others (Hamers et al. 1996) proposed the treatment of chemicals likely to have a narcotic mode of action as concentration- or dose-additive and those with more specific modes of action as response additive. An illustrative comparison of the Goedkoop and Spriensma (1999) and the Hamers et al. (1996) approaches is presented in Huijbregts et al. (2002).

Scenario 3 reflects an extreme in which the marginal change in PAF can be 0. This extreme may be unlikely in practice, although a significant overlap between the species affected by different chemicals is likely.

---

[1] It is statistically difficult to observe responses below such levels to justify adopting any one low dose–response curve shape for ecosystems. It is therefore justifiable to adopt the most straightforward measure from a low point of departure in the observable range. For this, we suggest the HC5, given that it considers a theoretically observable point, that it is an already widely adopted measure in a regulatory context, and that mixture interactions and site-dependent background compositions or concentrations do not need to be considered.

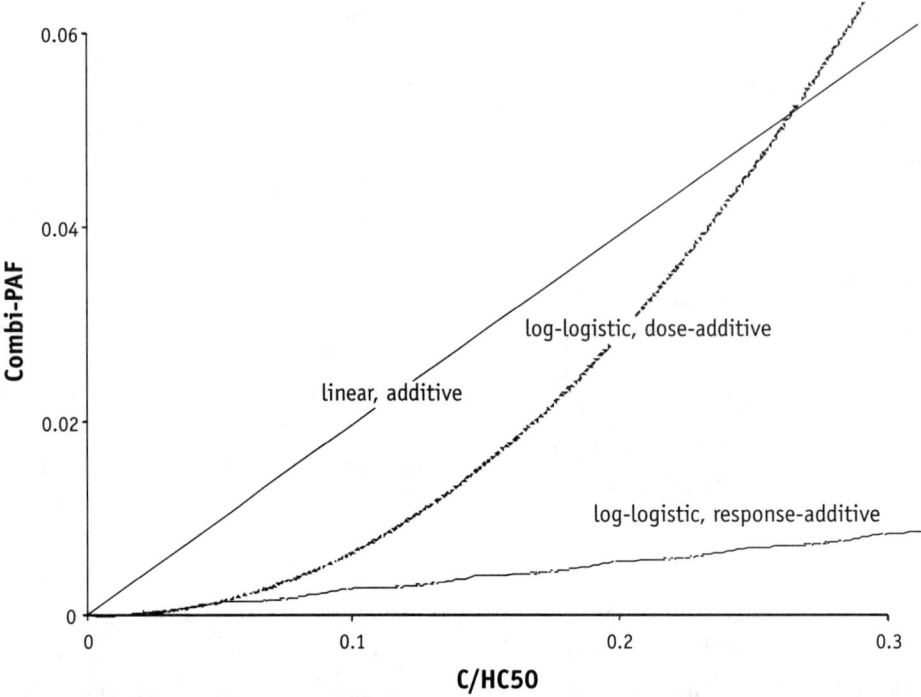

Figure 6-5 Comparison of dose-additive and response-additive curves for 2 mixtures of similar chemicals, assuming a straight line (characteristic of the secant gradient approach) and a log-logistic dose–response curve (characteristic of the tangential gradient approach). Each chemical is at an illustrative concentration of $C_i = 0.05/HC50_i$. The curve shape is represented by log SD = 0.36 for all chemicals. (The straight line represents the scenario in which the marginal gradient associated with all chemicals is assumed to be linear and is not a function of whether the chemical is dose or response additive.)

No one of these 3 scenarios will be generally applicable. A chemical may have dose-additive effects with some groups of chemicals and response-additive effects with other groups. Furthermore, a given chemical may not affect all species on an SSD by the same mode of action. It may be misleading to suggest that entire SSDs can be combined in terms of dose additivity or response additivity. A clear selection for best available measure for the marginal change in the number of species affected (or PAF) is therefore not currently feasible.

Ecosystem endpoint measure relevance: PDFs and PAFs

As described in 'Existing marginal ecosystem effect measures' (p 160) and 'Potentially affected fraction of species in mixtures' (p 162), the number or PAF of species is a common effect measure for ecosystems in current practice (either explicitly or implicitly). In the context of chronic exposures, a PAF usually represents the fraction of species that is stressed above each of their NOECs. This is described by an SSD curve. Hence, such SSDs describe the relationship between exposure concentration and the percentage of species exposed to levels above their NOEC. These SSDs reflect neither the extent to which each species is exposed beyond this NOEC level nor how this will affect their population (hence, biodiversity). Exposure above the NOEC does not necessarily indicate that the population cannot continue to function (or will not adapt). The degree of stress on each species and the extent of species loss (the PDF, which will influence biodiversity) are not reflected. For example, a chemical emission can result in an increase in the stress on a particular species but may not result in an increase in the number of species that are stressed, as illustrated in Scenario 3 in the preceding section.

Payet et al. (2001) propose the use of the chronic EC50 as one alternative to using NOECs. The advantages may be 2-fold. First, measures such as the EC50 have the possible advantage of introducing a more consistent measure than the NOEC (Newman et al. 2000). Second, although it is the subject of current research and debate, Payet et al. (2001) suggest that the measure may better reflect the concentration at which a species can no longer survive. Hence, SSDs based on EC50 measures could provide insights into the PDF of species at different exposure concentrations. The exact measures (ECx) used to define such an SSD are likely to be chemical and species specific.

Fundamental ecotoxicology research as well as further discussion in LCA circles are required to reach a consensus on the desired endpoint measure, or measures, for ecosystems.

Ecotoxicological research issues to be addressed include these:
- Are we measuring the right thing in laboratory tests? How do the effect measures (typically mortality or reduced growth and reproduction) that are observed in the ecotoxicological testing of chemicals predict disappearance of a species? Could more subtle effect measures (e.g., behavioural changes that reduce the species' ability to compete) that occur at lower exposure concentrations lead to better predictions?
- Are the test conditions relevant (e.g., low levels of particulate matter)?
- How do we account for the bioavailability of compounds predominantly bound in suspended particulate matter, and how do we account for biomagnification in food webs (increased exposure at higher trophic levels in food webs associated with ingestion)?

For the modelling of ecotoxic impacts in LCIA, additional aspects for further discussion and research include these:
- Which areas of protection (AoPs) are we addressing? It seems relevant to consider the impacts from ecotoxicity on the natural environment and on biotic natural resources, but what about the living parts of manmade environments (e.g., crop-fields, plantations, managed forests, and fish farms)? Traditionally, ecotoxicology addresses the toxic effects on the structure and function of natural ecosystems, but in principle, the same tools might be used to predict impacts on manmade or man-manipulated biological ecosystems, introducing a distinction between desired and undesired effects.
- If biodiversity is part of our concern, how relevant then is a measure like the PAF of species (i.e., the fraction of species exposed above their NOEC)? Obviously, the PDF predicts changes in the biodiversity of the exposed ecosystem, but how is the PDF modelled?
- Is it really the PDF we are interested in, or is the disappearance seen as more problematic for some species than for others? Do we want to give higher priority to species with a key role in the function of the ecosystem (if such can be identified) or to nearly extinct or economically important species? Is the loss of a species more important if the exposed ecosystem has a naturally low biodiversity (boreal ecosystems) than if the natural biodiversity is high (tropical ecosystems)?
- How do we scale the impacts of ecotoxicity against other impacts considered in LCIA? How many species have actually become extinct locally or globally as a consequence of exposure to manmade chemicals, versus other stressors, and how many are threatened with extinction as a consequence of our current activities?

- How can we interpret the effect indicator in terms of actual damage when, for a given substance, we apply the same indicator for all aquatic ecosystems based on the measured toxicity towards the same narrow group of standard test organisms, regardless of what organisms actually live in the exposed ecosystems?

Criteria to Evaluate Ecotoxicity Indicators

The criteria to evaluate ecotoxicity effect measures for their applicability as indicators in LCIA include the following aspects:

- Scientific validity — Is there a general acceptance of the approach, the assumptions it involves, and its interpretation in the relevant international community?
- Environmental relevance — Is it possible to interpret the result in terms of environmental impact or damage and to perform a subsequent quantitative weighting of the ecotoxicity impact score against the scores for the other impact categories such as climate change and acidification?
- Reproducibility — To what extent will different practitioners arrive at the same ecotoxicity effect indicator for the same substance? Is there a well-defined procedure for calculating the score and for collecting the necessary underlying data? Practitioners typically are environmental specialists but often are not ecotoxicological experts.
- Transparency — Is the procedure for calculating the ecotoxicity effect indicator transparent?
- Quantification of uncertainty — Is there an analysis of the uncertainty associated with the procedure for gathering data and calculating the indicator, and is it quantifiable?
- Complexity — Is the procedure for calculating the ecotoxicity effect indicator explainable to the practitioner?
- Feasibility — Is it feasible to calculate the ecotoxicity effect indicator for all the substances that potentially play a role in the overall ecotoxicity for the product life cycle? Many of these substances will not be identified as priority pollutants at a national level, and hence, they may not have been subject to detailed environmental studies.
- Data availability — What is the requirement for chronic data? Does the procedure allow the use of estimated data, and what is the need for data on the background situation of the exposed ecosystem? To a large extent, the feasibility is governed by the requirements for underlying ecotoxicity data.

Pre-selection of feasible indicators

The indicators are judged against the criteria set forward in the preceding section. The outcome is summarised in Table 6-3.

Here, we have chosen to focus on the discussion and evaluation of measures for ecosystems. However, we recognise that there may be interest in adopting measures for specific well-studied species. Similarly, we note that there may be a preference for initially ranking chemicals and emissions in terms of associated modes of action and potency, as performed in some chemical screening initiatives, in order to first identify chemical emissions likely to be of high concern to regulatory agencies and the public, for example. For many substances, we anticipate that it may not be possible to go beyond such a preliminary screening with currently available data.

Scientific validity
The scientific validity of the acute ecotoxicity measures is high: There is general acceptance of the measure, the assumptions underlying it, and its interpretation. The relevance of the measure as an ecotoxicity effect measure in LCIA is very low, but this is considered under a later criterion.

Table 6-3 Preliminary summary of the evaluation of the approaches for ecotoxicological effect assessment

Criterion	Average approach based on acute ecotoxicity data	Average approach based on PNEC[a], HC5[b], or HC50[c]	Marginal combi-PAF approach[d]
Scientific validity	+++	+	+
Environmental relevance	–	++	++
Reproducibility	+++	++	++
Transparency	+++	++	+
Quantification of uncertainty	–	–	–
Data availability	++	+	+

0 = Not applicable.
– = Insufficient.
+ = Acceptable.
++ = Good.
+++ = Excellent.

[a] Huijbregts 1999.
[b] Pennington 2000.
[c] Payet et al. 2001.
[d] Goedkoop and Spriensma 1999.

Regarding scientific validity, the methods for determining an SSD, the PNEC, or an HCx for exposed ecosystems are, as discussed earlier, encumbered with several problems:
- It is difficult to fulfil the condition that the species whose sensitivity to the substance is the basis for determining the sensitivity distribution are mutually independent random samples of the total number of species in the ecosystem.
- There are only empirical, but not particularly theoretical, arguments for selecting precisely the proposed, typically unimodal, statistical distribution types for the sensitivity of species in natural ecosystems, and it cannot be excluded that the sensitivity distribution will change if the ecosystem is simultaneously exposed to other forms of stress.
- It seems questionable, based on the existing set of data, to estimate one PNEC value that is assumed to be representative for all aquatic ecosystems in a region or the world.

The same reservations seem to apply to the determination of the PAF.

The determination of a PNEC through extrapolation with assessment factors, as applied in many risk assessment schemes, seems on average to overestimate the ecotoxicity of the substance. Empirically determined best-estimate assessment factors with uncertainty intervals have been proposed, but there is as yet no scientific consensus on an approach.

The average and marginal (or secant and tangential) gradient approaches differ theoretically but can both be used in an analogous way when assessing chemicals in isolation. Methods are being developed to account for dose and effect additivity within complex mixtures for ecosystems (reflecting the affected or the disappeared fraction of species), as well as accounting for physical and biological factors. The presence of multiple stressors, which was not addressed in this chapter, may even imply that the marginal gradient is not strongly chemical dependent and is purely a function of background conditions. Such theoretical assumptions behind current proposals remain the topic of significant debate. We note, however, that many of the current measures may provide similar insights in LCA and that some of the debates related to the theoretical underpinnings may ultimately be demonstrated to have limited consequences in practical terms.

Environmental relevance

Acute toxicity data are relevant only for short-term exposure situations, which will rarely be addressed in LCIA and cannot be used in conjunction with the time-integrated exposure concentrations calculated in current practice. We must therefore strive for measures of toxicity that represent long time exposures covering substantial parts of the organism's life or repeated exposures, that is, chronic toxicity measures. If chronic data are missing for some of the substances, we may have to extrapolate from acute data, but the aim must be to get as good, comprehensive, and relevant test data as possible for all the major contributors to ecotoxicity in the study, in order to increase the environmental relevance of the result.

The PNEC and HCx are related measures of chronic effects, usually calculated from NOEC data and possibly from an assumed SSD model. Such NOEC data are commonly considered to reflect the level at which the species is affected but not necessarily disappeared. Estimating measures such as HCx using ECx data may be one way to provide a measure that reflects the disappeared fraction of species. The significance of this measure when combined with time-integrated exposure concentrations requires further consideration.

Measures that attempt to take mixtures into account, as well as possibly physical and biological stresses, possibly step towards modelling damage with greater relevance to actual environmental situations. Marginal (or tangential) gradient measures can be considered to similarly improve relevance. However, without knowledge of the underlying model and scenario uncertainties, it is not clear that improvements in current practice are gained over measures sometimes perceived to be less relevant.

Reproducibility

The estimation of measures such as the PNEC or HCx fully depend on the identified ecotoxicity database used in their calculation. Because such data are often based on tests designed to be reproducible, and on documented procedures, these measures are typically reproducible. Problems are noted, however, when different practitioners use different data sources and different procedures. A code of best available practice and a single peer-reviewed database of ecotoxicological measures for LCIA are required.

Transparency

The use of acute ecotoxicity data as a basis for LCIA effect measures is usually transparent (although of lower relevance). However, differences in test protocols, for example, that result in differences in the measure may be unknown and can result in misleading conclusions. Other measures such as PNECs and HCxs, while requiring some additional procedural documentation, can also be calculated from such data in a transparent manner but will suffer from the same problems as acute measures. The use of measures that account for mixtures and the difficulty of understanding the underlying principles, even by named experts, result in additional complexity.

Quantification of uncertainty

The data or parameter uncertainty of a measure such as the PNEC or HCx for a substance can be reasonably well stated. In practice, it is rarely estimated or taken into account when LCA results are interpreted. Furthermore, uncertainty associated with the underlying toxicological data may not be stated. Estimation of model uncertainties and scenario uncertainties of available measures, which are important parts of whether a measure is accurate and environmentally relevant, is much more complex to determine.

In current practice, because unimodal distributions are assumed, all the ecosystem measures are mathematically related to the HC50 (median ecosystem response). The measure HC5 may require some extrapolation, but the parameter uncertainty may prove to be similar to that of the HC50 when only a limited number of data are available and both measures are in the theoretically observable range. Hence, the parameter uncertainty of the currently available measures for ecosystems is probably similar. Variations will depend on the exact data and procedures adopted.

The currently proposed measures for ecosystems have not been widely evaluated with measured data in the field. Many questions remain, such as how best to account for the bioavailability of compounds predominantly bound in suspended particulate matter[2] and how to account for biomagnification in food webs (increased exposure at higher trophic levels in food webs associated with ingestion). Theoretical differences and foundations are debated, but the inherent accuracy of the models is unknown. Similarly, the applicability of the scenarios represented by the models remains unknown. Nevertheless, such approaches are based on scientific assumptions and, probably, constitute the current best-available practice.

Data availability

Measured and predicted acute ecotoxicity data for selected individual species are most prevalent (see Annex, p 174). It is therefore a significant problem that chronic data are of greater relevance for use in LCA and that such data can be required for species belonging to several phyla. For this reason, possibilities exist to estimate PNECs or HCxs based on even small quantities of acute ecotoxicity data (measured and/or predicted). The resultant uncertainties can, however, be considerable.

Conclusions

An ecotoxicity indicator for LCIA will commonly be based on the extrapolation of measured and estimated data. In preliminary qualitative screening, such data can be used to identify chemicals likely to be of high toxicological concern. Comparison may be based on commonly tested single species. Traditionally in LCA, quantitative measures such as 1/PNEC have been used as comparative indicators of effects at the ecosystem level. Alternatives to account for mixtures and the disappeared, rather than the affected, fraction of species have recently been proposed, with the view of providing a more environmentally relevant basis for the measure. However, all approaches currently proposed are encumbered with significant problems, and it remains unclear which is the most scientifically justifiable. Model and scenario uncertainties, which are important components of the accuracy of a measure, are unknown. Parameter uncertainties are often not quantified. Yet, the parameter uncertainties are likely to be similar for many of the current methods, and the results of available models will probably yield consistent insights in many LCA studies.

We conclude that measures such as 0.05/PNEC or 0.05/HC5 can be seen as marginal measures, assuming linearity of the concentration–effect curve below PAF = 5%. Apart from terminology, differences may be primarily related to how extrapolation is conducted. Such measures may not, however, take into account the combined exposure to the complex chemical mixtures found in the environment. Here, so-called *combi-PAF approaches* offer a developing alternative, at least in terms of the fundamental principles. The resultant marginal measures from combi-PAF curves and the PNEC style of approaches may yield similar results in practice, and may have similar parameter and model uncertainties. The scenario uncertainty, and hence the overall uncertainty, or accuracy, of both types of approach is dependent on the environmental situation. This overall uncertainty may not

---

[2] Some current approaches assume 100% bioavailability and adopt the bulk water exposure concentration, while others, such as Huijbregts (1999), assume that none of a compound in suspended particulate matter is bioavailable. Both assumptions should be used in a sensitivity analysis in the absence of alternative insights. It may be extremely misleading to ignore contaminant concentrations in suspended particulates when we evaluate ecosystems, even if toxicity testing is often conducted in the absence of realistic particulate loadings.

be higher for the combi-PAF approaches than for PNEC approaches. We therefore acknowledge that the potential of a combi-PAF approach for representing mixture toxicity is interesting, particularly from the perspective of improved environmental relevance. However, the fundamental principles require further investigation before it can be considered for best practice in LCIA.

If concentrations of chemicals are low (PAF below, say, 0.05 for a chemical or group of dose-additive chemicals), then the best available practice for effect indicators in LCA may remain the 0.05/PNEC or 0.05/HC5. Research is encouraged to advance the state of the art by evaluating the validity of this straightforward measure or by justifying alternatives empirically.

PNEC and HC5 measures can differ in terms of the basis of extrapolation factors used in their derivation, as well as the data measurements or predictions adopted. We need to consider the merits of adopting best-estimate extrapolation factors with confidence intervals, rather than using factors of 10, for example.

Peer-reviewed extrapolation approaches and databases for use in LCA are a necessity to ensure transparency, reproducibility, quality, and accuracy. Parameter uncertainty should be quantified as far as possible. Other fundamental considerations include accounting for the bioavailability of chemicals bound in solid phases such as suspended matter (some current approaches assume 100% bioavailability, while others assume 0; we recommend that both should be used in a sensitivity analysis in the absence of alternative insights) and how to account for exposure by ingestion (biomagnification in food webs).

The underlying measure used in the calculation of PNEC, and sometimes for HC5, is the NOEC. NOECs for multiple species, sometimes represented in the form of SSDs, may be considered to provide a measure related to the PAF of species. From a decision-making perspective, it may be preferable to provide a comparison in terms of biodiversity-related measures such as the PDF of species. In addition to possibly improved consistency, some researchers promote the use of ECx data in place of NOEC to derive PDF measures. These differences need to be discussed and a consensus reached in LCA circles.

References

Aldenberg T, Slob W. 1993. Confidence limits for hazardous concentrations based on logistically distributed NOEC toxicity data. *Ecotoxicol Environ Saf* 25:48–63.

Ankley GT, Di Toro DM, Hansen DJ, Berry WJ. 1996. Technical basis and proposal for deriving sediment quality criteria for metals. *Environ Toxicol Chem* 15(12):2056–2066.

Bro-Rasmussen F, Calow P, Canton JH, Chambers PL, Fernandes AS, Hoffman L, Jouany J-M, Klein W, Persoone G, Scoullos M, Tarazona JV, Vighi M. 1994. EEC water quality objectives for chemicals dangerous to aquatic environments (List 1). *Rev Environ Contam Toxicol* 137:83–110.

Chapman PM, Fairbrother A, Brown D. 1998. A critical review of safety (uncertainty) factors for ecological risk assessment. *Environ Toxicol Chem* 17(1):99–108.

Clements RG, Nabholz JV, Zeeman MG, Osborn KC, Wedge R. 1993. Validation of structure activity relationships used by the USEPA's Office of Pollution Prevention and Toxics for the environmental hazard assessment of industrial chemicals. Philadelphia PA, USA: American Soc for Testing and Materials (ASTM). ASTM STP 1216.

De Zwart D, Sterkenburg A. 1999. Toxicity-based assessment of water quality. Poster presented at the Annual Meeting of the Society of Environmental Toxicology and Chemistry (SETAC); 1999 Nov 14–18; Philadelphia PA, USA. Bilthoven, NL: National Institute of Public Health and the Environment (RIVM).

Di Toro DM, Zarba CS, Hansen DJ, Berry WJ, Swartz RC, Cowan CE, Pavlou SP, Allen HE, Thomas NA, Paquin PR. 1991. Technical basis for establishing sediment quality criteria for non-ionic organic chemicals using equilibrium partitioning. *Environ Toxicol Chem* 10:1541–1583.

[ECETOC] European Centre for Ecotoxicology and Toxicology of Chemicals. 1995. Assessment factors in human health risk assessment. Brussels, B: ECETOC. Technical Report Nr. 68.

[ECETOC] European Centre for Ecotoxicology and Toxicology of Chemicals. 1997. The value of aquatic model ecosystem studies in ecotoxicology. Brussels, B: ECETOC. Technical Report Nr. 73.

[ECETOC] European Centre for Ecotoxicology and Toxicology of Chemicals. 1998. QSARs in the assessment of the environmental fate and effects of chemicals. Brussels, B: ECETOC. Technical Report Nr. 74.

Emans HJB, v.d. Plassche EJ, Canton JH, Okkerman PC, Sparenburg PM. 1993. Validation of some extrapolation methods used for effect assessment. *Environ Toxicol Chem* 12:2139–2154.

[EPPO] European and Mediterranean Plant Protection Organisation. 1993. Decision-making scheme for the environmental risk assessment of plant production products. *EPPO Bull* 23(1):151–157.

Erickson RJ, Stephan CE. 1988. Calculation of the final acute value for water quality criteria for aquatic organisms. Washington DC, USA: U.S. Environmental Protection Agency (USEPA) Office of Research and Development. Report Nr. PB88214994.

European Commission. 1996. Technical guidance documents in support of Directive 93/67/EEC on risk assessment of new notified substances and Regulation (EC) No. 1488/94 on risk assessment of existing substances (Parts I, II, III and IV). Luxembourg, L: Office for Official Publications of the European Community. EC catalogue numbers CR-48-96-001, 002, 003, 004-EN-C.

Finnveden G, Andersson-Sköld Y, Samuelsson M-O, Zetterberg L, Lindfors L-G. 1992. Classification (impact analysis) in connection with life cycle assessment: A preliminary study. In: Lindfors L-G, editor. Product life cycle assessment: Principles and methodology. Copenhagen, DK: Nordic Council of Ministers. Nord 1992:9. p 172–231.

Goedkoop M, Spriensma R. 1999. The Eco-indicator 99: A damage oriented method for life cycle impact assessment. Amersfoort, NL: PRé Consultants.

Guinée J, Heijungs R, van Oers L, van de Meent D, Vermeire T, Rikken M. 1996. LCA impact assessment of toxic releases. Generic modelling of fate, exposure and effect for ecosystems and human beings with data for about 100 chemicals. The Hague, NL: Ministry of Housing, Spatial Planning and the Environment (VROM). Report Nr. 1996/21.

Hamers T, Aldenberg T, Van de Meent D. 1996. Definition report: Indicator effects toxic substances (I_{tox}). Bilthoven, NL: National Institute of Public Health and the Environment (RIVM). Report Nr. 607128001.

Hauschild M, Wenzel H, Damborg A, Tørsløv J. 1998. Ecotoxicity as assessment criterion in life cycle assessment. In: Hauschild MZ, Wenzel H, editors. Environmental assessment of products. Volume 2, Scientific background. Boston MA, USA: Kluwer. ISBN 0412-80810-2.

Huijbregts MAJ. 1999. Priority assessment of toxic substances in the frame of LCA. Development and application of the multi-media fate, exposure and effect model USES-LCA. Amsterdam, NL: Universiteit van Amsterdam, Milieukunde.

Huijbregts M, Van de Meent D, Goedkoop M, Spriensma R. 2002. Ecotoxicological impacts in life cycle assessment. In: Posthuma L, Suter GW, Traas TP, editors. Species sensitivity distributions in ecotoxicology. Boca Raton FL, USA: Lewis.

[IRPTC/UNEP] International Register of Potentially Toxic Chemicals/United Nations Environment Programme.

1996. Inventory of critical reviews on chemicals. Geneva, CH: IRPTC/UNEP.

Jager T, Rikken MGJ, van der Poel P. 1997. Uncertainty analysis of EUSES: Improving risk management by probabilistic risk assessment. Bilthoven, NL: National Institute of Public Health and the Environment (RIVM).

Jolliet O, editor. Assies J, Bovy M, Finnveden G, Guinée J, Hauschild M, Heijungs R, Hofstetter P, Potting J, Udo de Haes HA, Wrisberg N. 1996. Impact assessment of human and ecotoxicity in life cycle assessment. In: Udo de Haes HA, editor. Towards a methodology for life cycle impact assessment. Brussels, B: Society of Environmental Toxicology and Chemistry (SETAC) Europe. p 49–61.

Jolliet O, Crettaz P. 1997. Critical surface-time 95. A life cycle impact assessment methodology including fate and exposure. Lausanne, CH: Ecole Polytechnique Fédérale de Lausanne (EPFL).

Jørgensen SE, Halling-Sørensen B, Mahler H. 1998. Handbook of estimation methods in ecotoxicity and environmental chemistry. Boca Raton FL, USA: Lewis.

Klepper O, Van de Meent D. 1997. Mapping the potentially affected fraction (PAF) of species as an indicator of generic toxic stress. Bilthoven, NL: National Institute of Public Health and the Environment (RIVM). RIVM Report Nr. 607504001.

Mead CD, Fairman R, Williams WP. 1997. Environmental risk assessment: approaches: Experiences and information sources. King's College, London, GB: European Environmental Agency.

Molak V. 1997. Fundamentals of risk analysis and risk management. Boca Raton FL, USA: Lewis.

Moore DRJ. 2000. A comparison of model performance for six QSAR packages that predict acute toxicity to fish. Hull, Quebec, CAN: Environment Canada, Chemicals Evaluation Division.

Nabholz JV, Clements RG, Johnson DW, Zeeman M. 1993. The use and application of QSARs in the Office of Toxic Substances for Ecological Hazard Assessment of New Chemicals. Philadelphia PA, USA: American Soc for Testing and Materials (ASTM). ASTM STP 1179.

Nendza M. 1998. Structure-activity relationships in environmental sciences. London, GB: Chapman and Hall.

Newman CM, Ownby DR, Mezin CA, Powell DC, Christensen TRL, Lerberg SB, Andersson BA. 2000. Applying species-sensitivity distributions in ecological risk assessment: Assumptions of distribution type and sufficient number of species. *Environ Toxicol Chem* 19:508–515.

[OECD] Organisation for Economic Co-operation and Development. 1992. Report of the OECD workshop on the extrapolation of laboratory aquatic toxicity data to the real environment. Paris, F: OECD. OECD Environment Monographs Nr. 59.

[OECD] Organisation for Economic Co-operation and Development. 1997. Users guide to hazardous substance data banks. (Available in OECD member countries.) Paris, F: OECD.

[OECD] Organisation for Economic Co-operation and Development. 1998. Harmonized integrated hazard classification system for human health and environmental effects of chemical substances. (As endorsed by the 28th Joint Meeting of the Chemicals Committee and the Working Party on Chemicals, Nov 1998.) Paris, F: OECD.

Payet J, Pennington DW, Jolliet O. 2001. The AMI method: A specific LCA tool for the assessment of toxic impacts on ecosystems. Draft report. Lausanne, CH: Ecole Polytechnique Fédérale de Lausanne (EPFL).

Pennington DW. 2000. Issues in the characterisation of toxicological impacts: Midpoints and endpoints. In: Bare JC, Hofstetter P, Pennington DW, Udo de Haes HA, organizers. Life Cycle Impact Assessment Workshop, Midpoints versus Endpoints: The Sacrifices and Benefits; 2000 May 25–26; Brighton, GB. Paris, F: United Nations Environment Programme (UNEP). Forthcoming as 'Evaluation of Environmental Impacts in Life Cycle Assessment'.

Pennington DW, Payet J. 2001. Toxicological extrapolation and uncertainty in relative comparison applications in the context of aquatic ecosystems. Draft report. Lausanne, CH: Ecole Polytechnique Fédérale de Lausanne (EPFL).

[RIZA] Institute for Inland Water Management and Waste Water Treatment. 1999. Effect factors for the aquatic environment in the framework of LCA. Haskoning, Nijmegan, National Institute of Public Health and the Environment (RIVM). Bilthoven, NL: Interfaculty Department of Environmental Science (Milieukunde), Univ of Amsterdam.

Smrchek JC, Zeeman M. 1997. Assessing risks to ecological systems from chemicals. In: Calow P, editor. Handbook of environmental risk assessment and management. London, GB: Blackwell Scientific.

Slooff W. 1992. Ecotoxicological effect assessment: deriving maximum tolerable concentrations (MTC) from single species toxicity data. Bilthoven, NL: National Institute of Public Health and the Environment (RIVM). Report Nr. 719102018.

Stephan CE, Mount DI, Hansen DJ, Gentile JH, Chapman GA, Brungs WA. 1985. Guidelines for deriving numerical national water quality criteria for the protection of aquatic organisms and their uses. Washington DC, USA: U.S. Environmental Protection Agency (USEPA). PB85-227049.

Suter GW. 1993. Ecological risk assessment. Boca Raton FL, USA: Lewis.

Suter GW. 1998. Ecotoxicological effects extrapolation models. In: Strojan CL, Newman MC, editors. Risk assessment: Logic and measurement. Chelsea MI, USA: Ann Arbor Pr. p 167–185.

Van de Meent D, Toet C. 1992. Dutch priority setting system for existing chemicals. The Hague, NL: Ministry of Housing, Spatial Planning and the Environment (VROM).

Udo de Haes HA, Jolliet O, Finnveden G, Hauschild M, Krewitt W, Müller-Wenk R. 1999. Best available practice regarding impact categories and category indicators in life cycle impact assessment. Background document for the second working group on life cycle impact assessment of SETAC Europe (WIA-2). *Int J LCA* 4(2):1–15.

[UNEP/ECETOC] United Nations Environment Programme/European Centre for Ecotoxicology and Toxicology of Chemicals. 1996. Inventory of information sources on chemicals: Intergovernmental organisations. Geneva, CH: International Register of Potentially Toxic Chemicals (IRPTC)/UNEP.

[USEPA] United States Environmental Protection Agency. 1984. Estimating concern levels for concentrations of chemical substances in the environment. Washington DC, USA: USEPA Environmental Effects Branch, Health and Environmental Review Division.

[USEPA] United States Environmental Protection Agency. 1994. Estimating concern levels for concentrations of chemical substances in the environment. Washington DC, USA: USEPA, Office of Pollution Prevention and Toxics (OPPT).

[USEPA] United States Environmental Protection Agency. 1997. Pollution prevention assessment framework. Washington DC, USA: USEPA, Office of Pollution Prevention and Toxics (OPPT).

[USEPA] United States Environmental Protection Agency. 1999. Water quality guidance for the Great Lakes system. Federal Register 40 CFR Part 142. USEPA.

[USEPA/EC] United States Environmental Protection Agency/European Community. 1994. Joint project on the evaluation of (quantitative) structure activity relationships. Washington DC, USA: USEPA/EC. Report Nr. EPA 743-R94-001.

Van de Meent D, Van Straalen NM. 1999. Two-way use of species sensitivity distributions in ecotoxicological risk assessment. Poster presented at the SETAC Annual Meeting; 1999 Nov 14–18; Philadelphia PA, USA.

Van Gestel CAM, Ma W. 1993. Development of QSARs in soil ecotoxicology: Earth worm toxicity and soil sorption of chlorophenols, chlorobenzenes and chloroanilines. *Water Air Soil Pollut* 69:265–276.

Van Leeuwen CJ, Hermens JLM. 1996. Risk assessment of chemicals: An introduction. Dordrecht, NL: Kluwer Academic.

Van Straalen NM, Dennemann CAJ. 1989. Ecotoxicological evaluation of soil quality criteria. *Ecotoxicol Environ Saf* 18:241–251.

Vermeire TG, Stevenson H, Pieters MN, et al. 1998. Assessment factors for human health risk assessment: A discussion paper. Bilthoven, NL: National Institute of Public Health and the Environment (RIVM) and Zeist, NL: Netherlands Organisation for Applied Scientific Research (TNO).

Wagner C, Løkke H. 1991. Estimation of ecotoxicological protection levels from NOEC toxicity data. *Water Res* 25:1237–1242.

Walz R, Herrchen M, Keller D, Stahl B. 1996. Impact category ecotoxicity and valuation procedure. Ecotoxicological impact assessment and the valuation step within LCA: Pragmatic approaches. *Int J LCA* 1(4):193–198.

Wenzel H, Hauschild MZ, Alting L. 1997. Environmental assessment of products. Volume 1, Methodology, tools and case studies in product development. London, GB: Chapman and Hall.

Zeeman MG. 1995. Ecotoxicity testing and estimation methods developed under Section 5 of the Toxic Substances Control Act (TSCA). In: Rand G, editor. Fundamentals of aquatic toxicology: Effects, environmental fate and risk assessment. Washington DC, USA: Taylor and Francis.

Annex: Data Sources

A number of national and international sources of measured toxicological data exist (e.g., International Register of Potentially Toxic Chemicals/ United Nations Environment Programme [IRPTC/UNEP] 1996; Mead et al. 1997; Organisation for Economic Co-operation and Development [OECD] 1997; United Nations Environment Programme/European Centre for Ecotoxicology and Toxicology of Chemicals [UNEP/ECETOC] 1996). The availability of toxicological data in the public domain is, however, significantly driven by regulatory requirements. Large-scale regulatory efforts include the screening of new and existing chemicals and the assessment of contaminated sites. In Europe, a number of directives and regulations dictate minimum or base datasets that must be submitted for new and existing chemicals to facilitate initial screening. However, the data required are typically from acute studies. As with most available toxicological data, the values submitted can be based on old test guidelines and quality assurance procedures. In the context of the U.S. Toxic Substances Control Act (TSCA), reliance on measured data is lower, and estimation techniques (quantitative structure-activity relationships [QSARs]; see below) are often used. Pesticides and pharmaceutical products are notable exceptions and undergo more detailed evaluations (although environmental concentrations of drugs are becoming a high concern).

In summary, regulatory activities are the most significant source of toxicity data but do not typically result in the availability of high-quality chronic measurements. This limited availability inhibits the ability to compare emissions or chemicals in terms of potential long-term impacts with high certainty.

Key sources of measured data

Directories of data sources include the UNEP/ ECETOC (1996), Mead et al. (1997), the OECD (1997), and the IRPTC/UNEP (1996). A number of international toxicological databases and national data sources are listed. These data sources can vary in the degree of quality assurance from listings of 'all' available values in the literature to highly peer-reviewed datasets. Data can be collected from a number of sources and sorted according to a preference hierarchy. These hierarchies may, however, reflect preference levels for use in screening. For example, the toxicological benchmarks (oral reference doses [RfDs], inhalation reference concentrations [RfCs], etc.) are often preferred. In the absence of benchmarks and chronic data, acute values and QSAR estimates are considered. Analogous approaches can be adopted in relative comparison studies, although underlying data from benchmarks and median test data should be used preferentially.

Data prediction: QSARs

To help corroborate measured data, but more importantly to estimate values for untested chemicals, QSARs are commonly used (Clements et al. 1993; ECETOC 1998; Jørgensen et al. 1998; Nendza 1998).

Quantitative structure-activity relationships are estimation models based on correlations among available data, hence QSAR predictions are predominantly for acute endpoints and only as good as the training sets used in their development. For aquatic species, toxicity is commonly correlated with the octanol–water partitioning coefficient (K_{OW}, a surrogate factor for partitioning between organic matter and water) or sometimes related descriptors (e.g., solubility). However, such descriptors should be used with care, particularly if the mode of action is unknown.

Typically, correlations between K_{OW} and toxicological effects data are linear. Bilinear relationships are sometimes proposed (e.g., for phenols), although such bilinear relationships may result from experimental error or simply indicate that steady-state levels of the more lipophilic chemical (log K_{OW} > 6) were not achieved in the species (body burdens) within the duration of the test. Alternatively, the lower toxicity at higher values may indicate pharmacokinetic inhibition (a limited ability of the chemical to transfer across fish membranes). QSAR estimates are therefore considered reliable only below a log K_{OW} of 6 (ECETOC 1998).

To avoid the problem of identifying mode of action and appropriate QSARs, a number of QSAR suites are readily available. For example, Ecotoxicity of Industrial Chemicals Based on Structure-Activity Relationships (ECOSAR) contains approximately 120 QSARs that facilitate estimation of aquatic toxicity for fish, invertebrates, and algae (USEPA 1997). For short-term and long-term fish tests, QSAR models can, however, provide the correct prediction of mechanism (reactive or nonreactive) for approximately 87% of chemicals (Nendza 1998). However, *reactive substructures* may not be identical for acute and chronic effects data for a given species. Furthermore, the correspondence across species of substructures considered to induce reactive modes of action is significantly limited (about 36% concordance between *Daphnia* and fish and about 24% between algae and fish). These factors will therefore introduce uncertainty in the prediction of long-term effects or mode of action from short-term tests and limit our ability to extrapolate amongst species.

Quantitative structure-activity relationship estimates are considered reliable only for narcosis (nonreactive) effects in general (ECETOC 1998). In general terms, QSAR estimates are sometimes considered to be within a factor of 10 of the actual value, providing the correlation is robust and applicability limits are observed (USEPA 1994). In relative comparison exercises, this factor should be considered an indicator of uncertainty and not incorporated in the calculation as a safety factor. However, the general applicability of this order of magnitude uncertainty factor is questionable, and the percentage of predictions expected to fall within this range is unknown. Evaluation of the ECOSAR QSARs, predominantly the acute correlations, have been performed with data for approximately 500 chemicals. Nabholz et al. (1993) presented a summary of the results for each correlation and chemical group. Approximately 85% of the predictions were within an order of magnitude of the measured results. However, later comparisons with European Union (EU) measured acute toxicity values for Daphnids indicated only a 71% agreement with an equal number of over- and underestimations (U.S. Environmental Protection Agency/European Community [USEPA/EC] 1994). With the exception of algae, the geometric mean of the predicted to the measured value ratios was conservative (0.24 for fish and 0.39 for Daphnids). These evaluation statistics provide a basis for estimation of the uncertainty of associated QSARs in a relative comparison context, as well as adjustments (extrapolation) to provide estimates of the most likely values. It should, however, be noted that the correlations in ECOSAR are continually being updated.

The EU technical guidance document for the assessment of new chemicals (European Commission 1996) recommends a number of QSARs for predicting nonpolar and polar narcosis endpoints. Acute and chronic endpoints are considered for fish, *Daphnia*, and algae. The QSARs are based on linear regression, valid for chemicals with a log K_{OW} in the range of 1 to 6 and predicted 95% of the values in the initial training set (used to develop the correlation) within a factor of 3. A comparison of the USEPA's ECOSAR and EU QSARs, limited to substances with a narcotic mode of action, indicated strong agreement (USEPA/EC 1994).

Moore (2000) compared 6 QSAR packages (ECOSAR, Assessment Tools for the Evaluation of Risk [ASTER], TOPKAT, OASIS, a probabilistic neural network [PNN], and a computational neural network [CNN]) using screened 96-h LC50 data from Aquatic Toxicity Information Retrieval (AQUIRE) database for fish species for 130 chemicals (including neutral organics, phenols, dinitrophenols, vinyl and allyl halides, esters, phosphate esters, aromatic amines, acrylates, hydrazines, imides, and others). These QSARs adopt significantly different statistical methods to describe the relationship between chemical descriptors. Although the tools primarily provide predictions for fathead minnow, no obvious improvement in correlation was generally observed when only the associated subset of data was considered. For each species with multiple toxicity test results, the geometric mean was calculated and then, using these as inputs, a grand substance geometric mean. Separate analyses were performed for different chemical classes and according to mode of toxic action.

Importantly, 3 of the QSAR packages (ECOSAR, TOPKAT, OASIS) provided warnings to indicate that toxicity predictions may be suspect for specific substances. ASTER does not provide predictions outside its range of applicability. TOPKAT was reported to have excellent performance within its optimal performance space (although only 37% of the substances could be considered). For the remaining substances, the use of the PNN was recommended. Based on the reported statistics, the 95th percentile of the ratio of the predicted and the measured values is 2709 for ASTER, 1330 for PNN, 1000 for ECOSAR (for chemicals without warnings and 2632 for the chemicals with warnings), 929 for OASIS (4552 with warnings), 884 for CNN, and 252 for TOPKAT (4476 with warnings).

It should be noted that the measured data were obtained using a variety of different testing protocols. As a result, some of the differences may be partially attributed to the measured toxicity values themselves. The extent of this uncertainty was not independently identified, although it was noted that a number of the significant outliers (differences >1000) had only one measured toxicity value. No recommendations could be made for chronic or nonmortality predictions.

Normalisation, Grouping, and Weighting in Life-Cycle Impact Assessment

Göran Finnveden, Patrick Hofstetter, Jane C. Bare, Lauren Basson, Andreas Ciroth, Thomas Mettier, Jyri Seppälä, Jessica Johansson, Greg Norris, Stephan Volkwein

Acknowledgements — Besides the authors, a number of persons have been involved in this work as members of the task group, taking part in meetings and/or providing written comments. Contributions have been provided by Magnus Bengtsson, Chalmers, Sweden; Norihiro Itsubo, JEMAI, Japan; Wolfram Krewitt, IER, Germany; Erwin Lindeijer, TNO, Netherlands; Helena Mälkki, VTT, Finland; and Yasunari Matsuno, MITI, Japan. We are also grateful for the comments provided by the reviewers: Mark Goedkoop, Ruedi Müller-Wenk, and Helias Udo de Haes.

Abstract — In this chapter, normalisation, grouping, and weighting are discussed as 3 optional elements in life-cycle impact assessment (LCIA). One starting point is an overview of methods that are available or could be developed. Another starting point is a discussion of requirements that can be put on weighting methods, resulting in a system of criteria. Based on these starting points, we discuss different methods. The relation and the congruence between the optional elements of LCIA deserve further attention. One important conclusion is that case-specific (internal) normalisation requires case-specific weighting. LCIA can rely on a number of different sciences and theories. Of special interest for the elements discussed here are environmental economics and decision analysis. Techniques, knowledge, and theories for the assessment problems developed in economics and decision analysis can be applied to the field of LCIA. From the evaluation of different weighting methods, an important conclusion is that distance-to-target (DtT) methods should not be used as weighting methods, unless it is explicitly assumed that all targets are of equal importance. Panel methods and monetisation methods are the 2 major groups of weighting methods that can be used in LCIA. Both consist of a large number of different approaches. Expressed preference methods can be used in both panel and monetisation methods; indeed, they have been used for a number of years, and there is some learning from this. One of the main outcomes of behavioural decision research during the last 2 decades is the view that people's preferences are often constructed during the elicitation process. Values for objects that are unfamiliar and complex are constructed rather than reported in the elicitation procedures. This constructive perspective has important implications for the use of methods based on expressed preferences in life-cycle assessment (LCA); for example, it is known that different types of biases occur during the elicitation process. Methods based on revealed preferences are all based on the assumption that preferences revealed in one context can be transferred to another context. A certain leap of faith is thus required. Another aspect is that we do not know which aspects were considered when decisions were taken and their relation to the weighting problems at hand. There are a number of different approaches for monetisation of environmental impacts (EIs). Since they measure different economic values (e.g., total economic value or just use values), an important conclusion is that monetary values may not always be comparable and additive just because they have the same units. The system of criteria for weighting methods suggested here is used on a generic level. We do, however, suggest that it also can be used on specific weighting methods and when choosing methods for a specific case study. There is no single weighting method that fulfils all relevant criteria and there probably will never be one. It is, however, important to note that weighting methods are widely used, and there is therefore a need to constantly review and improve weighting methods in order to provide decision-makers and other LCA users with the best available practice.

Introduction

The life-cycle impact assessment (LCIA) phase of life-cycle assessment (LCA) is divided into several elements according to the International Organization for Standardization (ISO 2000) standard. The elements normalisation, grouping, and weighting are described as optional elements that can be used depending on the goal and scope of the LCA. *Normalisation* relates the magnitude of impacts in different impact categories to reference values. *Grouping* is the assigning of impact categories into one or more sets; it may involve sorting and/or ranking. *Weighting* is the process of converting indicator results (i.e., results from characterisation or normalisation) by using numerical factors based on value choices; it may include aggregation across impact categories of the weighted results. According to this definition, weighting is a quantitative process.

It is significant that the last phase of an LCA is interpretation, in which conclusions are drawn and recommendations are made (ISO 1997). The results from the grouping and/or the weighting are thus not necessarily the final conclusions drawn from the overall LCA study, nor the only factors considered in a decision. There are several alternatives concerning the relations between the optional elements, as shown in these examples:
- Normalisation is performed without any subsequent grouping or weighting, and the results are fed into the interpretation.
- Normalisation is performed prior to a grouping exercise.
- Normalisation is performed prior to the weighting.
- Weighting is performed without any prior normalisation.

Although the ISO standards provide a framework and a terminology for LCA, it is clear that ongoing development presents different methodological choices. The scientific and methodological framework for LCIA is still being developed (ISO 2000) and has been under discussion in several international forums (e.g., Udo de Haes 1996; Udo de Haes and Wrisberg 1997; Bare et al. 1999, 2000; Udo de Haes et al. 1999).

The weighting phase in LCA literature is often called *valuation*. It is and has always been a controversial issue, partly because this element requires the incorporation of social, political, and ethical values (cf. Finnveden 1997). Values are involved not only in choosing weighting factors but also in choosing which type of weighting method to use, and in the choice of whether to use a weighting method at all (Finnveden 1997). The 2 latter choices are similar to choices made in other elements and phases of LCA (see 'On the Overall Structure and Methodology for LCIA', p 179), although these choices are less controversial. Despite the controversies, weighting is widely used (e.g., Hansen 1999). It is therefore important to critically review its methods and data. Evaluating weighting methods is difficult because the values involved are difficult to identify and evaluate. However, all weighting methods include some scientific parts, if we use a broad interpretation of the word *science* to include not only natural sciences but also social and behavioural sciences, economics, etc. These scientific parts can be evaluated, and the value choices can be identified and clarified. Ideally, we would like to highlight the value choices inherent in different weighting methods so that those who use them can ensure that these values are consistent with their own and that they can chose weighting methods accordingly.

In this chapter, we discuss normalisation, grouping, and weighting in LCIA, with special emphasis on weighting. We take as starting points the Society of Environmental Toxicology and Chemistry (SETAC) Europe First Working Group on LCIA (WIA-1) and its report on normalisation and weighting (Lindeijer 1996) and the ISO standards (ISO 1997, 2000). The starting premise for this chapter is that some people in some situations want to use weighting methods. This desire may be motivated by the need to further aggregate impact indicator results in order to facilitate decision-making or by the wish to identify the most important impact categories for the system under study, for example. The appropriateness of using weighting methods will not be discussed here. The main goal of this chapter is to provide guidance for people who are choosing methods on the basis of the inherent assumptions and limitations of different weighting methods, in order to promote a more informed selection of weighting methods. One starting point is an overview of methods that are available or could be developed (see 'Overview of Weighting Methods', p 185). Another starting point presented here is a discussion of requirements that can be put on a weighting method ('Criteria for the Evaluation of Weighting Methods', p 192). Based on these starting points, we discuss different methods (see 'Evaluation of Methods', p 195). We also discuss the overall framework of LCIA, normalisation, and grouping.

On the Overall Structure and Methodology for LCIA

Life-cycle impact assessment must rely on a number of different sciences and theories. Of course, natural sciences are the foundation for parts of LCIA, especially characterisation, but it is clear that natural sciences are not enough when it comes to the grouping and weighting elements. Other areas of knowledge and other theories are required for these elements. Environmental economics as described in textbooks (e.g., Turner et al. 1994; Tietenberg 1999) and a special issue of *Environmental Science and Technology* (2000) can be one basis for the weighting element (see 'Monetisation weighting methods', p 186). Another possible basis consists of the theories and methods of decision analysis. Seppälä et al. (2002) discuss the relationship between decision analysis and LCIA, some aspects of which are presented and further discussed in 'Panel weighting methods' (p 188).

Multiple criteria decision analysis (MCDA) is a set of methods developed in the operations research and management fields for the resolution of problems characterised by multiple objectives. MCDA methods developed for discrete decision problems (i.e., those in which a finite set of options is considered) are called *multiple attribute decision analysis* (MADA) methods. MCDA methods for continuous decision problems (in which an infinite number of options can be considered) are called *multiple objective optimisation* (MOO) methods. For a review of both classes of MCDA methods, see Stewart (1992). Examples of the application of MADA methods in LCA or similar applications include Miettinen and Hämäläinen (1997), Spengler et al. (1998), Seppälä (1999), Basson and Petrie (2000), and Basson et al. (2000). Examples of the application of MOO in LCA or similar applications include Azapagic and Clift (1998) and Stewart (1999).

It is interesting to note that LCIA methods can and have been usefully framed as instances of multiple attribute approaches to decision-making and applications of MCDA methodology. This means that findings and experiences from the MCDA field can be applied to LCIA, possibly forming a theoretical basis for elements of LCIA and helping to discern 'good' and 'bad' approaches to LCIA. A decision analytical approach is not restricted to weighting issues but offers a framework for all stages of LCIA (see Hertwich and Hammitt 2001; Seppälä et al. 2002). The large variety of MADA methods points to the potential variety that can be exhibited in LCIA methods.

There are several types of relations between weighting and the other phases and elements in LCA. Results from earlier elements are fed into the weighting, and the methodologies used for weighting should therefore be compatible with methodologies used for earlier elements. This also means that choices and assumptions made in earlier phases and elements, which may be described as value choices or assumptions, should be consistent with choices made in the weighting. In discussions on the relationships between weighting and other elements, the terms *bottom–up* and *top–down* are sometimes used. These terms are very general, and it is therefore necessary to define the bottom and the top. Therefore, an attempt to distinguish and describe 3 meanings of top–down is made here (Hofstetter 1999):

1) The first meaning assigns the bottom to the physical description of processes in terms of inputs and outputs of the product system, and the top to a clear description of the environment and of changes that are seen as constituting harm to the environment. Therefore, bottom–up could be labelled as *inventory-driven* and top–down as *relevance-driven* or *damage-driven*.

2) The second meaning sets the bottom on the level of available data and the analysts' or experts' view, while the top is at the level of the decisions to be taken and the decision-makers' view. Labels could therefore be *information-driven* versus *decision-driven*.

3) A third meaning concentrates on the modelling approach used. Bottom stands for the single processes and products, while top is situated at the level of political economies or their sectors. *Micro* for bottom–up and *macro* for top–down are often used as synonyms.

This chapter deals mainly with the first and second meanings.

Proponents of a top–down procedure (both first and second meanings) often claim that the LCA framework as suggested in, for example, SETAC documents and ISO standards typically are inventory and information driven (Müller-Wenk 1997; Hofstetter 1998, 1999), and this creates problems when the weighting step is added. Steen and Ryding (1992), and later Goedkoop (1995), were the first to look at LCA with some rigor in a damage-driven manner. Goedkoop (1995) norma-

tively defined safeguard subjects that represent the environment. Consequently, damage to the safeguard subjects became damage to the environment to be assessed. This line of thinking was further developed in Braunschweig et al. (1996) and Müller-Wenk (1997) and resulted in a new generation of damage-oriented methods (Goedkoop et al. 1998; Goedkoop and Spriensma 1999; Steen 1999a, 1999b). The methodology developed within the ExternE project for another context is also damage driven (ExternE 1995, 1999). However, although this new line of developments is an answer to the initially inventory-driven approaches, it is not (yet) necessarily relevance driven and may fail to be decision driven. When weighting is performed on a damage level, the indicators are defined on an endpoint level. In contrast, indicators can also be defined on a midpoint level, earlier in the environmental mechanism (Bare et al. 2000).

Several types of choices are made in the inventory analysis and in the selection of impact categories and the characterisation step of an LCIA: Some of these are either value choices, which should be compatible with the value choices used in the weighting, or assumptions with repercussions for the weighting step. One example is time used as a possible system boundary in both the inventory analysis (e.g., in modelling landfills [Finnveden 1999a]) and the impact assessment (Udo de Haes et al. 1999). Another example concerns the level in the environmental mechanism that is considered appropriate for defining indicators for the characterisation. This depends partly on views on what possibility science has to predict environmental impacts (EIs; e.g., Finnveden et al. 1992; Finnveden 1997; Bare et al. 2000). In order to make the choices as transparent as possible, we suggest this procedure (Hofstetter 1999):

1) Address value choices explicitly, rather than hiding them or shifting them into another phase. Where appropriate, discuss choices with respect to ethics and societal value settings.

With that, the set of reasonable value settings is made transparent.

2) Check whether the LCA tool provides answers to the choices already made in the goal and scope definition. For example, LCA studies environmental interventions and potential impacts that are spread in place and time. Distance or time discounting (meaning that a lower importance is given to impacts occurring at a distant place or time just because they are distant) — widely practised in today's decision-making — is therefore excluded by definition (although a cutoff in time often is used in current LCA practice, and there often is also a focus on impacts in the region of the foreground processes).

3) Design the model in such a way that choices and assumptions are treated like input parameters and therefore accessible for re-evaluation.

4) Suggest or quantify and complement one or several settings with information on value-independent uncertainties such as variability, statistical variation, and approximation.

The selection of impact categories, indicators for the categories, and models to quantify the contributions of different resources and emissions to the impact categories are choices that may depend on societal values and worldviews. Different societal values and worldviews may require different models for characterisation. Different weighting methods may require different characterisation methods. The choice of what is considered best available practice for characterisation methods may therefore depend on the weighting method. However, the relationship should work both ways, so the choice of weighting method may therefore depend on the choice of characterisation methods.

Normalisation

Normalisation is an element within LCIA that is used mostly as input to a grouping or a weighting element but sometimes also directly as input to the interpretation phase. Here, the focus is on normalisation as a step prior to weighting. Currently, there are many different ways in which normalisation is being conducted worldwide, but no single standardised method exists. Conducting normalisation in several different ways can point out the significance of various choices. Two categories of normalisation techniques are presented:

1) external normalisation or determination of relative contribution and
2) case-specific or internal normalisation.

In either case, the following equation is used:

$$N_i = S_i/R_i \qquad (7\text{-}1),$$

where i is the environmental problem (or impact category), N is the normalised results, S is the product system results prior to normalisation, and R is the reference value.

External normalisation

In an external normalisation, the reference value in Equation 7-1 is selected on the basis of a reference system. The goal of this approach is to place a set of case-specific LCIA results into wider context. As with all approaches to normalisation, this approach also has the effect of adjusting the characterisation results to a common unit. Relative contribution approaches are considered *external normalisation* because they require an external database of the other contributors within the temporal and spatial region.

As a special case of external normalisation, normalisation using externally determined targets could be identified. In this case, the reference value in Equation 7-1 is a target value. This type of normalisation often has been regarded as a weighting method and is therefore discussed in later sections.

Selection of a reference system

The choice of a reference system for external normalisation can be framed as the result of choices on each of several dimensions of the reference situation. These dimensions include the system basis, system treatment, spatial scaling, temporal scaling, and magnitude scaling. Options for each of these dimensions are listed below. For each dimension, the option that has been most commonly selected by LCA practitioners to date is indicated with boldface type (Bare et al. 2002).

- System basis — **a region**, an economic sector or group of sectors, or a product type or group.
- System treatment — **all in-system activities** (e.g., **in-region activities** or in-sector activities) or all activities in the system's supply chain (e.g., all activities supplying an area's final demand, or a sector and its full supply chain, or a reference product and its life cycle). Choices may include impact reference system, environmental intervention reference system, marginal damages, and average damages.
- Spatial scaling — **nation**, continent, or world.
- Temporal scaling — **per year** or per day.
- Additional magnitude scaling — **none**, per capita, or per dollar of output.

Reference system considerations

First it should be noted that there is no a priori requirement that the impact reference systems selected for different impact categories should be identical (including having the same geographic scale or temporal scale). Thus, valuation could conceivably elicit tradeoff weights between the U.S. impacts of year 2000 acid deposition versus the global impacts of ozone depletion during the 21st century. At issue is the appropriate reference system (especially as it relates to the temporal and spatial scales) for LCIA normalisation. Bare et al. (2002) propose that the following 4 conditions should be met in the selection of a reference system:

1) The normalisation and weighting reference systems must be consistent within each impact category.
2) For panel methods, the reference system and its related impacts and/or environmental interventions must be clearly comprehensible to the panellists who do the weighting.
3) An appropriate accounting must be made, relating the impact and environmental intervention reference systems (temporally and spatially, as well as considerations such as whether marginal or average changes are used). This accounting will require the appropriate use of scientific scale of the environmental mechanism. That is, projections of the environmental interventions causing impacts must be appropriate temporal and spatial links to the impacts caused. This may make the normalisation quite complex by taking into account residence time of the environmental intervention, impact delay time, recovery time, and spatial distribution of environmental interventions over time.
4) The emissions and extractions for the reference system should be estimable with as much accuracy as necessary, using empirical data.

It should be noted that while Conditions 1 and 2 are nonnegotiable, the selection of a reference system will require striking the best possible balance between Conditions 3 and 4. First, consider the proposed conditions for selection of a reference system, starting with the first 2 requirements for congruence with the weighting process. These conditions are consistent with a recent presentation by Hofstetter (2000a), which recommended that the normalisation step be designed to reflect each individual's perspective reference system that would be used within the weighting process. If Hofstetter is correct, then there are 2 possible options. One option is to attempt to alter the individual's perspective by communicating a normalisation-consistent (impact) reference system while preparing the panel for the weighting process. This will require one adjustment calculation for each impact category, relating the data for the emissions reference system to the specified impact reference system for each impact category. This may be inconsistent with concerns Hofstetter (2000b) raised about the influence of media or communication with panellists upon the results. The second option is to allow the normalisation impact reference systems to be those governed by each individual's perspective. This would then require identification of each individual's perspective for each impact category (no small task), followed by impact-emissions adjustment calculations tailored to each panellist, based on the relationships between the emissions reference systems and each individual's impact reference systems.

Conditions 3 and 4 address the practicality and availability of data and the role played in driving the selection of the reference systems' temporal and spatial scales in many cases. For example, it has become acceptable to conduct normalisation and weighting on the basis of political boundaries because this division is most often used when emissions (and thus normalisation) databases are compiled. But it must also be recognised that the scientific spatial and temporal scale of the impact category has been considered and recommended by ISO 14042 (2000) for selecting a reference state. This is most important for midpoint categories that require some knowledge of the relationship of impacts to characterised information. Disregard for the scientific scale on a midpoint level would complicate the weighting process not only by requiring a knowledge of the scientific relationships of midpoints and endpoints but also by requiring a budgeting exercise to allocate impacts to various environmental interventions.

Case-specific normalisation

In a case-specific (or internal) normalisation, data from 2 different options within the same LCA study are referenced relative to each other (Equation 7-1). Thus, no external data are required. Proponents of the internal comparison approach to normalisation address primarily the problem of noncommensurate units. In this view, normalisation is primarily (or exclusively) seen as an operational prerequisite to weighting. LCA practitioners sometimes have employed this approach before taking a simple, additive, weighted-sum approach to weighting. In using this approach, practitioners have attempted to make consistent units after the characterisation element by using a comparison to baseline. In this technique, one of the options may be selected as the baseline and the other options' impact categories are listed as a percent increase or decrease when compared to baseline. Caution should be exercised when internal normalisation is used because it poses unsolved problems in the weighting.

Comparison between external and case-specific normalisation

The 2 major types of normalisation techniques are displayed in Table 7-1.

One key concern for using normalisation is the congruence of the methodology in principle and practice with the weighting practice. When any method of case-specific normalisation is used in

Table 7-1 Viewpoints and methods for normalisation

Viewpoint regarding purpose	Explanation of viewpoint	Class of methods	General description of methods
Case-specific	Normalisation is needed to resolve noncommensurate units prior to weighting.	Internal or case-specific	Division of scores in each category by some function of the case's values in that category, e.g., maximum value, sum, values for selected alternative
External: determination of relative contribution	Normalisation is needed to assess relative contribution of results across categories.	External or generic	Division of scores in each category by an estimate of the total impacts in that category for a chosen system or region over a chosen period of time

conjunction with case-independent weights, the results are arbitrary. This occurs because there is no linkage or congruence between the weighting step and normalised results. The weights reflect the importance of each impact category in general, while the normalisation results reflect the relative performance of the studies. If or when a case-specific normalisation is used, a case-specific set of weighting factors must also be used, and a very involved communication of normalisation results must be communicated to the panel.

The congruence between normalisation and weighting is also important in the case of external normalisation. In addition, in this case, the normalisation and weighting systems must be consistent. For example, spatial and temporal scaling must be consistent and comprehensible for panellists if a panel method is used. This is important because the weighting factors have to reflect the tradeoffs between the impacts caused by the reference values, according to theories from decision analysis (cf. Braunschweig et al. 1996; Lindeijer 1996; Seppälä and Hämäläinen 2001).

When is normalisation needed?

The term *normalisation* is ambiguous. In a broad sense, normalisation could mean any process of converting different numbers into other numbers with a common unit. In a more narrow sense, normalisation converts the results from the characterisation into numbers with a common unit (Equation 7-1). In the ISO standard (2000) and in this chapter, the term is used in the narrow sense. If normalisation is used in the broader sense, it can be claimed that some sort of normalisation is a necessary step prior to weighting for most weighting methods (Bare et al. 2002). However, in the more narrow meaning, normalisation is not necessary for all weighting methods. For panel methods (discussed below), normalisation makes the elicitation of weights easier. From a decision analysis point of view, the aim of normalisation is to convert the different scales of category indicator (CI) results into the same range (e.g., [0,1]), which is an essential stage before the weighting in some MADA methods. Different methods for decision analysis use different techniques for normalisation. In the context of a decision analysis framework, it is possible to conduct both external and case-specific normalisation (Seppälä et al. 2002).

Grouping

The term *grouping* was introduced in the ISO process and has previously not been used in the LCA literature. According to the ISO documents, weighting is a quantitative process because numerical factors are used. In contrast, grouping is a qualitative or semiquantitative process that involves sorting and/or ranking. It acknowledges that some decision-makers may find it useful to group impact categories that they consider conceptually related or that they value in a similar manner. Examples of these include grouping impact categories that relate to damage to different areas of protection (AoPs) or impacts that are significant at a global, regional, and local level. This is the grouping step, and its intent is to assist in the overall evaluation of environmental performance and, where relevant, to propose a 'best' or preferred alternative. Further to the initial grouping, the ISO standard (ISO 2000) suggests that the impact indicators may also be ranked on an ordinal scale (e.g., indicating high, medium, or low priority). Such a qualitative ranking could be used to select or screen a set of alternatives; for example, if we decide that global effects are significantly more important than local effects, then we may select only those that perform well in global-scale impact categories for further consideration.

Several methods do not rely on an explicit weighting of impact categories but still aim to enable the overall evaluation of the environmental performance of the alternatives. One such method is the 'Umweltbundesamt ranking-method' as described by Schmitz and Paulini (1999), which is a development of the 'Verbal-argumentative approach' described by Giegrich and Schmitz (1996). Another is a ranking method originally developed by Volkwein et al. (1996) with further development and updating by Klöpffer et al. (2000), Bundesministerium für Forschung und Technologie (BMBF 2001), and Fleischer et al. (2001). In the version published by BMBF (2001), the method is called *uncertainty-oriented ranking* (UNORRA), highlighting the role of a data quality analysis. Parts of these methods may be described as grouping methods. However, they also include other elements of LCIA, and it can be argued that they include methods for interpretation.

Overview of Weighting Methods

Among recent overviews of weighting methods are Lindeijer (1996), Hertwich et al. (1997), Powell et al. (1997), Finnveden (1999b), and Bengtsson (2000). All weighting methods discussed here result in weighting factors, V_i, which express the contribution to the total potential EI from the intervention or impact category i. The total potential EI can then be calculated as

$$EI = \sum V_i S_i \text{ or}$$
$$EI = \sum V_i N_i \qquad (7\text{-}2),$$

where S is the product system results prior to normalisation, and N is the normalised result. In principle, other types of equations could be anticipated but have so far not been used for LCA. The weighting can be performed at either a midpoint level or an endpoint level (cf. Bare et al. 2000).

Based on Braunschweig et al. (1996), Lindeijer (1996) classified existing weighting methods and approaches into 5 main groups:
1) proxy,
2) technology,
3) monetisation,
4) panels, and
5) distance to target (DtT; authorised targets or standards).

These 5 groups are discussed below. The first two are evaluated here whereas the others, which have been more discussed in the LCA community, are evaluated in 'Evaluation of Methods' (p 195).

Proxy weighting methods

Proxy approaches use one or a few quantitative measures, stated to be indicative for the total EI (Lindeijer 1996). Examples include energy requirements or total mass displacement (material intensity per unit service [MIPS]). Clearly, the extent to which environmental problems such as ecotoxicity and ozone depletion can be adequately covered by these indicators is problematic (Lindeijer 1996). Instead of attempting a weighting between different types of environmental problems, the proxy approaches pick one or a few important ones. Therefore, no intereffect weighting is included, an inherent limitation of the proxy approaches. Another limitation is that these approaches do not give a comprehensive picture of the EIs. Therefore, proxy approaches have limitations as weighting methods for LCA. They may, however, be used as characterisation methods and may provide input to the interpretation.

It has been suggested that expressing impacts in their real market prices can be regarded as a proxy method as well (Udo de Haes 1999). Here, however, it is treated as a monetisation method.

Technology abatement weighting methods

Technology abatement approaches consider the requirements for the abatement of impacts and in most cases are combined with some other measure, for example, costs to reduce the burdens. In the latter case, technology approaches can also be described as monetisation methods. One exception that may be described as a technology method is the Ecological Footprint, a concept for estimating the biologically productive area necessary to support a studied system, which also can be used as a weighting method for LCA (Wackernagel and Rees 1996). In this approach, emissions are transformed into area by calculating the area needed for assimilation of the emission. This calculation is then dependent on the technology used for the assimilation, and this is the reason why it is treated as a technology method here. In its current version, the Ecological Footprint can handle only very few pollutants, among them CO_2 and nitrogen emissions (Holmberg et al. 1999). There are thus large gaps in the types of impacts that can be covered and, therefore, limitations in the approach as a weighting method for LCA.

Monetisation weighting methods

Approaches to monetisation

Monetisation method is used here as an umbrella term for all methods that have a monetary measure as the unit for weighting factors. There are a large number of different approaches for monetising EIs. There are also a large number of ways of classifying the different approaches, which sometimes lead to a somewhat confused discussion. The classification here is from Finnveden (1999a) and is based on but not identical to those of other sources (e.g., Turner et al. 1994; White et al. 1996; Bockstael et al. 2000; System of Integrated Environmental and Economic Accounting [SEEA] 2000):
1. Methods based on willingness to pay [WtP]
   1.1 Individual's revealed preferences
   1.2 Individual's stated preferences
   1.3 Society's WtP
2. Methods not based on WtP.

The first distinction is between methods that are based on WtP and methods that are not (no distinction is made here between WtP and 'willingness-to-accept'). Methods based on WtP measure an 'economic value' (Bockstael et al. 2000). Environmental economists distinguish between different types of economic values that relate to natural environments. Again, the terminology is not completely agreed upon; the following, however, is based on Turner et al. (1994). The first distinction is between *use values* and *non-use values*. The use values include both direct and indirect use values. An example of a direct use value is the timber value of a forest. The indirect use value includes the recreation value of the forest, the value of carbon fixation, etc. Non-use values are noninstrumental values that are attributed to objects without the direct intention of actually using them. Such values include the concern for, sympathy with, and respect for the rights and welfare of nonhuman beings (*existence values*). We may also attach a value to knowing that future generations will be able to enjoy use and non-use values (*bequest value*). Also the option to use an object, even if there is no intention of using it at the moment, may be attributed a value (*option use value*). The total economic value is the sum of the use and non-use values (Turner et al. (1994).

Methods based on an individual's revealed preferences assume that people reveal their preferences in market behaviour. The revealed preferences are normally related only to the use values, and sometimes only to the direct use value. Direct use values can often be derived from actual market prices, for example, the market price of timber. In addition, some indirect use values may be derived from market values, though often indirectly. The travel cost method and hedonic pricing methods are examples of methods for evaluating use values, which may include some, but typically not all, indirect use values. The travel cost method is a revealed preference method that can be used to estimate demand curves for recreation sites and thereby value those sites. These values are then derived from people's travel costs. Hedonic pricing methods also attempt to evaluate environmental services by studying their influence on certain market prices. One example is the price of a house, which is determined by a number of factors, including environmental aspects. Another example is wages that may vary according to the risks associated with different types of jobs.

Non-use values can normally not be derived from revealed preferences (Turner et al. 1994). The Contingent Valuation Method (CVM) bypasses the need to refer to market prices by asking individuals explicitly to place values upon environmental assets (Turner et al. 1994). Because of this, the CVM is often referred to as an *expressed* or *stated preference method* (Turner et al. 1994). There are of course some similarities between CVMs and the panel methods discussed below. CVMs have been used extensively, and there are guidelines developed for them (e.g., Arrow et al. 1993; Carson 2000). As an alternative method to the use of open-ended bidding questions or dichotomous choices of referendum questions, conjoint analysis has been developed (Farber and Griner 2000).

A society's WtP may be derived from political and governmental decisions. In these methods, it is assumed that meaningful information on environmental values can be derived from political and governmental decisions. One way of deriving a 'societal price' or a collectively revealed preference is to study society's efforts to avoid impacts. An example may be the costs of reducing emissions to a decided emission limit. The marginal cost for removing the pollutant to the emission limit can be seen as the monetary value the society puts on the pollutant. These costs may be called *prevention costs* or *abatement costs*. Yet another way of deriving a societal price is to look at 'green taxes'. If there are taxes on emissions, these taxes may be seen as the society's WtP (or rather, willingness-to-accept) for that specific pollutant.

There are also a number of monetisation methods that are not based on WtP. Methods that are not based on a WtP do not measure the economic value (Bockstael et al. 2000). They are often based on an estimation of a cost to do something; however, if it is not clear that somebody is willing to pay this cost, it is not a measure of a WtP. A first example of such a method is a further development of one of the approaches mentioned above for evaluating a society's WtP. In this approach, the marginal cost for removing the pollutant to an emission limit is calculated. If the emission limit is a future target value, for example, a critical load value (if such are available), it is no longer a WtP that is evaluated (since it is not clear whether somebody is actually willing to pay), but another type of cost. Another example may be the cost for remediation of damage. This approach is useful only if remediation is possible at all.

Because different monetisation methods cover different types of economic values (i.e., different use and non-use values), different methods should result in different results, and they do. For example, the total economic value, as measured by the CVM, is in some cases an order of magnitude larger than the economic value derived from market valuations (Konjunkturinstitutet [KI] 1998). This may be explained by the non-use values that are included by CVM. Just because something is expressed in monetary terms does not make it immediately comparable or additive to another measure in the same unit that uses a different method (cf. Bockstael et al. 2000). If a monetisation weighting method is used, the same method should therefore ideally be used to derive all economic values within the method. Care should also be taken when the results from a monetisation weighting method are compared to other types of costs.

One difference between the approaches concerns whose values are considered and how societal values are assessed. The neoclassical approach in economics suggests that the societal values are the sums of individual values. From this perspective, it is reasonable to look at the behaviour of individuals or ask for their values. However, it may also be claimed that there are societal values that are different from the sums of individuals' values. From this perspective, we may assume that societal values are expressed through decisions taken by the society, for example, by governments. If so, it is reasonable to look at the behaviour of governments or governmental organisations and/or ask them for their values. It may also be claimed that a government represents a society as a whole, and that taking the values of a government is relevant regardless of whether these values are more than the sum of individuals' values.

In monetised valuation methods, future impacts are often discounted. An explicit discount is not necessary for some types of economic valuations. For example, if the economic valuation is derived from a society's WtP, an implicit discount has already been made. However, if damage costs are to be estimated, a choice has to be made concerning how to handle costs in the future (e.g., either a discount must be used or a cutoff must be made after a certain time). In the latter case, all impacts occurring after the cutoff are neglected.

Monetisation methods used in LCIA

Several monetisation methods have been developed for weighting in LCA. The Environmental Priority Strategies (EPS) system (Steen 1999a, 1999b) primarily aims at estimating damage costs. The same is true for the method developed for the ExternE project (ExternE 1995, 1999), resulting in several suggestions for application within LCA (Dobson 1998a, 1998b; van Beukering et al. 1998; Spadaro and Rabl 1999; Krewitt, Trukenmüller et al. 2001). In addition, the Explicit LCA (XLCA) method developed by Newell (1998) aims at estimating damage costs. In practice, all these methods use a mix of different types of economic valuations. Although data from contingent valuation (CV) studies sometimes are preferred (e.g., van Beukering et al. 1998), in practice, the data are often a mixture of CV studies, hedonic pricing methods, abatement costs, and market values (Finnveden 1999b). The Tellus system (Tellus Institute 1992; Zuckerman and Ackerman 1994) used measures of costs associated with decided emission limits and environmental taxes to estimate society's WtP. The Ecotax '98 method (Johansson 1999) is based on environmental taxes as estimates of a society's WtP for deriving weighting factors. In the 'Virtual Pollution Prevention Costs 99', marginal prevention costs are used as estimates for society's WtP (Vogtlander and Bijma 2000). These methods have published weighting factors. In addition, a number of studies describe different approaches (e.g., Krozer 1992; Huppes et al. 1997) or apply an economic valuation to LCA on a more ad hoc basis (e.g., Craighill and Powell 1995; Carlsson 1997; Sonesson et al. 2000).

Panel weighting methods

Panel approaches is used as a heading for a number of different approaches that have one thing in common: It is assumed that the relative importance of damages, impact categories, or interventions (weighting factors) can be derived from an individual or a group of people by elicitation. *Elicitation* is the process of gathering judgments concerning the problem through specially designed methods of verbal and/or written communication (Meyer and Booker 1990). Panel methods differ, for example, according to the following aspects (e.g., Meyer and Booker 1990; Brunner 1998; Seppälä 1999):

- Size of a panel and type of panellists — environmental experts, experts from other sciences, stakeholders, lay people, or a representative mix
- Elicitation situation — questionnaires, interviews, interactive group, and Delphi (see Dalkey 1969); one-round procedure or multi-round procedure with or without feedback
- Question format
- Presentation of background information
- Response modes — ranks, ratings, pairwise comparisons, ranges, etc.
- Type of aggregation — a consensus, use of mathematical methods to combine multiple panellists' data into a single estimate or single distribution of estimates.

In LCA, most panel methods have been developed on an ad hoc basis in connection with a specific case study or impact assessment methodology. Examples of studies in which quantitative panel methods have been used are Anonymous 1991; Kortman et al. 1994; Wilson and Jones 1994; Nagata et al. 1995; Poulamaa et al. 1996; Huppes et al. 1997; Lindeijer 1997; Sangle et al. 1999; Seppälä 1999; Harada et al. 2000; Mettier and Baumgartner 2000. Some of them have developed sets of generic weighting factors, for example, the Index of Environmental Friendliness (Poulamaa et al. 1996) and Eco-indicator 99 (Goedkoop and Spriensma 1999). An example of country-specific weighting factors is presented in Seppälä (1999).

The bullet points that classify different panel approaches above are also the major unresolved research questions that await further exploration:

- Because LCA can support decisions by private industry, consumers, nongovernmental groups, and governments or regulators, there is no single panel composition that fulfils all requirements. Previous panel procedures used primarily environmental experts from the LCA field and sometimes a wider stakeholder group (Kortman et al. 1994). This choice was primarily guided by the ease of access to panellists and their potential to understand the issues at stake. Although there is no evidence that panel results would differ if lay people were asked, there also is no evidence that they would be similar.
- Elicitation situations, question formats, and the background information provided do bias panellists' responses. This bias is potentially large because panellists are asked to state preferences about a largely unobservable and nonperceivable good at a level of detail that goes beyond their preexisting preferences. Therefore, panellists form their preferences during the panel procedure, which explains its large influence. Such a bias was directly shown in unpublished results of the study by Mettier and Baumgartner (2000) and was suggested as interpretation in the study by Wilson and Jones (1994). Hofstetter presented results of a review of LCA panel methods (see Bare et al. 2000), which show that, in most cases, the difference between the most and least important impact categories was a factor of 3. He suggested that this narrow range of weighting factors is an artefact and that anchoring effects biased these studies.
- As already suggested in 'Normalisation' (p 181), the different temporal and spatial dimensions of environmental problems, including scenario-dependent estimates for future damages, complicate the weighting task for panellists to an extent that cognitive limits hamper the possibilities to elicit 'true' preferences. These cognitive limits may also lead to a low sensitivity towards the scope of the objects to be valued, which has been found in many CV studies. This implies problems for normalisation because it is unclear whether panellists really understand the scope of estimated damages or impacts; that is, it is uncertain whether their answers really refer to the provided reference.

In the available LCA literature, no elicitation procedure was tested for all sources of biases that may be relevant, and more research is ongoing and needed if we are to understand the relevance of the different elements of the procedure.

Potential contributions to panel methods in LCA from multiple attribute decision analysis

The way in which judgement has been elicited has differed widely within LCIA applications because there is a wide range of possibilities for elicitation processes. There is presently no established procedure for panel methods in the LCA context. There are 2 major sources for knowledge and experience for elicitation techniques: CVMs (discussed above in 'Monetisation weighting methods', p 186) and methods for MADA presented in 'On the Overall Structure and Methodology for LCIA' (p 179) and further discussed below.

In general, weighting elicitation procedures used in MADA offer techniques, knowledge, and experience for evaluating the weighting factors in LCIA because one of the most important methodological issues in many MADA methods is to quantify tradeoffs among attributes (CI results or interventions in the context of LCA; see Seppälä et al. 2002) as importance weights (weighting factors). However, these methods vary according to their methodological bases (see, e.g., Guitouni and Martel 1998), which give different requirements for elicitation of weighting factors in the context of each MADA method.

One helpful distinction amongst MADA methods is whether they are based on Multi-Attribute Utility/Multi-Attribute Value Theory (MAUT/MAVT, where MAVT can be considered as a theory under MAUT). MAUT/MAVT is a major theory of MADA methods, which are aimed at producing a complete ranking of the alternatives. It provides a clear axiomatic foundation for rational decision-making under multiple objectives (e.g., Keeney and Raiffa 1976; French 1988). There are also non-MAUT/MAVT methods such as Analytical Hierarchy Process (AHP) and so-called *outranking methods* (e.g., ELECTRE and PROMETHEE) that screen or select the most preferred alternatives.

MAUT/MAVT is especially interesting for the LCA community because the typical calculation rule for the total potential EI used in LCIA (Equation 7-2) corresponds to a simple additive weighted model derived from MAVT (Seppälä 1999). The outranking methods require establishing thresholds and absolute performance standards, which provide substantial methodological challenges for LCA, which bases its evaluation on indicators.

In MAUT/MAVT, there are numerous procedures for determining weighting factors (see, e.g., von Winterfeldt and Edwards 1986). The tradeoff procedure has a strong theoretical foundation (Keeney and Raiffa 1976; Weber and Borcherding 1993), but it is rather difficult to use. Easier methods such as the ratio estimation and the swing procedure are therefore widely used (von Winterfeldt and Edwards 1986). For example, in the LCA application of the Finnish forest industry, the ratio estimation was used to elicit weighting factors from 58 experts working with environmental issues (Seppälä 1999). The ratio method requires the panellist to first rank the relevant impact categories according to their importance. The least important impact category is assigned a weighting factor of 10 and all others are judged as multiples of 10. The resulting raw weighting factors are normalised to sum to 1. According to MAVT, the question format must be adjusted in the context of Equation 7-2 so that panellists express their opinions about the importance of different impacts caused by reference values R_i (see Seppälä and Hämäläinen 2001).

If in Equation 7-2, the weighting factors are defined by ratio estimation, the whole procedure corresponds to a method called *Simple Multi-Attribute Rating Technique* (SMART; Edwards 1977). Equation 7-2 can be constructed by using theoretical elements of SMART. However, there are variations of SMART in which weighting procedures can vary (see von Winterfeldt and Edwards 1986). SMART is one of the most commonly used techniques under MAVT.

In Equation 7-2, weighting factors can also be defined by the AHP developed by Saaty (1980). MAVT has close connections with AHP (see, e.g., Salo and Hämäläinen 1997), although they have different evaluation scales to determine weighting factors. In AHP, weighting factors are evaluated by a pairwise comparison process. In AHP, it is possible to check consistency of judgements on the basis of an index obtained from the calculation method.

In decision analysis, there are also so-called *ranking methods* that can be used if the panellists are able to rank the criteria (impact categories) in order of importance. Ranking methods such as the Expected Value Method (Rietveld 1984a, 1984b), the Extreme Value Method (Paelinck 1974, 1977), and the Random Value Method (Voogd 1983) have different calculation procedures with assumptions in order to produce weighting factors on the basis of the information on orderings.

If the LCIA application is constructed according to MADA methods other than the additive weighted methods (such as SMART and AHP), we may get different aggregation models, compared to Equation 7-2. Examples of these other methods are the outranking methods (e.g., ELECTRE and PROMETHEE; see Roy and Vanderpooten 1996). Different methodological bases of MADA methods also mean that each method has its own requirements for elicitation of weighting factors. Thus, the practitioners must understand relationships between weighting factors and other aggregation elements in the method used when the elicitation procedure is chosen (Seppälä et al. 2002).

Distance-to-target weighting methods

Several weighting methods relate the weighting factors to some sort of target (Lindeijer 1996). These methods are conveniently called *distance-to-target* (DtT) *methods*, although this name in some cases may be somewhat misleading. The major differences between methods are

1) the precise shape of the equation relating the targets to the valuation weighting factors (discussed in Ahbe et al. 1990),
2) the choice of targets, and
3) whether inventory or characterisation data are used in the weighting, and if so, which type.

The simplest type of equation is

$$V_i = 1/T_i \qquad (7\text{-}3),$$

where V is the weighting factor and T is the target (expressed in units related to the target). If Equation 7-3 is inserted in Equation 7-2, the result is

$$EI = \sum S_i/T_i \qquad (7\text{-}4).$$

In this equation, every term of the sum is identical to the normalisation equation (Equation 7-1), with the reference value as the target value and the EI as the normalised result. More complicated equations have also been used, involving the current level in addition to the target level.

This group includes a large number of different methods. Among them are the Ecoscarcity Method (Ahbe et al. 1990), with updates (Bundesamt für Umwelt, Wald und Landschaft [BUWAL] 1998) and national adoptions (e.g., Lindfors et al. 1995). Examples of other methods include the Effect Category Method by Baumann et al. (1993) and Baumann and Rydberg (1994), the MET-Point Method (Kalisvaart and Remerswaal 1994), and methods by Corten et al. (1994), Kortman et al. (1994), and Schaltegger and Sturm (1991). All these methods use as targets either political or administrative target levels or 'critical' or 'sustainable' levels. The targets are thus set externally to the method. A common feature is that the targets are always assumed to be equally important. This assumption is further discussed in 'Evaluation of Methods' (p 195).

A slightly different approach is represented by the Eco-indicator 95 (Goedkoop 1995), with updates by Frischknecht (1998). In this approach, the target levels are set internally within the method development, with the explicit aim that the targets all should be equally important. The procedure for determining the targets is not clearly described.

In a recent paper, Seppälä and Hämäläinen (2001) discuss the theoretical foundations of DtTs on the basis of decision analysis theories. Their conclusion is that Equation 7-4 is consistent with the additive weighted model derived from MAVT if targets are set with the explicit aim of being equal and if linear damage functions passing through the origin are assumed. If the assumption concerning equally important targets does not hold, every term in Equation 7-4 can be considered an externally normalised result. Then the normalised results must be multiplied by the corresponding weighting factors before summing. Thus, in this case, an additional weighting step is needed in order to calculate the total potential EI. This weighting task can in principle be conducted using weighting techniques and rules applied in panel methods.

Other classifications

There are alternative classifications of weighting methods: methods based on expressed or stated preferences and methods based on revealed preferences. The methods based on stated preferences include panel methods as well as methods based on CV studies and conjoint analysis. Methods based on revealed preferences include monetisation methods based on individuals' revealed preferences, society's WtP, and methods not based on WtP. Yet another classification of weighting methods is between endpoint methods based on damage assessments and midpoint methods that weight interventions, impacts, or threats rather than damages (Bare et al. 2000). Both these groups of methods may include panel and monetisation methods.

Criteria for the Evaluation of Weighting Methods

Weighting methods differ. Criteria, being 'standards on which judgments or decisions may be based' (Britannica Compact Disc [BCD] 1998) are often used for describing and evaluating weighting methods (Giegrich et al. 1995; Braunschweig et al. 1996; Lindeijer et al. 1996). With more than a few criteria, it is vital to have an underlying structure, a *system of criteria*. Such a system of criteria for weighting methods can be useful in several applications. It can be used to evaluate weighting methods on a general level; that is the primary aim here. It can also be useful when weighting methods are chosen for a particular case study and when methods are being developed and improved.

To come to a target system for weighting methods, it is useful to recall that, basically, every weighting method produces an output, the result, on the basis of some input data (Figure 7-1).

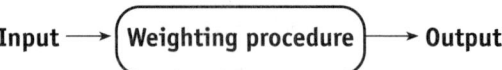

Figure 7-1 Input, procedure, and output as 3 principal elements of a weighting method

Based on this, the criteria for weighting methods may be divided into
1) result- or outcome-related criteria,
2) procedure-related criteria, and
3) input-related criteria.

Not all criteria are equally important. We have distinguished between 3 classes of criteria:
1) must criteria,
2) must candidates, and
3) nice-to-have criteria.

Must criteria are criteria that are measurable and that a specific weighting method must fulfil in order to be acceptable. If a weighting method fails on a must criterion, it is regarded as an unacceptable method. Every weighting method that is acceptable must score above a certain threshold in each must criterion. Therefore, it is essential for must criteria to be measurable. In this chapter, the measurement of the fulfilment of a criterion and thresholds related to criteria were not worked out to the extent that we feel comfortable prescribing must criteria. Instead, a number of criteria that have the nature of must criteria have been identified; these are called *must candidates*. *Nice-to-have criteria* are criteria without thresholds. This means that a method that fails completely in a nice-to-have criterion may be accepted anyway as a useful weighting method.

A distinction can also be made regarding the application level. Some criteria are applicable on a general level, whereas others are applicable only on a case-specific level.

The system of criteria for weighting methods is presented in Figure 7-2.

Input-related criteria

The first input-related criterion is *low requirements on data* (I1), which is further divided into the *necessary low amount of data* (I11) and the demanded quality of data, *low requirements on the quality of data* (I12). If everything else is constant, a method seems preferable if it needs a lower amount of data than another method, and a method is preferable if it accepts a lower quality of data. Another criterion is *good availability of required data* (I2). A method is considered preferable if the data necessary are more easily available. Finally, a method is considered preferable if it has *low requirements on technical skills* [I3] of the persons involved. *Technical skills* refer to the knowledge necessary for the method.

Procedure-related criteria

A procedure must be based on good practice in sciences (*science-based* [P1]); this improves the acceptance of the result of the method. This criterion can be further divided into *high transparency of the procedure* (P11), a *clear discernment of objective* (unbiased) *and subjective elements* (P12), and a *systematical approach* leading to a general applicability of the methods (P13). A method is transparent if it presents its results in an open, comprehensive, and understandable way (ISO 1997). That means, for example, that it is possible to see, if we look closely at the method, why a certain outcome is the outcome of the method. The clear discernment of objective and subjective elements means that a method may comprise both types of elements but that it is possible to identify both types separately (and also their influence on the result of the method). *Acceptable scientific practice in the science used* (P14) is the last criterion relating to the science-based criterion, where the term *science* is broadly defined to include not only natural sciences but other fields as well. In some cases, the weighting method used in the LCA context is taken from other fields. This criterion allows LCIA to profit from experience gained in other fields, providing the methods are consistent with what is regarded as acceptable practice in those fields.

Another point is a *low effort necessary for the execution* of the method (P2). A method is preferable if it has *high flexibility* (P3). High flexibility may be further divided into the ease by which *new environmental problems can be included* (P31), *new value choices can be included* (P32), and *new characterisation methods can be included* (P33). *Completeness*, that is, including all relevant environmental problems, is another procedure-related criterion (P4).

Adequate representation of values (P5) consists of *adequate representation of values in the choice of the weighting procedure* (P51) and *adequate representation of values in the choice of the weighting factors* (P52). Criteria P51 and P52 ensure that the values underlying the weighting procedure and the weights (weighting factors) themselves are consistent with the values underlying the specific LCA study and also with the values that the persons performing or commissioning the LCA want to have represented. The meta-criterion P6 is further divided into *reflecting the subjectivity of weighting* (P61), and *including an intereffect weighting* (P62). These criteria check that the method under discussion includes value choices and that impacts are weighted against each other, either directly or indirectly, by a common measure (e.g., a monetary measure). Both aspects are regarded as necessary requirements for a weighting method. (It can be noted that criterion P61 asks for subjective elements, criterion P12 asks for a clear separation of subjective and objective elements, and criterion P5 asks whether the subjective elements have an adequate representation of values.)

Criterion P7, *consistent with the application*, is divided into P71, *Is the method based on weighting of marginal or average impacts and is this consistent*

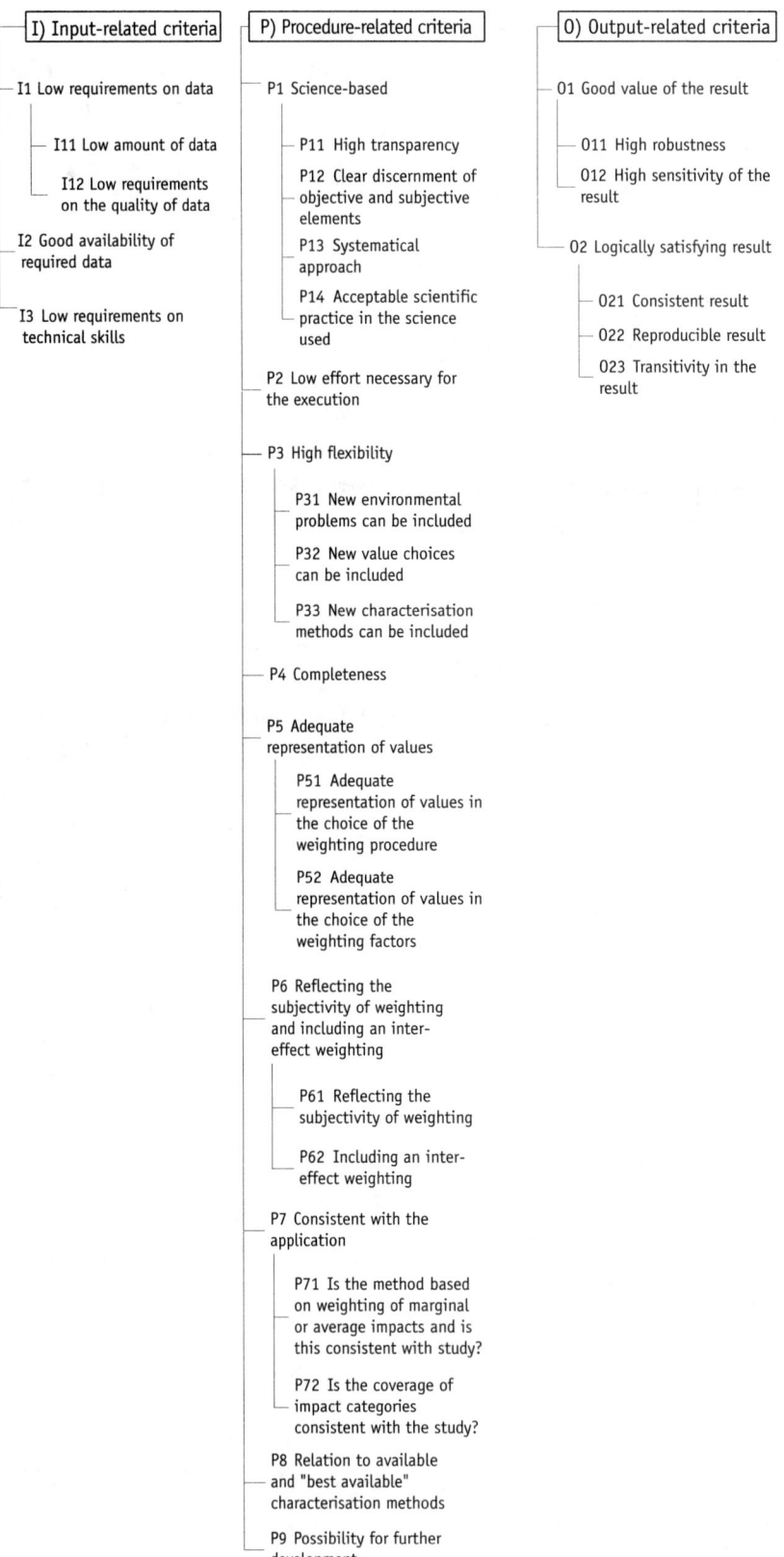

Figure 7-2 System of criteria for weighting methods

with the study? and P72, *Is the coverage of impact categories consistent with the study?* Discussions concerning average versus marginal data have earlier mostly been confined to the inventory analysis (e.g., Clift et al. 1999) and the characterisation (Udo de Haes et al. 1999) but are also of relevance for the weighting. The criterion P8, *relation to available and best available characterisation methods*, reflects that a necessary condition for a weighting method is a link or an interface to the results of the characterisation used. Finally, for the procedure-related criteria, there is P9, *possibility for further development*. This criterion is a meta-criterion; it checks whether an undesirable result achieved for a criterion by a specific method may be changed by further development of the method.

Output-related criteria

For the output-related criteria, the first main criterion is *good value of the result* (O1), being further divided into *high robustness* (O11) and *high sensitivity of the result* (O12). A result is robust if it is not qualitatively influenced by uncertainties (in input data or occurring during the procedure). A result is sensitive if significant changes in input data are reflected in changes in the result (see, e.g., Tomovic and Vukobratovic 1972). The second main criterion is a *logically satisfying result* (O2). This criterion is further divided into 3 points: *consistent result* (O21), *reproducible result* (O22), and *transitivity in the result* (O23). A result is regarded as consistent if it does not show any logical contradictions. It is regarded as reproducible if a similar result is achieved when the weighting is performed under similar conditions (the meaning of this is further discussed in the next section). The meaning of transitivity is best shown by example: If *a* has more weight than *b* and *b* more weight than *c*, then *a* should also have more weight than *c*, according to the weighting procedure (Eisenführ and Weber 1993).

The following are regarded as must candidates: P1, P5, P6, P7, and O2. Completeness (P4) is an important criterion but not regarded here as a must candidate because P72 is included as a case-specific must candidate. All criteria are useful on both a general and a case-specific level, except criteria P5 and P7, which are applicable only on a case-specific level.

Evaluation of Methods

In this section, the set of evaluation criteria developed in 'Criteria for the Evaluation of Weighting Methods' (p 192) is applied to different weighting methods. The following types of methods are evaluated:

- Panel methods are evaluated as a group. This is difficult because the group consists of a number of different methods. The procedure discussed is based on an endpoint evaluation in which the weighting factors are elicited by questionnaire. This procedure is thus similar to the Eco-indicator 99 (Goedkoop and Sprensma 1999).
- Distance to target methods are divided into 3 types:
  1) Distance to critical (or sustainable) target (DtcT), where the targets are based on critical or sustainable loads (e.g., Baumann et al. 1993; Kortman et al. 1994)
  2) Distance to political target (DtpT), where the targets are based on political values (e.g., Ecoscarcity Method [Ahbe et al. 1990])
  3) Distance to equally important target (DteT), where the targets are decided with the explicit aim that they should be equally important. This probably is accomplished with a panel method when the DteT is being developed (e.g., Eco-indicator 95 [Goedkoop 1995]).
- Monetisation methods are discussed according to the following types:
  1) Willingness to pay based on damage costs (WtPDam) (e.g., Steen 1999a, 1999b;

XLCA from Newell 1998; and methods from ExternE 1995, 1999). In practice, these methods mix different types of data, including both stated and revealed preferences.

2) Willingness to pay based on environmental taxes (WtPTax) (e.g., Johansson 1999).
3) Willingness to pay based on marginal prevention costs (WtPPrev) (e.g., Vogtlander and Bima 2000). In this approach, the marginal prevention costs are calculated for a political target situation.

Evaluation

Table 7-2 provides a summary of a tentative evaluation using the criteria list in 'Evaluation of Methods' (p 195). Each criterion is also discussed in the text below.

Input-related criteria

I1) Low requirements on amount of data. DtcT, WtPDam, and WtPPrev have fairly high demands in terms of scientific or technical data. This is also true for Panel and DteT because they must create a context that enables the panel members to express their preferences. Because Panel and DteT both require large numbers of judgements, they require a fairly large number of preference data.

I2) Good availability of data. Panel methods score well because panel members are readily available. However, for WtPDam, data on damages are available for some impacts but not for all relevant impacts. Because of the lack of data, WtPDam methods typically mix different types of environmental valuations, at the risk of introducing inconsistencies. The DtcT method requires critical or sustainable targets, which are available for some impact categories, but it is not possible to calculate such targets for all types of environmental problems (e.g., Chadwick and Nilsson 1993). DtpT, WtPTax, and WtPPrev all require data on either political targets or taxes. Such data are readily available in some cases but not in all. Because targets may be decided by different actors and with different aims, available data may not always be compatible with each other, and there is a risk of inconsistencies. Data for DteT are not readily available but could be developed using panel methods.

I3) Low requirements on technical skills. Panel methods and WtPDam methods that use CV or similar methods require fairly high cognitive skills among the respondents. The same methods and DteT also require fairly high technical skills among the practitioners who construct the procedure. WtPPrev probably requires relatively low (interdisciplinary) skills if data on prevention costs are available; if not, calculations require fairly high technical skills. WtPTax and DtpT require low skills if, and only if, the taxes and political targets are available for WtPTax and DtpT.

Procedure-related criteria

P11) High transparency. Transparency can be interpreted in several ways. According to one interpretation, transparency is not an inherent property of a type of method but instead depends on how detailed procedures and calculations are documented and published for the specific method. The chances of having good transparency increases with simpler procedures and calculations. Transparency can also refer to the arguments behind the values. This type of transparency is typically low for stated preference methods (panel methods and WtPDam based on CVMs) because panellists typically are asked about their preferences but not their arguments. This type of transparency, on the other hand, can be higher for methods based on revealed preferences from the political arena (DtpT, WtPTax, and WtPPrev) because reasons and arguments are published principally in background documents for the decisions. Methods based on revealed preferences for individuals (as are often used in WtPDam) do not have this advantage.

Transparency can also refer to the procedure for eliciting the preferences. If so, methods based

7: Normalisation, grouping, and weighting in life-cycle impact assessment

Table 7-2 Tentative evaluation of weighting methods

Criteria	Panel	DtcT	DtpT	DteT	WtPDam	WtPTax	WtPPrev
I11) Low requirements on amount of data			☺			☺	
I2) Good availability of data	☺	☹		(☹)			
I3) Low requirements on technical skills			☺?			☺?	☺ ☺
P11) High transparency[a]		n.a.		n.a.			
P12) Clear discernment of objective and subjective elements[a]	☺?	n.a.	?	n.a.	?	?	?
P13) Systematic approach[a]				?			
P14) Acceptable scientific practice in the sciences used[a]		☹	☹			?	?
P2) Low efforts necessary for the execution		☹☺	☹☺	n.a.	☹☺	☹☺	☹☺
P31) New environmental problems can be included				n.a.			
P32/33) New value choices and characterisation methods can be included			☹?	n.a.			
P4) Completeness				n.a.			
P5) Adequate representation of values[a]	?	☹	?	?	?	?	?
P61) Reflecting the subjectivity of weighting[a]	☺	☹	☺	☺	☺	☺	☺
P62) Including an inter-effect weighting[a]	☺	☹	☹	☺	☺	?	?
P8) Relation to available and best available characterisation methods	?	?	?	?	?	?	?
P9) Possibilities for further developments		☹	☹				
O1) Robustness or sensitivity	☺				☺		☺
O21/23) Consistent results and transitivity[a]							
O22) Reproducibility[a]	☺?	?	☺?	?	☺?	☺?	☺?

☺ = Method scores well for the criterion.
☹ = Method scores badly for the criterion.
☺ or a blank cell = Method scores neither particularly well nor badly for the criterion.
n.a. = Method is not analysed for the criterion.
? = Method is difficult to evaluate.
[a] Criteria suggested as 'must candidates'.

on stated preferences can have a higher transparency because the questions asked and the answers given can be documented. For methods based on revealed preferences, this is not possible.

P12) Clear discernment of objective and subjective elements. The separation of values and facts is not always clear cut, as discussed, for example, by Finnveden (1998), Hofstetter (1998), Tukker (1999), and Hertwich et al. (2000). This criterion can be interpreted in different ways:

1) It can refer to the process of arriving at preferences. If this interpretation is used, the criterion is difficult to evaluate because it is not known how panellists arrive at their answers

and to what extent they mix objective and subjective elements. Similarly, in methods based on revealed preferences, it is not known to what extent objective and subjective elements have been mixed in the decisions that are used as revealed preferences.

2) According to another interpretation, the stated or revealed preferences may be regarded as the subjective elements, and the procedures that are used and the calculations that are performed as the objective elements. According to this interpretation, there is a fairly good discernment between subjective and objective elements.

3) According to a third interpretation, there is for all methods a weak distinction between subjective and objective elements because all methods include different types of choices concerning which facts to use and present, what procedures to choose, etc.

Yet another point is that methods based on revealed preferences include extra subjectivity by requiring an extra step when the revealed preferences are interpreted and transferred into weighting factors. This step is not required for methods based on stated preferences.

P13) Systematic approach. Because most methods struggle with the large array of environmental effects, and because there are data gaps, most current methods may need improvement on this criterion. DteT so far has no operationalisation.

P14) Acceptable scientific practice in the sciences used. For both panel methods and monetisation methods, there is a scientific tradition, although there is a lack of formal requirements in many areas. One exception is CV studies, which may be used in WtPDam methods for which there are published lists of quality criteria (e.g., Arrow et al. 1993; Carson 2000). In practice, however, WtPDam methods usually mix different methods, based on stated and revealed preferences. WtPPrev is based on regulators' revealed preferences approaches, which are empirical means of establishing WtP that is used in different applications (e.g., White et al. 1996; SEEA 2000). Many economists would, however, use this approach as a second-best option, compared to damage assessments (SEEA 2000). WtPTax is difficult to evaluate in many countries but is considered established practice in Sweden (it is, e.g., a recommended approach in guidelines for cost–benefit analysis of infrastructure investments [Statens Institut för Kommunikations Analys (SIKA) 1999]). If the taxes are Pigouvian taxes, that is, taxes corresponding to the external costs, WtPTax theoretically would be acceptable. Different evaluations of this criterion can be explained by different viewpoints as to whether it can be reasonably assumed that the taxes are Pigouvian and whether the approach also can be acceptable in other cases. For both panel and monetisation methods, the evaluation of this criterion may be different for specific methods. It is unclear what sciences the DtcT and DtpT are based on, and the methods therefore score badly for this criterion. Natural scientists calculating critical loads would normally disagree that a high DtcT implies a higher importance than a lower DtcT. If the targets are explicitly assumed to be equally important, that is, the DteT approach, the method using Equation 7-2 can be consistent with calculation rules derived from MAVT (Seppälä and Hämäläinen 2001). A DtcT approach, however, would require a different calculation rule according to MAVT (Seppälä and Hämäläinen 2001).

P2) Low efforts necessary for the execution. For all methods, the efforts for execution are low if generic weighting factors have been established. The critical aspect, therefore, is the efforts necessary for establishing new weighting factors. For panel methods, the efforts necessary are fairly high. This also is the case for WtPDam and WtPPrev, unless there are studies available that can be used. In the latter case, the efforts could be fairly low. For the methods DtcT, DtpT, WtPTax, and WtPPrev, the efforts are low if data on targets and taxes are available and very high if they are not.

P31) **New environmental problems can be included.** This criterion includes both recently discovered environmental problems and well-known problems that are not (yet) included in LCA practice. Different evaluations may be necessary for these 2 aspects. All methods score fairly low, but for different reasons. Panel methods will normally require that the whole procedure be redone if new problems are to be included. Because regulated revealed preference methods (including DtpT, WtPTax, and WtPPrev) rely on a slow political process, the possibility of including new environmental problems is usually fairly low. The amount of scientific knowledge required to establish critical levels is normally very high, resulting in a low score for DtcT on this criterion. Please note that this evaluation refers to the weighting step; the possibilities of including new environmental problems in the characterisation step are not evaluated here.

P32/33) **New value choices and characterisation methods can be included.** If DtpT relies on targets set on impact rather than emission level, then it may be difficult or impossible to adjust for new characterisation methods. In all other cases and for new value choices, recalculations are necessary and possible (at the effort specified in I1 to I3). For methods based on stated preferences, it may be necessary to redo the process if the respondents' answers are likely to depend on the characterisation method.

P4) **Completeness.** The evaluation of the completeness criterion largely follows from criteria I2 and P31. One additional point is that resources are sometimes excluded from monetisation methods on theoretical grounds because it may be argued that resource 'impacts' are internalised in resource prices already and should not be double counted (e.g., Newell 1998). Again, note that this evaluation concerns the weighting element and not the characterisation. If different weighting methods require different characterisation methods, an evaluation of the combined characterisation and weighting elements may be significantly different.

P51/52) **Adequate representation of values in the choice of weighting procedure and weighting factors.** As noted in 'Evaluation of Methods' (p 195), this criterion cannot be evaluated on a general level. This is because what is regarded as adequate is influenced by ethical and ideological values (e.g., Finnveden 1997), which cannot be decided here. Some general comments can be made, however. One question is this: Whose values are relevant? Are they the values of the society as a whole or of only a subset of the population, and if so, which subset? If societal values are of interest, some would argue that decisions taken by elected governments reflect societal values, whereas others argue that values revealed by individuals in markets are more adequate; yet others would argue that only in a panel with adequate representation can the societal values be reflected. Another question is whether values revealed in one context can be used as valuations in another context, which is done in methods based on revealed preferences, or if only stated preferences offer an adequate basis. A possible problem with methods based on revealed preferences is that somebody is making an assumption that the revealed preference can be transferred to another context. Behind this assumption are values, and it is not clear to what extent these are representative. If only stated preferences are regarded as adequate, yet another question is whether it is adequate to develop generic weighting sets where preferences stated in one context are used in another. The DtcT and DtpT methods do not include any weighting between different targets (see below), and the values are therefore not present at all (DtcT) or only partly present (DtpT).

P6) **Reflecting the subjectivity of weighting and including an intereffect weighting.** Stated preference methods allow the reflection of both subjectivity and intereffect weighting. In theory, DtcT lacks the subjectivity of weighting because critical targets should be based on natural science alone, and it does not include an intereffect weighting because different targets are not weighted against each other. DtpT includes subjective elements because the setting of targets is a process that includes subjective values. However, there is no explicit intereffect weighting because the impact categories are not weighted against each other when the targets are set. This is an inherent limitation of DtpT and DtcT methods because there is no requirement in the target setting that the targets should be set in relation to each other. Costs, however, are normally considered (explicitly or implicitly) when targets are set. The basic assumption behind the WtPPrev methods is that the costs associated with the politically set targets reflect the societal valuation. As in all monetisation methods (exemplified here by WtPDam, WtPPrev, and WtPTax), this assumption is based on the notions that prices reflect values, and by introducing a common unit (i.e., money), an intereffect weighting is taking place. Differences between different monetisation methods reflect different types of prices and different types of values. A major difference between DtpT on the one hand and WtPPrev and WtPTax on the other hand is that the latter two have a common unit that can be interpreted as a measure of value. However, some do not accept that the common unit implies an intereffect weighting, explaining the question mark in Table 7-2.

P8) Relation to available and best available characterisation methods. Today, there is no general consensus on what constitutes best available characterisation methods, making it difficult to evaluate this criterion. However, a few general observations can be made. Some methods are flexible and can use essentially any characterisation methods that are available. On the other hand, methods that are based on damage assessments require characterisation methods that are based on an endpoint level, which may be regarded as a drawback for such methods because they are more limited in their choice of characterisation method. Another observation is that characterisation methods earlier in the environmental mechanism may be more readily available than later. For example,

the SETAC Europe task group on human toxicity concludes that current knowledge does not enable the elaboration of effect indicators for all types of toxic effects (Chapter 5). They suggest that toxicological potency indicators are pre-selected (Chapter 5). This may be regarded as a limitation of the methods based on damage modelling (some panel methods and WtPDam) because they require methods for effect indicators.

P9) Possibilities for further developments. The major limitation of the DtcT and DtpT methods is their lack of intereffect weighting. This limitation can be overcome only by changing the method into a DteT method, in which the targets are set with the explicit aim of being equal, or by combining them with monetary information, as in WtPPrev. For the other methods, there is room for further development.

Output-related criteria

O1) Robustness or sensitivity. There is some evidence that the results from both panel studies and CV studies are not influenced by changes in the specified environmental problems. In the literature, this is discussed as the scoping or embedding effects. For CV studies, there is a debate whether this is an inherent shortcoming or a shortcoming of specific studies (e.g., Hanemann 1996; Carson 2000). Another aspect is that, in some panel studies, the weighting factors are allowed to vary only between 1 or 2 orders of magnitude, decreasing the sensitivity (e.g., when AHP is used). WtPPrev is based on 2 types of data: targets and marginal pollution prevention costs. The latter are often nonlinear, growing exponentially at high targets. This affects robustness and makes the results less reliable.

O21/23 Consistent results and transitivity. Results from panel procedures are, in most cases, consistent and do not show any logical contradictions, provided the procedure is conducted well and inconsistencies are dealt with through reexamination. Possible inconsistencies in the data used are discussed in relation to criteria I1 and I2. Besides these aspects, logical inconsistencies should not be inherent properties of the different weighting methods. However, in practice, logical errors cannot be excluded, and this is therefore an important criterion when specific methods with generic sets of weighting factors are evaluated. All methods follow the structure of an additive calculation rule. Therefore, intransitive results occur only if the normalisation step is incongruent to the weights.

O22) Reproducibility. The question marks in Table 7-2 indicate that our experience may not be sufficient to judge this criterion. Some results suggest that panel methods show good reproducibility (Lindeijer 1997; Harada et al. 2000), but this could also be an artefact of insensitivity. WtPTax may be reproducible but only if the procedure is clearly described. For DtpT, the interpretation necessary to derive a large set of political targets may involve steps that are not reproducible. For WtPDam, the assessment of damage costs may be reproducible but give an unacceptably large range of values. For WtPPrev, the selection of prevention technologies and the allocation of multiprevention technologies may lead to nonreproducible results.

Criteria P71 and P72 are application dependent and not addressed in Table 7-2. Some comments can be made regarding criteria P71, however (Is the method based on weighting of marginal or average impacts and is this consistent with the study?). If marginal analysis is chosen in the inventory and characterisation modelling, then the weighting should reflect values at the margin as well. This is normally the case for monetisation methods, which generally are based on marginal changes. A difference between CV studies and panel methods is that, in typical CV studies, one aspect is considered and valued while everything else is constant. In panel studies, everything is typically valued at the same time, but preferences about marginal changes can be elicited, although marginal procedures may be more complex than average ones.

Concluding remarks

One important conclusion from this evaluation is that the DtTs fail on several criteria described as must candidates. Both DtpT and DtcT lack an intereffect weighting and do not use acceptable scientific practice in the sciences used. DtcT also lacks adequate representation of values because it does not adequately reflect the subjectivity of weighting. These problems are inherent to DtTs, unless they transform into either panel or monetisation methods or base the methods on targets that are set with the explicit aim of being equally important. The lack of intereffect weighting occurs because, when targets are set, there is no requirement that they should be of equal importance. When the targets are used in a DtT, however, this is assumed, and this is a limitation of these methods.

Distance to target methods that use externally determined targets have many similarities with normalisation methods. The targets may be either authoritative targets or critical or sustainable targets. Mathematically, the expressions are similar to targets (or functions of targets) as reference values. In parallel to other normalisation techniques, they can be used to place a set of case-specific LCA results into a wider context, and they also have the effect of adjusting the characterisation results to a common unit. However, in parallel to the results of both external and case-specific normalisation, the results are not necessarily additive just because they have a common unit. If it is explicitly assumed that all targets are equally important, then the addition of results from normalisation that uses externally determined targets can be acceptable. In cases where this assumption is not justified, which is the normal case because targets are normally not set with the aim of setting equally important targets, these methods should not be used as weighting methods because they do not include an intereffect weighting, that is, the different impact categories are not weighted against each other. However, the information provided by target values may well be used in the interpretation phase of LCA.

The results in Table 7-2 also illustrate that the evaluation of weighting methods against different criteria is not straightforward. Different persons interpret and evaluate criteria differently. Thus, some of the question marks in Table 7-2 illustrate that different persons can reach different conclusions. This is, for example, the case for monetisation methods based on revealed preferences for important criteria such as the science base, the representation of values, and the intereffect weighting, where for some people these methods score well but for others they are not acceptable.

The system of criteria is applied here on a general level. We expect that it can be used to evaluate specific weighting methods, possibly giving other types of results. We suggest that it can be used on combinations of characterisation and weighting methods in order to evaluate a larger part of the LCIA methodology.

Conclusions

In this chapter, normalisation, grouping, and weighting are discussed as 3 optional elements in LCIA. The relation and the congruence among them deserve further attention. One important conclusion is that case-specific (internal) normalisation requires case-specific weighting. If generic weighting factors are applied to results after a case-specific normalisation, the results are arbitrary. This is a limitation of case-specific normalisation if generic weighting factors are to be used.

Life-cycle impact assessment can rely on a number of different sciences and theories. Of special interest for the elements discussed here are environmental economics and decision analysis. Techniques, knowledge, and theories for the assessment problems developed in economics and decision analysis can be applied to the field of LCIA. Weighting methods are used to a large extent, including for comparative assertions as illustrated by presentations at the LCA Case Study Symposium (SETAC Europe 2000). This is in contrast to the ISO standard (ISO 2000). It would therefore be interesting if future updates of the ISO standard could address weighting and the role of value choices in all elements and phases of an LCA.

From the evaluation of different weighting methods an important conclusion is that DtTs should not be used as weighting methods, unless it is explicitly assumed that all targets are of equal importance. Panel methods and monetisation methods are the 2 major groups of weighting methods that can be used in LCIA. Both consist of a large number of different approaches.

Expressed preference methods, especially CV studies, have been used for a number of years, and some learning has resulted. One of the main outcomes of behavioural decision research during the last 2 decades is the view that peoples' preferences are often constructed during the elicitation process (Slovic 1995). According to this perspective, people have beliefs and values for the objects they are questioned about, but these values are not numerically quantified prior to the elicitation process. Following this *constructive perspective*, values for objects that are unfamiliar and complex are constructed rather than reported in the elicitation procedures. This constructive perspective has important implications for the use of methods based on expressed preferences in LCA. From descriptive decision analysis, we know that different types of biases occur during the elicitation process. For example, results are dependent on how questions are phrased and what information is provided but are sometimes insensitive towards the scope of the provided information, which causes a problem for normalisation. It is not clear how the procedures should be performed to avoid these biases and produce reliable results. There is guidance on how to overcome some biases (e.g., von Winterfeldt and Edwards 1986; Gregory et al. 1993; Carson 2000). However, some of the problems remain, and an important area for research is therefore to develop procedures that minimise biases and increase reliability. The constructive perspective may also have implications for the possibilities of applying weighting factors derived in one context to another context. That is, it has implications for the use of generic weighting sets.

Methods based on revealed preferences all assume that preferences revealed in one context can be transferred to another context. A certain leap of faith is thus required. This leap will be larger if the preferences are not expressed on the level of impact categories. Another concern is that the aspects considered when decisions were taken and their relation to the weighting problems at hand are not known.

There are a number of different approaches for monetisation of EIs. Because they measure different economic values (e.g., total economic value or just use values), an important conclusion is that monetary values may not always be comparable and additive just because they have the same units. One limitation of currently available damage-oriented monetisation methods is the mixture of different monetisation methods. Because some impacts and damages have market prices and some do not, there is the risk of an unbalanced evaluation. Other inherent limitations of this group of methods are the uncertainties in handling future impacts and the choices related to discounting and/or using a cutoff.

Endpoint methods require that damages be modelled quantitatively. The importance of this limitation varies between different impact categories and also depends on worldviews. Hofstetter (1998) has suggested using an indicator for unknown damages to partly circumvent this limitation. This is a suggestion that should be further explored. At the same time, it can be noted that quantifying the unknown is, of course, impossible. Another approach is to seek robust knowledge that all stakeholders can agree on (cf. Tukker 1999).

This would probably imply that the weighting should be based on impacts earlier in environmental mechanisms rather than later (cf. Bare et al. 2000).

The system of criteria for weighting methods suggested here is used on a generic level. However, we suggest that it also can be used on specific weighting methods. In that case, some criteria that are difficult to evaluate on a general level, for example, concerning logical consistencies, may turn out to be useful. The system of criteria may also be useful when methods for a specific case study are chosen. In that case, the application-dependent criteria may turn out to be useful.

It should be clear from this chapter that there are is no single weighting method that fulfils all relevant criteria, and there probably never will be one. It is important, however, to note that weighting methods are widely used, and there is therefore a need to constantly review and improve weighting methods in order to provide decision-makers and other LCA users with the best available practice. In all cases, an attempt should be made to use the best possible science in weighting methods. Methods should be applied with an understanding of their assumptions, limitations, and inherent value choices. Efforts should be made to ensure that these are consistent with the values of the decision-makers and stakeholders.

References

Ahbe S, Braunschweig A, Müller-Wenk R. 1990. Methodik für Oekobilanzen auf der Basis Ökologischer Optimierung. Bern, CH: Bundesamt für Umwelt, Wald und Landschaft (BUWAL). Schriftenreihe Umwelt Nr. 133.

Anonymous. 1991. Integrated substance chain management. Bilthoven, NL: Assoc of the Dutch Chemical Industry (VNCI).

Arrow K, Solow R, Portney PR, Learner EF, Radner R, Schuman H. 1993. Report to the NOAA Panel on Contingent Valuation. *Federal Register* 58:4601.

Azapagic A, Clift R. 1998. Linear programming as a tool for life cycle assessment. *Int J LCA* 3:305–316.

Bare JC, Hofstetter P, Pennington DW, Udo de Haes HA. 2000. Life cycle impact assessment workshop summary; Midpoints versus endpoints: The sacrifices and benefits. *Int J LCA* 5(6):319–326.

Bare JC, Norris G, Hofstetter P. 2001. Normalisation in life cycle impact assessment. Forthcoming.

Bare JC, Pennington DW, Udo de Haes HA. 1999. Life cycle impact assessment sophistication. *Int J LCA* 4(5):299–306.

Basson L, Perkins AR, Petrie JG. 2000. The evaluation of pollution prevention alternatives using non-compensatory multiple criteria decision analysis methods. Annual Meeting of the American Institute of Chemical Engineers (AIChE); 2000 Nov; Los Angeles CA, USA. New York NY, USA: AIChE Manuscript Center. Presentation Record 230c.

Basson L, Petrie JG. 2000. The development of a decision support framework for fossil fuel based power generation. Annual Meeting of the American Institute of Chemical Engineers (AIChE); 2000 Nov; Los Angeles CA, USA. New York NY, USA: AIChE Manuscript Center. Presentation Record 225c.

Baumann H, Ekvall T, Eriksson E, Kullman M, Rydberg T, Ryding S-O, Steen B, Svensson G. 1993. Miljömässiga skillnader mellan återvinning/återanvändning och förbränning/deponering. Malmö, S: Stiftelsen Reforsk. FoU Nr. 79. (In Swedish.)

Baumann H, Rydberg T. 1994. Life cycle assessment. A comparison of three methods for impact analysis and evaluation. *J Cleaner Prod* 2:13–20.

[BCD] Britannica Compact Disc. 1998. Encyclopaedia Britannica 98 CD. Chicago IL, USA: BCD.

Bengtsson M. 2000. Environmental valuation and life cycle assessment [thesis]. Göteborg, S: Chalmers Univ of Technology, Department of Environmental Systems Analysis.

[BMBF] Bundesministerium für Forschung und Technologie. 2001. Systematische Auswahlkriterien für die Entwicklung von Verbundwerkstoffen unter Beachtung ökologisher Erfordernisse–Abschlussphase. Bonn, D: BMBF.

Bockstael NE, Freeman III AM, Kopp RJ, Portney PR, Smith VK. 2000. On measuring economic values for nature. *Environ Sci Technol* 34:1384–1389.

Braunschweig A, Förster R, Hofstetter P, Müller-Wenk R. 1996. Developments in LCA valuation. St-Gallen, CH: Institut für Wirtschaft und Ökologies an der Hochschule St. Gallen (IWÖ-HSG). IWÖ-Diskussionsbeitrag Nr 32.

Brunner S. 1998. Panel methods and their application for weighting. Zürich, CH: Swiss Federal Institute of Technology (ETH).

[BUWAL] Bundesamt für Umwelt, Wald und Landschaft. 1998. Bewertung in Ökobilanzen mit der Methode der Ökologischen Knappheit. Ökofaktoren 1997. Bern, CH: BUWAL. Schriftenreihe Umwelt Nr 297. (In German.)

Carlsson M. 1997. Economics in ORWARE: A welfare analysis of organic waste management. Uppsala, S: Swedish Univ of Agriculture, Dept of Economics. Report 114.

Carson R. 2000. Contingent valuation: A user's guide. *Environ Sci Technol* 34:1413–1418.

Chadwick MJ, Nilsson J. 1993. Environmental quality objectives: Assimilative capacity and critical load concepts in environmental management. In: Jackson T, editor. Clean production strategies. Boca Raton FL, USA: Lewis. p 29–39.

Clift R, Frischknecht R Huppes G, Tillman A-M, Weidema B. 1999. A summary of the results of the working group on inventory enhancement. *SETAC Europe News* 10(3):14–20.

Corten FGP, Haspel B, v.d. Kreuzberg GJ, Sas HJW, de Wit G. 1994. Weighting environmental problems for product policy. Phase 1. Delft, NL: (In Dutch). As cited by Lindeijer (1996).

Craighill AL, Powell JC. 1995. Lifecycle assessment and economic valuation of recycling: A case study. Norwich, GB: Centre for Social and Economic Research on the Global Environment (CSERGE). CSERGE Working Paper WM 95-05.

Dalkey NC. 1969. An experimental study of group opinion: The Delphi method. *Futures* 1:403–406.

Dobson P. 1998a. The multiple pathway method. A guide to the application of the methodology. Leatherhead, Surrey, GB: Pira International.

Dobson P. 1998b. The implementation of the multiple pathway method. Case study: A waste management study. Leatherhead, Surrey, GB: Pira International.

Edwards W. 1977. How to use multiattribute utility measurement for social decision making. *IEEE Trans Sys Man Cybernet* SMC-7:326–340.

Eisenführ F, Weber M. 1993. Rationales Entscheiden. Berlin, D: Springer Verlag.

Environmental Science and Technology. 2000. Special issue on economic valuation. *Environ Sci Technol* 34(8).

ExternE. 1995. Externalities of energy. Volumes 1–6. Brussels, B: European Commission DG XII.

ExternE. 1999. Externalities of energy. Volumes 7–10. Brussels, B: European Commission DG XII.

Farber S, Griner B. 2000. Using conjoint analysis to value ecosystem change. *Environ Sci Technol* 34:1407–1412.

Finnveden G. 1997. Valuation methods within LCA: Where are the values? *Int J LCA* 2:163–169.

Finnveden G. 1998. On the possibilities of life-cycle assessment [dissertation]. Stockholm, S: Department of Systems Ecology, Stockholm Univ.

Finnveden G. 1999a. Methodological aspects of life cycle assessment of integrated solid waste management systems. *Resour Conserv Recycl* 26:173–187.

Finnveden G. 1999b. A critical review of operational valuation/weighting methods for life cycle assessment. Stockholm, S: Avfallsforskningsrådet (AFR), Swedish EPA. AFR Report 253.

Finnveden G, Andersson-Sköld Y, Samuelsson M-O, Zetterberg L, Lindfors L-G. 1992. Classification (impact analysis) in connection with life cycle assessment: A preliminary study. In: Product life cycle assessment: principles and methodology. Copenhagen, DK: Nordic Council of Ministers. Nord 1992:9. p 172–231.

Fleischer G, Ciroth A, Gerner K, Kunst. 2001. Nachhaltiges Witschaften am Beispiel von Schienenfahrzeugen (Bahnkreis). Berlin, D: Technical Univ of Berlin. Project funded by German Ministry of Education and Research (BMBF), 1998–2000. Final report available at http://edok01.tib.uni-hannover.de/edoks/e01fb01/330464833.pdf. Accessed 8 April 2002.

French S. 1988. Decision theory: An introduction to the mathematics of rationality. Chichester, GB: Ellis Horwood.

Frischknecht R. 1998. Life-cycle inventory analysis for decision-making [PhD thesis]. Zürich, CH: Swiss Federal Institute of Technology (ETH). ETH Nr 12599.

Giegrich J, Mampel U, Duscha M, Zazcyk R, Osorio-Peters S, Schmidt T. 1995. Bilanzbewertung in produktbezogenen Ökobilanzen. Evaluation von Bewertungsmethoden, Perspektiven. In: Methodik der produktbezogenen Ökobilanzen: Wirkungsbilanz und Bewertung. Berlin, D: Umweltbundesamt. Texte 23/95. (In German).

Giegrich J, Schmitz S. 1996. Valuation as a step in impact assessment: Methods and case study. In: Curran MA, editor. Environmental life-cycle assessment. New York NY, USA: McGraw-Hill.

Goedkoop M. 1995. The Eco-indicator 95: Final report and manual for designers. Amersfoort, NL: PRé Consultants.

Goedkoop M, Hofstetter P, Müller-Wenk R, Spriensma R. 1998. The Eco-indicator 98 explained. *Int J LCA* 3:352–360.

Goedkoop M, Spriensmaa R. 1999. The Eco-indicator 99: A damage oriented method for life cycle impact assessment. Amersfort, NL: PRé Consultants.

Gregory R, Lichtenstein S, Slovic P 1993. Valuing environmental resources: A constructive approach. *J Risk Uncert* 7:177–197.

Guitouni A, Martel JM. 1998. Tentative guidelines to help choosing an appropriate MCDA method. *Eur J Operation Res* 109:501–521.

Hanemann WM. 1996. Theory versus data in the contingent valuation debate. In: Bjornstad DJ, Kahn JR, editors. The contingent valuation of environmental resources: Methodological issues and research needs. Cheltenham, GB: Edward Elgar.

Hansen OJ. 1999. Status of life cycle assessment (LCA) activities in the Nordic region. *Int J LCA* 4:315–320.

Harada T, Fuji Y, Nagata K, Inaba A, Mettier T. 2000. Panel test for Japanese experts aiming to weight safeguard subjects. In: Proceedings of the Fourth International Conference on Ecobalances; 2000 Oct 31–Nov 2; Tsukuba, Japan. Tokyo, J: Society of Non-Traditional Technology.

Hertwich EG, Hammitt JK. 2001. A decision-analytic framework for impact assessment. Part I: LCA and decision analysis. *Int J LCA* 6:5–12.

Hertwich EG, Hammitt JK, Pease WS. 2000. A theoretical foundation for life-cycle assessment. Recognizing the role of values in environmental decision making. *J Ind Ecol* 4(1):13–28.

Hertwich EG, Pease WS, Koshland CP. 1997. Evaluating the environmental impact of products and production processes: A comparison of six methods. *Sci Total Environ* 196:13–29.

Hofstetter P. 1998. Perspectives in life cycle impact assessment: A structured approach to combine models of the technosphere, ecosphere and valuesphere. Boston MA, USA: Kluwer Academic.

Hofstetter P. 1999. Top-down: Arguments for a goal-oriented assessment structure. Landsberg, D: LCA Global Village, Ecomed. http//www.ecomed.de/journals. Accessed 15 Mar 2002.

Hofstetter P. 2000a. Consequences of the damage approach on the valuation step and LCA in general. In: Proceedings of the USEPA conference InLCA; 2000 Apr 25-27; Washington DC/Arlington USA. Forthcoming.

Hofstetter P. 2000b. Looking at the full picture: Implications associated with valuation. In: Bare J, Hofstetter P, Pennington D, Udo de Haes HA. 2000. Midpoints versus endpoints: The sacrifices and benefits. Proceedings of the UNEP/USEPA/CML Expert Workshop; 2000 May 25–26; Brighton, GB. Forthcoming.

Holmberg J, Lundqvist U, Robért K-H, Wackernagel M. 1999. The ecological footprint from a systems perspective of sustainability. *Int J Sustain Dev World Ecol* 6:17–33.

Huppes G, Sas H, de Haan E, Kuyper J. 1997. Efficient environmental investments. Paper presented at the SENSE International Workshop; 1997 Feb 20; Amsterdam, NL. Leiden, NL: Leiden Univ Centre for Environmental Science.

[ISO] International Organization for Standardization. 1997. Environmental management—Life cycle assessment—Principles and framework. International Standard ISO 14040. Geneva, CH: ISO.

[ISO] International Organization for Standardization. 2000. Environmental management—Life cycle assessment—Life cycle impact assessment. Geneva, CH: ISO. International Standard ISO 14042.

Johansson J. 1999. A monetary valuation weighing method for life cycle assessment based on environmental taxes and fees [Masters thesis]. Stockholm, S: Stockholm Univ, Department of Systems Ecology.

Kalisvaart SH, Remmerswaal JAM. 1994. The MET-points method: A new single figure environmental performance indicator. In: Udo de Haes HA, Jensen AA, Klöppfer W, Lindfors L-G, editors. 1994. Integrating impact assessment into LCA. Brussels, B: Society of Environmental Toxicology and Chemistry (SETAC) Europe. p 143–148.

Keeney RL, Raiffa H. 1976. Decisions with multiple objectives: Preferences and value tradeoffs. New York NY, USA: J Wiley.

[KI] Konjunkturinstitutet. 1998. En utvärdering av FNs miljöräkenskapsuppställningar. Värderingsstudier. Stockholm, S: KI. Miljöräkenskaper Rapport 1998:9. (In Swedish).

Klöpffer W, Schmidt W-P, Volwein S. 2000. Modul Umwelt. In: Fleischer G, Becker J, Braunmiller U, Klocke F, Klöpffer W, Michaeli W, editors. Eco-Design. Berlin, D: Springer. p 88-102.

Kortman JGM, Lindeijer EW, Sas H, Sprengers M. 1994. Towards a single indicator for emissions: An exercise in aggregating environmental effects. Amsterdam, NL: Univ of Amsterdam, Interfaculty Department of Environmental Sciences.

Krewitt W, Trukenmüller A, Bachmann TM, Heck T. 2001. Country specific damage factors for air pollutants: A step towards site dependent life cycle impact assessment. *Int J LCA* 6(4):199–210.

Krozer J. 1992. Decision model for Environmental Strategies of Corporations (DESC). The Hague, NL: Institute for Applied Economics (TME).

Lindeijer E. 1996. Normalisation and valuation. In Udo de Haes, editor. Towards a methodology for life cycle impact assessment. Brussels, B: Society of Environmental Toxicology and Chemistry (SETAC) Europe. 75–93.

Lindeijer E. 1997. Results try-out Japanese/Dutch LCA valuation questionnaire 1996. IVAM ER. Amsterdam, NL: Univ of Amsterdam.

Lindfors L-G, Christiansen K, Hoffman L, Virtanen Y, Juntilla V, Hanssen OJ, Rønning A, Ekvall T, Finnveden G. 1995. Nordic guidelines on life-cycle assessment. Copenhagen, DK: Nordic Council of Ministers. Nord 1995:20.

Mettier T, Baumgartner T. 2000. A non-monetary approach to weight environmental damages in life cycle assessment: panel methods compared to contingent valuation and other monetary approaches. Paper Nr. 26 presented at ESEE 2000, 3rd Biennial Conference of the European Society for Ecological Economics; 2000 May 3–6; Vienna, A.

Meyer MA, Booker JM. 1990. Eliciting and analyzing expert judgement. Washington DC, USA: U.S. Nuclear Regulatory Commission, Los Alamos National Laboratory, Los Alamos NM, USA. NUREG/CR-5424, LA-11667-MS.

Miettinen P, Hämäläinen RP. 1997. How to benefit from decision analysis methods in environmental life cycle assessment (LCA). *Eur J Operation Res* 102:279–294.

Müller-Wenk R. 1997. Safeguard subjects and damage functions as core elements of life-cycle impact assessment. St Gallen, CH: Institute für Wirtschaft und Ökologie (IWÖ). IWÖ-Diskussionsbeitrag Nr. 42.

Nagata K, Fuji Y, Ishikawa M. 1995. Proposing a valuation method based on panel data, preliminary report, Tokyo. In: Brunner S. 1998. Panel methods and their application for weighting. Zürich, CH: Swiss Federal Institute of Technology (ETH).

Newell SA. 1998. Strategic evaluation of environmental metrics: Making use of life cycle inventories [thesis]. Cambridge MA, USA: Massachusetts Institute of Technology, Department of Materials, Science and Engineering.

Paelinck JHP. 1974. Qualitative multiple criteria analysis, environmental protection and multi-regional development. *Pap Reg Sci Assn* 36:59–74.

Paelinck JHP. 1977. Qualitative multiple criteria analysis: An application to airport location. *Environ Plan* 9:883–895.

Poulamaa M, Kaplas M, Reinikainen T. 1996. Index of environmental friendliness: A methodological study. Helsinki, FS: Statistics Finland, Environment 1996:13.

Powell JC, Pearce DW, Craighill AL. 1997. Approaches to valuation in LCA impact assessment. *Int J LCA* 2:11–15.

Rietveld P. 1984a. Public choice theory and qualitative discrete multicriteria evaluation. In: Bahrenberg G, editor. Recent developments in spatial data analysis. Aldershot, GB: Gower. p 409–426.

Rietveld P. 1984b. The use of qualitative information in macro economic policy analysis. In: Despontin M, Nijkamp P, Spronk J, editors. Macro economic planning with conflicting goals. Berlin, D: Springer. p 263–280.

Roy B, Vanderpooten D. 1996. The European School of MCDA: Basic features and current works. *J Multi-Criteria Decision Anal* 5:(1) 22–38.

Saaty TL. 1980. The analytical hierarchy process. New York NY, USA: McGraw-Hill.

Salo A, Hämäläinen RP. 1997. On the measurements of preferences in the analytical hierarchy process. *J Multi-Criteria Decision Anal* 6:309–319.

Schaltegger S, Sturm A. 1991. Methodik der ökologischen Rechnungslegung in Unternehen. Basel, CH: Wirtschaftswissenschaftliches Zentrum der Universität Basel. WWZ-Studien Nr. 33. (In German).

Schmitz S, Paulini I. 1999. Bewertung in Ökobilanzen. Methode des Umweltbundesamtes zur Normierung von Wirkungsindikatoren, Ordnung (Rangbildung) von Wirkungskategorien und zur Auswertung nach ISO 14042 und 14043. Version '99. Berlin, D: Umweltbundesamt (UBA). UBA-Texte 92/99.

[SEEA] System of Integrated Environmental and Economic Accounting. 2000. Extending the monetary accounts to include valuation of degradation. Draft Chapter 5 of revised SEEA handbook. Forthcoming from UN, World Bank, OECD, and Eurostat. http://ww2.statcan.ca/citygrp/london/london.htm. Accessed 15 Mar 2002.

Seppälä J. 1999. Decision analysis as a tool for life cycle impact assessment. In: Klöpffer W, Hutzinger O, editors. LCA documents, Volume 4. Bayreuth, D: Ecoinforma Pr.

Seppälä J, Basson L, Norris G. 2002. Decision analysis frameworks for life cycle impact assessment. *J Ind Ecol*. Forthcoming.

Seppälä J, Hämäläinen RP. 2001. On the meaning of the distance-to-target methods and normalisation in life cycle impact assessment. *Int J LCA* 6:211–218.

[SETAC Europe] Society of Environmental Toxicology and Chemistry Europe. 2000. Presentation summaries. 8th LCA Case Studies Symposium; 2000 Nov 30; Brussels, B. Brussels, B: SETAC Europe.

[SIKA] Statens Institut för Kommunikations Analys. 1999. Översyn av samhällsekonomiska kalkyprinciper och kalkylvärden på transportområdet. Stockholm, S: SIKA. SIKA Rapport 1999:6.

Sonesson U, Björklund A, Carlsson M, Dalemo M. 2000. Environmental and economical analysis of management of biodegradable waste. *Resour Conserv Recycl* 28:29–53.

Spadaro JV, Rabl A. 1999. Estimates of real damage from air pollution: Site dependence and simple impact indices for LCA. *Int J LCA* 4(4):229–243.

Spengler T, Gelderman J, Hähre S, Sieverdingbeck A, Rentz O. 1998. Development of a multiple criteria based decision support systems for environmental assessment of recycling measures in the iron and steel making industry. *J Cleaner Prod* 6:37–52.

Steen B. 1999a. A systematic approach to environmental priority strategies in product development (EPS). Version 2000: General system characteristics. Göteborg, S: Chalmers Univ of Technology, Centre for Environmental Assessment of Product and Material Systems (CPM). CPM Report 1999:4.

Steen B. 1999b. A systematic approach to environmental priority strategies in product development (EPS). Version 2000: Models and data of the default method. Göteborg, S: Chalmers Univ of Technology, Centre for Environmental Assessment of Product and Material Systems (CPM). CPM Report 1999:5.

Steen B, Ryding S-O. 1992. The EPS environ-accounting method. Göteborg, S: Swedish Environmental Research Institute (IVL). Report Nr. B1080.

Stewart M. 1999. Environmental life cycle assessment for design related decision making in minerals processing [PhD dissertation]. Cape Town, South Africa: Univ of Cape Town.

Stewart TJ. 1992. A critical survey on the status of multiple criteria decision making theory and practice. *Omega Int J Manage Sci* 20:569–586.

Tellus Institute. 1992. The Tellus packaging study. Boston MA, USA: Tellus Institute.

Tietenberg T. 1999. Environmental and natural resource economics. 5th ed. New York NY, USA: Addison-Wesley.

Tomovic R, Vukobratovic M. 1972. General sensitivity theory. New York NY, USA: American Elsevier.

Tukker A. 1999. Frames in the toxicity controversy. Dordrecht, NL: Kluwer Academic.

Turner RK, Pearce D, Bateman I. 1994. Environmental economics, an elementary introduction. Hemel Hempsted, GB: Harvester Wheatsheaf.

Udo de Haes HA, editor. 1996. Towards a methodology for life cycle impact assessment. Brussels, B: Society of Environmental Toxicology and Chemistry (SETAC) Europe.

Udo de Haes HA. 1999. Weighting in life-cycle assessment. Is there a coherent perspective? *J Ind Ecol* 3(4):3–7.

Udo de Haes HA, Jolliet O, Finnveden G, Hauschild M, Krewitt W, Müller-Wenk R. 1999. Best available practice regarding impact categories and category indicators in life cycle impact assessment. *Int J LCA* 4:66–74 and 167–174.

Udo de Haes HA, Wrisberg N, editors. 1997. Life cycle assessment: State of the art and research needs. LCA Documents, Volume 1. Bayreuth, D: Ecoinforma.

van Beukering F, Oosterhuis F, Spaninks F. 1998. Economic valuation in life cycle assessment. Amsterdam, NL: Vrije Universiteit, Institute for Environmental Studies. Working Paper W98/02.

Vogtlander JG, Bijma A. 2000. The 'Virtual Pollution Prevention Costs '99'. *Int J LCA* 5:113–124.

Volkwein S, Gihr R, Klöpffer W. 1996. The valuation step within LCA. Part II: A formalized method of prioritization by expert panels. *Int J LCA* 1:182–192.

von Winterfeldt D, Edwards W. 1986. Decision analysis and behavioral research. Cambridge, GB: Cambridge Univ Pr.

Voogd H. 1983. Multicriteria evaluation for urban and regional planning. London, GB: Pion.

Wackernagel M, Rees W. 1996. Our ecological footprint. Gabriola Island, BC, CAN: New Society.

Weber M, Borcherding K. 1993. Behavioral influences on weight judgments in multiattribute decision making. *Eur J Operation Res* 67(1):1–12.

White AL, Savage D, Shapiro K. 1996. Life-cycle costing: Concepts and applications. In: Curran MA, editor. Environmental life-cycle assessment. Chapter 7. New York NY, USA: McGraw Hill.

Wilson B, Jones B. 1994. The phosphate report. London, GB: Landbank Environmental Research and Consulting.

Zuckerman B, Ackerman F. 1994. The 1994 update of the Tellus Institute packaging study impact assessment method. Boston MA, USA: Tellus Institute.

The Conceptual Structure of Life-Cycle Impact Assessment

Helias A. Udo de Haes, Erwin Lindeijer

Acknowledgements — This chapter is based on a broad discussion, which started within the Task Group on Resources and Land Use, but later involved the full working group and outside participants. Written contributions have been received from Jane C. Bare, Göran Finnveden, Edgar G. Hertwich, Patrick Hofstetter, Gjalt Huppes, Norihiro Itsubo, Olivier Jolliet, Walter Klöpffer, Wolfram Krewitt, Ruedi Müller-Wenk, Willie Owens, David W. Pennington, Bengt Steen, Arnold Tukker, Stephan Volkwein, and Bo Weidema. The text, where possible, aims at consensus, but on a number of critical issues, differing viewpoints have been included as a basis for further discussion. On such open issues, the authors indicate where they stand.

Introduction

This book reports on new developments in life-cycle impact assessments (LCIAs) and aims to contribute to the development of best available practice in this field. These new developments concern not only technical knowledge about the fate and effect of substances but also conceptual issues. This last chapter focuses on the latter, addressing terminology and the definitions of impact categories, category indicators (CIs), and so-called *areas of protection* (AoPs), that is, classes of endpoints that society wants to protect. We do not pretend to be able to identify best available practice in LCIA yet. Instead, the goal of this chapter is to identify new issues, to clarify differing viewpoints, but also to make our own suggestions. Just as the publication of Udo de Haes et al. (1999) marked the start of the Society of Environmental Toxicology and Chemistry (SETAC) Europe Second Working Group on Life-Cycle Impact Assessment (WIA-2) and laid the groundwork for what is reported here, this chapter aims to build a bridge to the next stage, that is, the definition of best practice in the United Nations Environment Programme (UNEP)–SETAC Life-Cycle Initiative.

Trains of thought differ regarding the identification of best available practice on conceptual issues. Leaving intermediate standpoints out of consideration for the moment, one train of thought is that all choices in LCIA should be dependent on the goal and scope of the study. The other is that best available practice should and can be identified. We follow the latter train of thought, but at the same time we realise that practitioners must have considerable freedom to make specific choices in line with the goal and scope of a particular life-cycle assessment (LCA) study. In the UNEP–SETAC Life-Cycle Initiative, a choice has been made to identify best available practice with generic application dependency. This means that different types of applications may well require different types of best practice, while remaining part of one consistent framework. It is our hope that we can reach a similar choice for the conceptual issues discussed here. Without a common conceptual framework, it is hardly possible to define a best-practice set (or sets) of impact categories and their CIs, and it is precisely these that should be the basis for the work on LCIA in the coming UNEP–SETAC Life-Cycle Initiative. Within such an overall framework, it must then be possible to make further choices regarding impact categories, CIs, average versus marginal modelling, spatial differentiation, etc.

The primary aim of this chapter is to clarify questions regarding the framework for LCIA. One question is what the implications of such a framework would be for the preceding life-cycle inventory (LCI) analysis phase: Different impact categories and CIs require different types of data from LCI results. Regardless of how true this is, it is not a guiding principle of this chapter. Consequently, the implication may be that identifying best practice in the field of LCIA will set LCI requirements that at present cannot be or cannot fully be fulfilled. It will be the task of the UNEP–SETAC Life-Cycle Initiative to define best practice in LCIA in such a way that it can be implemented.

The structure of this chapter is as follows. 'LCIA Terminology' (below) deals with the terminology regarding the environmental mechanism of an impact category. We present some options for adapting this terminology, but for the rest of the chapter, we use the terminology of the first report of the WIA-2 (see Udo de Haes et al. 1999). 'Viewpoints and Approaches in LCIA' (p 213) presents differing viewpoints on 3 critical issues in LCIA:
1) LCA as a tool for decision-making versus LCA as an example of systems analysis,
2) the scope of LCIA, and
3) the comparison of bottom–up versus top–down approaches in LCIA.

'Classification of AoPs in LCIA' (p 216) explores 2 different ways to structure AoPs in LCIA. In 'Overview of the AoPs' (p 220), a visual representation of these 2 approaches is presented and discussed.

LCIA Terminology

In this section, we discuss and define a number of terms related to LCIA. In the rest of the chapter, the terminology, for continuity reasons, will follow that of the first report of the WIA-2 (Udo de Haes et al. 1999). Where alternatives are provided, the possibility to define new terms is in the hands of the UNEP–SETAC Life-Cycle Initiative, and later in a possible revision of International Organization for Standardization (ISO) 14042. The terms we discuss here are *environment*, *environmental mechanism*, *environmental intervention*, *midpoint*, *endpoint*, and *areas of protection*.

Environment

In ISO 14001, the term *environment* is defined as 'surroundings in which an organisation operates, including air, water, land, natural resources, flora, fauna, humans and their interrelation'. This definition is in line with systems analysis, which defines that a universe can be subdivided into a system and the system environment. We will follow this general requirement. But we can define the system here at 2 relevant levels: 1) the level of a given product system, and 2) the level of the total of product systems in a given area (or on earth). The total of product systems will be called the *economy* and is further elaborated in 'Decision support versus systems analysis' (p 213).

There are alternatives for the term *environment*, such as *nature* or *ecosphere*. However, the term *environment* is used in the 14000 series of the ISO standard and in environmental management in general; another term with roughly the same meaning would introduce unnecessary confusion.

Environmental mechanism

The *environmental mechanism* of an impact category is, according to ISO 14042, the total of environmental processes that link environmental interventions to the endpoints for that given impact category. The concept in itself is clear. However, in practice the term *mechanism* causes confusion because it seems to refer to a single environmental process (for instance, the degradation of a compound or the response of vegetation to eutrophication). A possible alternative is *impact chain*. The term *chain* does not show the branching of the interlinked processes, but this confusion seems small compared to *environmental mechanism*, which seems hardly to be used. Yet another option is *environmental pathway*, as used in environmental pathway analysis (Spadaro and Rabl 1999). We therefore propose replacing the term *environmental mechanism* with *impact chain* or *environmental pathway*. (But for the remainder of this chapter, we will use ISO 14042 terminology.)

Environmental intervention

Environmental interventions (or, possibly, *anthropogenic interventions*) are the physical elements that cross the border between the product system and the environment (Udo de Haes et al. 1999). First, environmental interventions consist of the extraction of natural resources that enter the product system and the emission of hazardous substances (or other types of physical elements such as radiation) that leave the product system and enter the environment. But they also involve other elements, that is, physical changes in the environment caused by activities in the product system (in particular, different types of land use, cutting of trees, shooting of animals, lowering of groundwater table, etc.). ISO 14042 uses the term *elementary flows* in this context. These elementary flows cover the extraction of resources and the emissions of substances, but they do not cover land use. In addition, this term seems to focus on chemical elements and not on compounds and is therefore rather confusing. Another term that is used in this context is *exchanges* with the environment, or *environmental exchanges* (Wenzel et al. 1997). In itself, this is probably a very good term, which covers all 3 types of elements mentioned above. A point to consider is that the term *exchange* is used in LCI to refer to the relationship between LCI databases; for instance, there is the term *exchange format*, which has nothing to do with passing the system boundary between product system and environment. A last possibility concerns the term *stressor*, as was originally coined in the Sandestin, Florida, USA workshop on LCIA (Fava et al. 1993). This term is a good candidate, but it is limited to elements that pass the boundary of the product system and does not include elements at midpoint level (see the next section), as was originally the case. In our opinion, the term *stressor* may present the least misunderstanding, and it links up well with the history of LCIA. (However, we will use the term *environmental interventions* for this chapter.)

Midpoint and endpoint

Midpoints concern all elements in an environmental mechanism of an impact category that fall between environmental interventions and endpoints. This leads to the necessity to define endpoints. We define *endpoints* as those elements of an environmental mechanism that are themselves of value to society. ISO 14042 mentions forests and coral reefs as examples, in contrast to ambient concentrations of hazardous substances. Other examples are physical aspects of human health such as lifetime or bodily functions, plant or animal species, or natural resources such as fossil fuels and mineral ores.

Regarding the terms *midpoint* and *endpoint*, a number of issues must be clarified. It is important to note that the meaning of *endpoint* as used by ISO 14042 differs from its meaning in environmental risk assessment. There, an endpoint is any physical element in the analysis that is regarded as the dependent variable. In ISO–LCA language, category endpoints are at the physical damage level, that is, the level of elements that are of value to society. But perhaps the gap is smaller than indicated here. In risk assessment, a distinction often is made between measurement endpoints and assessment endpoints. Then measurement endpoints link up with CIs, and assessment endpoints are the same as the category endpoints defined by ISO. We write *physical damage level* as distinct from *monetary damage level* because we must recognise that the term *damage* is used at these 2 different levels. A further point regarding the term *endpoint* concerns the distinction between an element itself and an effect on an element, for instance, a forest versus damage to a forest. We advise using ISO terminology, meaning that endpoints are the physical elements themselves, not their changes caused by environmental interventions.

Another question is whether elements like disability-adjusted life years (DALYs) or quality-adjusted life years (QALYs) are to be regarded as endpoints in LCIA terminology. Like monetary damage, these do not involve physical elements but are value constructs. As close as possible to the ISO 14042 definition, the term *endpoint* will be used here for the physical elements themselves; then DALYs and QALYs are derived from endpoints, or weighted endpoints, but are not the endpoints themselves.

A last point regarding endpoints concerns the phrase *of value to society*; this can refer both to intrinsic values such as elements related to human life, biodiversity, or works of art and to functional values, that is, values related to elements that are a means to an end, such as abiotic resources. Although this distinction is not absolute (we can look at works of art, and thus they also have a function), by *functional values* we mean physical or practical use.

Environmental relevance

The term *environmental relevance* is explained in ISO 14042 but not strictly defined there. Based on ISO 14042, we define environmental relevance as the degree of linkage between the CI and the category endpoints of a given impact category.

Areas of protection

The last term in this context is *area of protection*. In the first report of the WIA-2 (Udo de Haes et al. 1999), an AoP is defined as a class of endpoints. In ISO 14042, three such classes are mentioned, be it in a rather implicit way: 'human health', 'natural environment', and 'natural resources'. In this chapter, another classification of AoPs will be discussed, as well as 2 possibly additional AoPs: 'manmade environment' (first introduced by Udo de Haes et al. 1999) and 'life support functions' (LSFs; first proposed in LCIA context by Udo de Haes; see Bare et al. 2000). Another term that describes classes of endpoints is the expressive term *safeguard subject*, introduced by Steen and Ryding (1992) and used in the Swiss literature (Müller-Wenk 1997). It is important to note that these 2 terms convey exactly the same message: They relate to the endpoints as physical elements, not to the underlying societal values. So, following this terminology, the human right to life or economic welfare cannot be an AoP or a safeguard subject; neither can respect for nature or cultural values. We propose to keep the term *AoP* because it is more in line with the discussion in ISO 14042 (although not strictly defined there) and because this term is used in the earlier publication of the WIA-2.

The above terminology rather closely connects with terminology used in decision analysis (cf. Keeney 1992; Hertwich and Hammitt 2001), which starts from the values that are affected. In the framework of these authors, the terms *stressor, insult, stress, consequence,* and *value lost* are used, as are the terms *attribute* and *means–ends objective network*. In Table 8-1, we indicate how these terms correspond to the terms used here. It seems fruitful to analyse more precisely whether the terminology used here corresponds with the given decision analysis terminology, and whether it may be fruitful to adopt terms from this field.

Table 8-1 Correspondence between terms from decision analysis and terminology used in LCIA (this chapter)[a]

Terms used in decision analysis	Corresponding terms used in LCIA and this chapter
Stressor	Environmental intervention
Insult	Midpoint
Stress	Midpoint
Consequence	Endpoint
Value lost	Area of protection
(Not specified)	Societal value
Attribute	Category indicator
Means–ends objective network	Environmental relevance

[a] cf. Keeney 1992; Hertwich and Hammitt 2001.

Viewpoints and Approaches in LCIA

In this chapter, we discuss 3 contrasting viewpoints on LCIA, which are relevant for the definition of impact categories and CIs. These viewpoints have been expressed in the last few years and have not been dealt with in ISO 14042:
1) The role of LCA as a decision support tool versus LCA as example of systems analysis
2) The types of problems that should be addressed in LCA and in LCIA in particular
3) The distinction between bottom–up and top–down approaches in LCIA.

Decision support versus systems analysis

Life-cycle assessment is an analytical tool for decision support. At the same time, it is often seen as an example of systems analysis, a scientific discipline that structures the analysis of complex systems. There can be a friction between these 2 statements. We may argue that systems analysis and decision support set different requirements. Thus, we can perform a systems analysis study that has no relevance for decision support. And, conversely, we can provide decision support regarding a complex system like a product system, without doing a systems analysis study. Is this a real friction or can it be resolved? In our opinion, the latter is the case. If we select relevant input variables that can be influenced by decision-making, particularly including the selection of alternative products to provide a predefined function, and output variables that are regarded as good indicators of environmental damage and benefits, then a systems analysis study can become very relevant as a means of decision support. As long as LCA meets these requirements, we argue that it can be directed at decision support, and at the same time, follow the requirements of the environmental systems analysis framework. Note that LCA as a decision support tool always has to start with a system definition in the goal and scope definition phase.

Taking this as a starting point, it is important to define what is the system, what are the surroundings of the system, and consequently, what is the boundary between the system and its surroundings or its environment. In line with what we said in 'LCIA Terminology' (p 210), the system can be defined at 2 relevant levels: 1) the level of the given product system and 2) the level of the total of product systems, that is, the economy. In the latter case, the environment of the system (i.e., the environment of the economy) coincides with what is called *environment* in current language. It also relates well to the definition of environment in ISO 14001, with the enumeration of elements in the environment mentioned in 'LCIA Terminology' (p 210). In the former case, the environment of the product system includes more elements, particularly unit processes that are part of other product systems. In this chapter, we will use the term *environment* mainly in the sense of *environment of the economy* (see Figure 8-1). If we use the terms otherwise, we will so indicate.

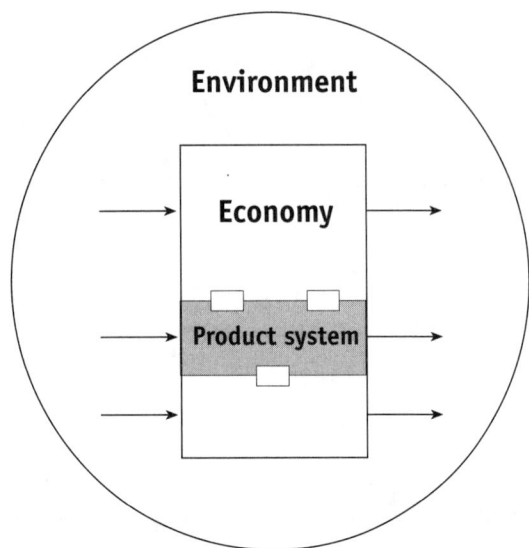

Figure 8-1 Product system and economy: Two levels of system definition. (Small rectangles indicate multiple processes, which are shared by the product system under study and other product systems, i.e., the rest of the economy.)

At the level of the product system, the relationships with other product systems (thus, inside the economy) are the fields of multiple processes and of allocation. The relationships with the environment consist of the environmental interventions as described in 'LCIA Terminology' (p 210). At the higher level — the economy and the environment — other terms are possible. Instead of *economy*, we can use the term *technosphere*, as proposed by Frische et al. (1982) with the same meaning. The term *technosphere* is also used by Hofstetter (1998), but with a slightly different meaning. *Technosphere*, as used by Hofstetter, is opposed to *ecosphere* (the environment) and *valuesphere*, suggesting that only the physical processes themselves constitute the technosphere. In our view, the modelling of processes that together constitute the economy and that are to be analysed in LCA (particularly LCI), may well include principles that surpass the physical level. For instance, the function that is provided by a product is of a nonphysical nature. In addition, allocation based on system extension explicitly includes the analysis of market processes, just as economic allocation involves monetary values. The physical level must be the basis for the analysis of processes, but the analysis itself may go well beyond that.

One more remark must be made regarding the boundary between system (in the sense of the economy) and environment: where to place the humans? Are they part of the economy or part of the environment? Suppose that we had taken human society as the system; then the answer should have been 'inside the system'. However, at the higher level, the system has been defined as the total of product systems (the economy). Although this may include market mechanisms and prices, this system does not include the humans themselves. So, contrary to common sense, according to this definition, the humans and aspects such as human health are regarded as part of the environment, which in fact is in line with the description of the term *environment* given in ISO 14001.

The scope of LCIA

Life-cycle impact assessment can include any type of problem that can be related to the functional unit of a product system. The choice of the problems to be included in a particular LCA study, then, can be made in the goal and scope definition phase. The question, however, is this: What types of problems are to be included in environmental LCA, that is, LCA as the environmental analytical tool of the ISO 14000 series? This question is particularly relevant for the UNEP–SETAC Life-Cycle Initiative. If we aim to identify best practice in a given area, we should know what is and what is not part of this area.

The most straightforward answer is this: problems concerning damage to elements in the environment (like the elements mentioned in the ISO 14001 definition of the environment) and problems related to elements that are damaged by environmental processes. There is a difference between these 2 categories of damage. The latter includes damage to elements within the economy, which is due to an environmental pathway. Examples include acid rain damage to buildings or climate change damage to crops. This is further discussed in 'Exploration of manmade environment in relation to system boundary' (p 218).

The above guideline excludes from LCIA damage to elements within the economy that do not involve environmental processes. In fact, these types of impact are part of the product system itself. A product system therefore not only fulfils a function but also can lead to internal damage within the product system itself without any involvement of processes in the environment. An example is the material damage caused by car accidents. We would like to stress again that LCA also can include the analysis of these types of impacts, but that these are additional to the environmental impacts that are part of the scope of an environmental management tool. Note that, given the fact that the above definition of the economy includes human health as part of the environment, in principle all human health effects related to a product system are to be included.

Midpoint versus endpoint approaches

Life-cycle impact assessment deals with damage to endpoints, either in the environment or in the economy, caused by environmental interventions. This does not necessarily imply that the full chain of processes from interventions up to the endpoints needs to be modelled quantitatively. In this context, 2 approaches can be distinguished: 1) a midpoint approach and 2) an endpoint approach. The midpoint approach starts from the environmental interventions and takes these as input to models that bring us further along the environmental mechanism of accepted impact categories. Given the acceptance of such categories in decision-making, the results, expressed in terms of midpoint variables, can be regarded as relevant for decision-making. On the other hand, the endpoint approach starts at the other end of the environmental mechanisms: the endpoints that directly matter to society and the societal values behind them. The CIs are then chosen at this endpoint level, and models that link the interventions with these endpoint indicators must be constructed.

Historically, the midpoint approaches have set the scene in LCIA, including prominent examples such as the critical volume approach (Basler and Hofmann 1974), the Nordic LCA guide (Lindfors et al. 1995), the Centre of Environmental Science (CML) thematic approach (Heijungs et al. 1992), the Environmental Design of Industrial Products (EDIP) model (Wenzel et al. 1997), and the Sandestin workshop on LCIA (Fava et al. 1993). They also have structured the way of thinking and the examples chosen in ISO 14042.

Since the mid-1990s, the endpoint approach was set on the LCIA agenda. It already had a long history, particularly in the Environmental Priority Strategies (EPS) approach from Steen and Ryding (Steen and Ryding 1992; Steen 1999), but it received strong impetus from Switzerland (Müller-Wenk 1997), from the ExternE project (Spadaro and Rabl 1999), and again from the Netherlands in the Eco-indicator approach (Goedkoop et al. 1998; Goedkoop and Spriensma 1999). At the moment, in Japan, impact assessment models are being developed according to this approach (Itsubo 2000). The endpoint approach starts from the main values in society, connected with AoPs, or *safeguard subjects*, groups of endpoints that are linked to these societal values. From these values and connected endpoints, the modelling goes back to the environmental interventions. In line with ISO 14042, three major AoPs are generally identified in these approaches:
1) human health,
2) natural environment (particularly involving nonhuman life or biodiversity), and
3) natural resources.

The first real discussion of the midpoint and endpoint approaches took place at a workshop in Brighton, UK in 2000, organised under the auspices of the UNEP Division of Technology, Industry and Environment (UNEP-DTIE; Bare et

al. 2000). The conclusions of this workshop can be summarised as follows:
- Both types of approaches have specific value.
- Midpoint approaches give results that are relatively certain (although sometimes still quite uncertain) but less environmentally relevant because they focus on variables that generally are far removed from the endpoints that matter to society.
- Endpoint approaches, on the other hand, give results that are expressed in very relevant terms but are relatively (to extremely) uncertain.
- The development of one encompassing framework would be an important step forward, particularly if the framework includes the most important variables of both types of approaches, thus enabling modelling according to the 2 approaches and allowing comparison of the results.

The SETAC Europe First Working Group on Life-Cycle Impact Assessment (WIA-1) clearly focused its work on midpoint approaches (Udo de Haes 1996). The WIA-2 took clear steps to incorporate both types of approaches in one encompassing framework. In the preparation of the UNEP–SETAC Life-Cycle Initiative, the latter conclusion has recently been incorporated as one of the leading principles for the LCIA programme. The required encompassing framework must include, on the one hand, a broad picture of the environmental interventions concerned and, on the other hand, a clear picture of the relevant types of endpoints and their underlying societal values. This second requirement is the basis for the next sections about AoPs.

Classification of AoPs in LCIA

As described in 'LCIA Terminology' (p 210), AoPs are classes of endpoints. They enable a clear link with the societal values, which are the basis for the protection of the endpoints concerned. Table 8-2 gives an overview of the AoPs with underlying societal values, as presented in the first report of the WIA-2. These societal values can be substantiated by referring to international agreements, such as conventions, declarations, and protocols. Important agreements in this context are the Universal Declaration of Human Rights of 1948; the Additional Protocol of the American Convention on Human Rights in the Area of Economic, Social and Cultural Rights of 1988 from the Organisation of American States; the Rio Declaration of 1992; and the Rio Convention on Biological Diversity, also of 1992 (see Volkwein and Klöpffer 1996).

Table 8-2 Areas of protection and underlying societal values[a]

AoP	Societal values
Human health	Intrinsic value of human life, economic value
Natural environment	Intrinsic value of nature (ecosystems, species), economic value of LSFs
Natural resources	Economic and intrinsic values
Manmade environment	Cultural, economic, and intrinsic values

[a] Udo de Haes et al. 1999. Reprinted by courtesy of *Int J LCA* 4(2):66–74 and 4(3):167–174.

Because AoPs are the basis for determining relevant endpoints, their definition implies value choices. Thus, there is not one correct way to define a set of AoPs. In this chapter, we will discuss 2 different possibilities. The first one starts from a physical (or ecological) classification of endpoints and links these afterwards with societal values. The other starts from the societal values and links these afterwards with endpoints in the environment. Diagrams of both approaches are presented in 'Overview of the AoPs' (p 220).

Classification of AoPs according to physical characteristics

There are different possibilities for classifying endpoints according to physical characteristics. One possible classification is this:
- Atmosphere
- Hydrosphere
- Pedosphere
- Geosphere
- Biosphere
  - Plants
  - Animals (including man).

Such an ecological classification of endpoints is quite straightforward. However, there are problems: The link with the economy is unclear, and it is very difficult to link directly to societal values.

Another possible classification of endpoints according to physical characteristics is this:
- Abiotic environment
- Biotic environment
- Human health
- LSFs.

A further subdivision is possible in this classification. For instance, within the abiotic environment we can distinguish between abiotic resources (such as fossil fuels, minerals, and water and soil [in part]), natural structures (such as glaciers and mountains), manmade structures (such as materials, buildings, and works of art), and possibly land. Likewise, within the biotic environment we can distinguish between biodiversity, biotic resources, and a manmade structure such as crops. In addition, we can distinguish a separate AoP, dealing with the LSFs, that is, the total of regulating mechanisms in the environment that support life on earth. This concerns a combination of processes within the abiotic and the biotic environments. This AoP will be discussed in more detail in the next section.

The advantages of this second classification are that it can be linked rather well with the economy and that the link with the societal values is clearer, particularly at the level of the subdivisions. But still, the latter AoPs are quite heterogeneous regarding their connection with societal values.

Classification of AoPs according to societal values

The 3 AoPs that are included in ISO 14042, that is, human health, natural resources, and natural environment, in fact are classified according to societal values, as we elaborate in this section. Two dichotomies will be taken as guiding principles. First, there is the distinction between human (or manmade) on the one hand and natural on the other. Second, there is the distinction between intrinsic values and functional values. This leads to some changes in the original setup of ISO 14042. Following the report of the WIA-1, we introduce a new AoP, manmade environment (see 'Exploration of manmade environment in relation to system boundary', p 218), which contains, on the one hand, endpoints of intrinsic value and, on the other hand, endpoints of functional value. And we suggest splitting the elements of the AoP natural environment into 'biodiversity' and 'natural landscapes' with intrinsic value and the LSFs (see 'Splitting up the AoP natural environment', p 219). Then one step further is to regard natural resources as a sub-AoP of natural environment. This setup of AoPs according to societal values is presented in Table 8-3.

We will target 2 issues regarding the AoPs in this value-based line. First, there is the manmade environment, which was already proposed in the first report of the WIA-2 (see 'Exploration of manmade environment in relation to system boundary', p 218). Second, we will devote attention to the splitting of the AoP natural environment into 3 sub-AoPs (see 'Splitting up the AoP natural environment', p 219).

Table 8-3 Classification of AoPs according to societal values

Societal values	AoPs Human or manmade	Natural
Intrinsic values	• Human health • Manmade environment (landscapes, monuments, works of art)	• Natural environment (biodiversity and natural landscapes)
Functional values	• Manmade environment (materials, buildings, crops, livestock)	• Natural environment (natural resources) • Natural environment (LSFs)

Exploration of manmade environment in relation to system boundary

In the first report of the WIA-2, Udo de Haes et al. (1999) suggested including manmade environment as a fourth AoP, covering crops, production forests, buildings, and materials. The reason was that, without this AoP, we would not have a basis for including impacts in the analysis of acid rain on silviculture or on monuments, or of tropospheric ozone on agricultural crops. The inclusion of this AoP, logical as it may seem, is not without problems. The main problem is that, for consistency reasons, the inclusion of this AoP may open up the analysis to all kinds of impacts that traditionally did not belong to LCIA and that may divert it from its original aim, that is, a focus on unintended negative impacts on the natural environment. Examples include impacts on commercial elements such as buildings or materials, positive impacts on agriculture from river dams, and economic benefits from reclamation of natural areas. This may lead to a shift in focus from unintended (negative) impacts towards intended (positive) impacts, and thus render LCA a tool for the promotion of economic activities despite their damage to what we formerly called the *environment*.

We will explore 4 different options that may set limits to and avoid such an undesirable expansion of the scope of LCIA. The first concerns the position of the system boundary. In 'The scope of LCIA' (p 214), we argued that environmental LCIA should focus on damage to elements in the environment and on elements in the economy, which are affected through environmental processes. This means that elements within the economy, which are affected by processes within the product system, are not part of the LCIA phase but are part of the LCI phase. An earlier example concerned material damage caused by car accidents.

A second option for limiting the scope in a generic way concerns the distinction between intrinsic and functional values. Thus, we may choose to include monuments but to exclude materials or commercial buildings. However, in the AoP natural environment, we propose regarding elements with both functional and intrinsic values (see Table 8-3). So, limiting the scope of the AoP manmade environment would be rather inconsistent. But it remains advisable to split the AoP manmade environment precisely according to this criterion. Crops, materials, and buildings fall predominantly under the functional values; monuments, works of art, and manmade landscapes fall predominantly under the intrinsic values.

A third option for limiting the scope of LCIA concerns a possible distinction between damage and benefits. We may argue that LCIA should focus on damage and should exclude benefits. If so, the positive economic impacts of a dam would not be included in environmental LCIA. But there are strong arguments against such a limitation. Why only include the acidifying impacts from SO_2 and exclude its mitigation of climate change? Why exclude possibly positive impacts of NO_X emission on agricultural crops? So in fact, we think that the exclusion of positive impacts would lead to biased results.

A fourth option may lie in the distinction between intended and unintended impacts. Intended impacts should be dealt with as part of the function of the system, and consequently should be part of the LCI phase and not of LCIA. If we choose to limit LCIA to unintended impacts, then the positive economic impacts of river dams or of measures aimed at enhancing biodiversity should not be part of LCIA but should be included in the definition of the function of the product system. This line of reasoning seems well founded.

Such reasoning leads to a sharper focus, and helps to avoid a shift away from traditional environmental impacts; the fourth option may be particularly effective in this respect. But no final decision should be taken here because it is not clear whether a workable solution can be found. If the resulting scope of LCIA still is seen as too large, a last resort will be to define positive and negative lists of types of impact that should and should not be part of LCIA, as suggested by Müller-Wenk (personal communication). So we should be able to say 'strictly speaking, damage to commercial buildings should be regarded as part of LCIA, but given the attention which is given to this issue in other contexts, LCIA should not include this type of impact in its scope'. Taken one step further, this focus can be application dependent or even case dependent. In the latter case, the limitation would be part of the goal and scope definition phase, and therefore outside the realm of this discussion.

Splitting up the AoP natural environment

Starting from the distinction between intrinsic and functional values, we propose to differentiate between biodiversity and natural landscapes and LSFs within the AoP natural environment. Now, there are 2 ways to deal with this. One is to define 2 new AoPs instead of one, each with its own indicators. The other is to continue to regard natural environment as one AoP, within which different indicators are defined in relation to the different societal values. Because we see the latter as the most generally applicable structure, and to keep the result simple, we follow this line of reasoning here. But there is indeed little reason to keep natural resources as a separate AoP. Rather, it can be seen as a third subcategory within the AoP natural environment. The links between these 3 subcategories are quite strong, and the boundaries are not sharp.

First sub-AoP: Natural resources

Natural resources are those elements that are extracted physically for human use. They comprise both abiotic resources such as fossil fuels and mineral ores and biotic resources such as wood and fish. These have predominantly a functional value for society.

Second sub-AoP: Biodiversity

The biodiversity part of the AoP natural environment is defined here according to the Biodiversity Convention and includes genetic diversity, species diversity, and ecosystem diversity. True, these classes of endpoints can also have functional values. But if these are predominant in a given case, the given elements can instead be grouped under natural resources. The natural landscapes are in fact ecosystems on a higher level. They have, just as the abovementioned manmade landscapes, predominantly intrinsic values related to recreation and aesthetics.

Third sub-AoP: Life support functions

The LSFs concern the major regulating functions of the natural environment, which enable life on earth (both human and nonhuman). These particularly include regulation of the earth's climate, hydrological cycles, soil fertility, and biogeochemical cycles of substances. Like the natural resources, the LSFs are of functional value to society. From a value perspective, these two are therefore fundamentally different from the AoPs with intrinsic value to society, particularly those connected with human health, biodiversity, and works of art.

The question of whether to include the LSFs as a separate sub-AoP is not only of academic significance. Suppose we want to choose the CIs at endpoint level (i.e., at the level of physical damage) directly in relationship to societal values. If we want to include only AoPs that have intrinsic value to society, then it would suffice to select indicators for human health and for biodiversity, just as is done in the Eco-indicator approach (Goedkoop et al. 1998; Goedkoop and Spriensma 1999). But if we also want to include functional values, then it becomes relevant to include indicators for the natural resources and the LSFs.

A few more remarks are needed about the possible sub-AoP of LSFs. First, it should be clear that it is composed of a number of subclasses, which cannot easily be represented by one indicator, thus giving further shape to a hierarchical setup. To be more precise, this sub-AoP may well cover more impact categories, each with its own CI. Second, it is interesting to compare this sub-AoP with *unknown damage*, as introduced by Hofstetter (1998). Although there is resemblance, there are also differences. Hofstetter's unknown damage is based on a negative definition; it shrinks with increasing knowledge. Here, a positive description is used, based on the natural regulation functions and of comparable significance to the natural resources. Third, it should be recognised that the inclusion of an LSF sub-AoP implies that the elements involved are to be regarded as endpoints. Climate regulation and soil fertility, midpoints in the context of other AoPs and sub-AoPs, are regarded as endpoints here.

Comparison of two types of classification of AoPs

Both ways of defining AoPs — according to physical characteristics or according to value characteristics — have their merits. A classification according to physical characteristics will appeal to ecologists; a classification according to value characteristics will appeal to environmental scientists. In principle, we prefer the latter option, taking into account that LCA is a decision support tool and must consequently choose indicators that are relevant to societal decision-making. From our point of view, choosing relevant indicators can be performed in a more transparent way if we start with a value-based definition of AoPs. Cows and pigs should be treated differently than wildlife, not to mention humans, although from an ecological point of view, such a distinction makes little sense. On the other hand, this also can be coped with in the physical classification approach by defining sub-AoPs, which implies that this option is also possible. The 2 types of AoP classification are presented graphically in the next section.

Overview of the AoPs

In the preceding section, a structuring of AoPs was presented, following 2 different approaches in LCIA: 1) an approach starting from the physical characteristics and 2) an approach starting from societal values. In this section, we will elaborate both in a more visual way. Diagrams, as presented in Figures 8-2 and 8-3, can be regarded as underlying paradigms for LCIA. They represent a realistic model, which serves as guidance for the modelling in LCIA.

Visual representation of the classification of AoPs according to physical characteristics

Figure 8-2 presents a classification of the AoPs according to physical characteristics, resulting in a well-ordered picture. Apart from the AoPs themselves, possible subdivisions are included. An advantage of this way of structuring is that there is a clear link with traditional disciplines. But this setup also has a number of disadvantages and limitations. One disadvantage is that the link with societal values is less clear. To some extent, this point can be resolved at the subdivision level: The protection of the elements under the heading 'Biotic environment' takes place on the basis of different underlying societal values. But there are limitations here. LSFs are included, but this gives some friction with the other AoPs, as it involves both abiotic and biotic processes. This AoP therefore implies the need for a sharp distinction between elements (to be put in the other AoPs) and processes (to be put under the LSF). A comparable problem holds true for biodiversity, which in its definition according to the Convention on Biodiversity also includes ecosystems that consist of both

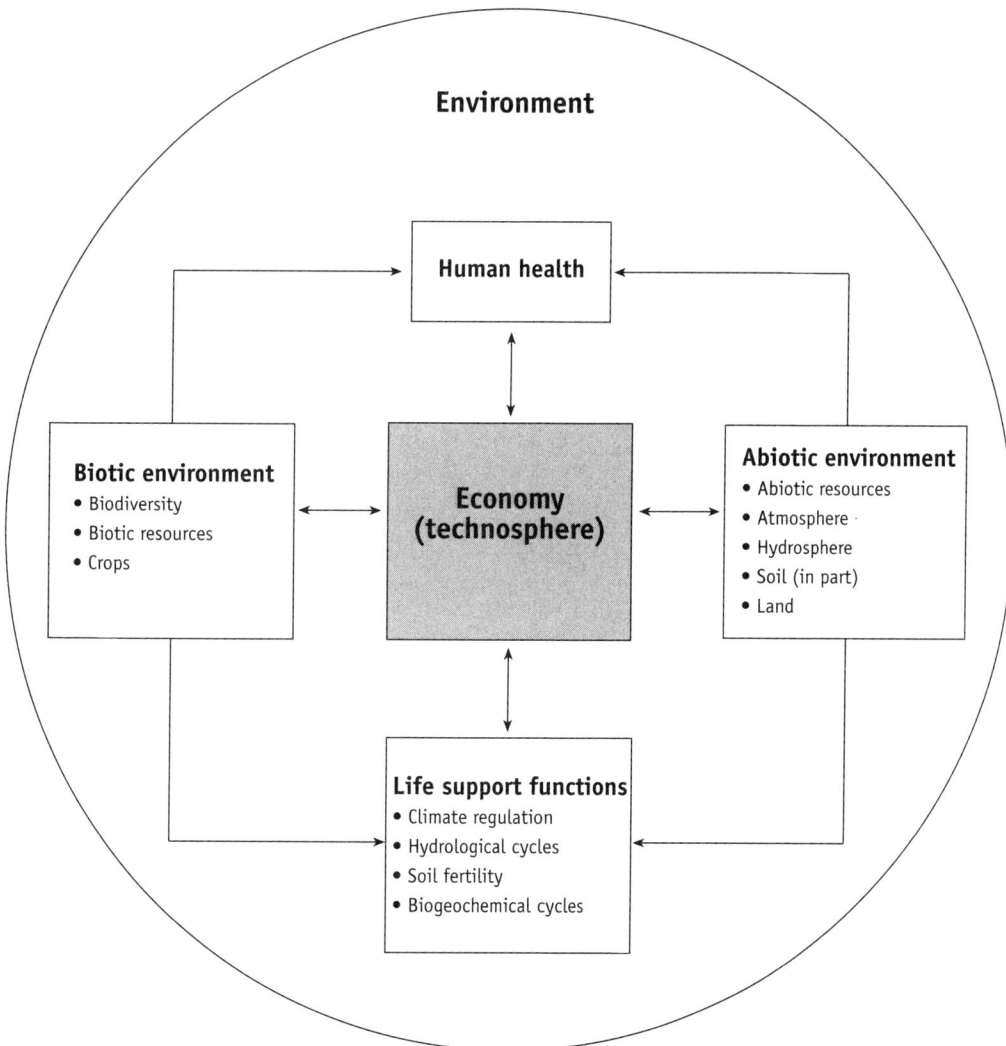

Figure 8-2 Classification of AoPs according to physical characteristics. (Arrows pointing both ways express interactions between the economy and AoPs. Other arrows indicate main relationships between AoPs.)

biotic and abiotic elements; if we accept this definition of biodiversity it would be wrongly placed under 'Biotic environment'. And the same holds true for landscapes, which also include abiotic and biotic elements. Also, there is no clear room for the distinction between natural and manmade; thus, there is no distinct place for the AoP manmade environment.

Visual representation of the classification of AoPs according to societal values

Figure 8-3 presents a classification of AoPs according to societal values. Here, the main divisions concern the distinctions between manmade and natural and between intrinsic values and functional values. Compared to the physical classification, the advantage is that it will be easier to adopt this classification for weighting purposes. The disadvantage is that the distinction between intrinsic and functional values may not be fully clear.

The next step would be to change the original pathway diagram in the first report of the WIA-2 (Udo de Haes et al. 1999), according to the classification presented above. However, doing so is beyond the scope of this chapter.

Conclusions

The discussions in this chapter led to the following conclusions.
1) Not all experts agree on the need to identify best available practice for LCIA at a conceptual level. However, the authors and most participants in the discussion regard best practice at the conceptual level as a prerequisite for the identification of best practice at a more specific level, that is, the level of impact categories, CIs, and characterisation factors.
2) The identification of best available practice in LCIA at a conceptual level first of all pertains to the definition of terminology and to a general technical framework for LCIA. At a more specific level, this framework may well be dependent on the type of application, or even on the goal and scope of the LCA study at hand.
3) Some alternatives to ISO 14042 terminology should be considered.
4) Main topics for the framework are the definitions of *system* and *environment* and the structure of the environmental mechanism of impact categories (the environmental impact chain or impact pathway).
5) System and environment can be defined at 2 relevant levels: the product system with its environment, and the economy (the total of all product systems) with its environment. The term *environment* is further used at this higher level.
6) Life-cycle assessment is regarded as part of systems analysis and, at the same time, as support for decision-making. This is seen by some as a contradiction. However, if relevant input and output variables are selected, LCA, and consequently LCIA, can be an analytical tool in the field of systems analysis, which aims to support societal decision-making. We should investigate whether it is useful to adopt terminology used in the field of decision analysis.
7) The problems analysed in LCIA are environmental problems; this means that the endpoints of impact chains either must lie in the environment or must be affected through environmental processes. Damage within the product system itself must be coped with in the LCI phase.
8) Two main approaches to LCIA have large consequences for further characterisation modelling: midpoint approaches and endpoint approaches. We propose developing an encompassing framework of environmental pathways, which would incorporate both midpoint and endpoint approaches.

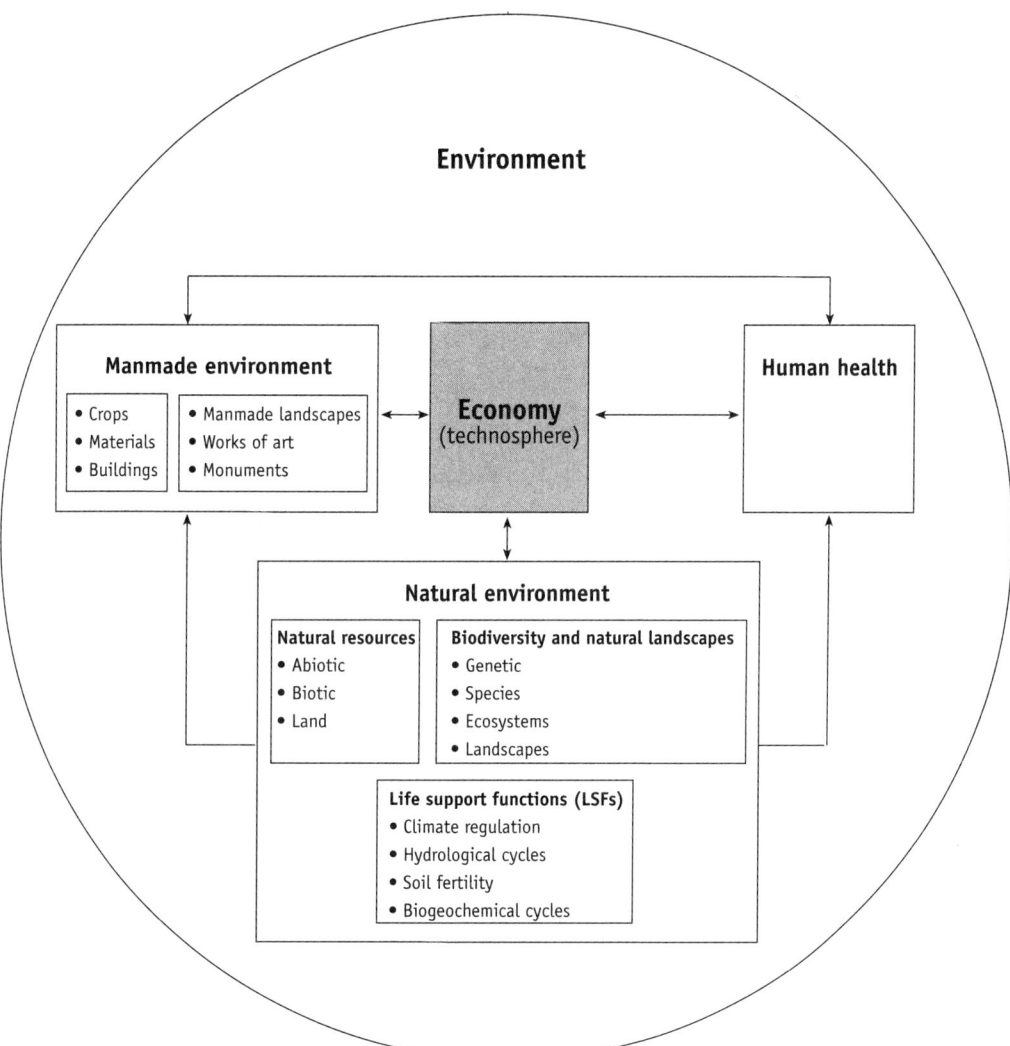

Figure 8-3 Classification of AoPs according to societal values

9) The main topic of this chapter concerns the definition of AoPs, that is, classes of endpoints for LCIA. These AoPs can be classified according to physical (or ecological) characteristics or societal values. Both types of classification have been elaborated. Further discussion is needed before best practice regarding the definition of AoPs can be established.

10) The classification according to physical characteristics offers a good link to various scientific disciplines; however, it does not offer transparent links to societal values. On the other hand, a classification according to societal values can be linked well to decision-making but is less self-evident from an ecological point of view. We prefer a classification according to societal values.

11) In both types of classification, a new AoP or sub-AoP concerns the LSFs. These concern the main regulation functions in the environment, which support life on earth. The definition of these functions as a separate AoP of sub-AoP is relevant, because this leads to the need, in both midpoint and endpoint approaches, to define CIs that reflect impacts on these LSFs.

12) In the classification according to societal values, the manmade environment is defined as a separate AoP. Although such a definition seems quite logical, this AoP brings about a considerable risk that the scope of LCIA will be extended unduly. We have identified and discussed options for coping with such a risk.
13) The chapter concludes with the visual representation of the results of the 2 approaches to AoP classification. If possible, such simple figures should be further developed, with the goal of achieving broad support within the LCIA community and thus acting as a paradigm for further research in this scientific area.
14) The most logical place for this further work is the UNEP–SETAC Life-Cycle Initiative.

References

Bare JC, Hofstetter P, Pennington DW, Udo de Haes HA. 2000. Life cycle impact assessment workshop summary; Midpoints versus endpoints: The sacrifices and benefits. *Int J LCA* 5(6):319–326.

Basler and Hofmann [consultancy]. 1974. Vergleich der Umweltbelastung von Behältern aus PVC, Glas, Blech und Karton. Bern, CH: Eidgenössisches Amt für Umweltschutz.

Fava J, Consoli F, Denison R, Dickson K, Mohin T, Vigon B, editors. 1993. A conceptual framework for life-cycle impact assessment. Pensacola FL, USA: Society of Environmental Toxicology and Chemistry (SETAC).

Frische R, Esser G, Schönborn W, Klöpffer W. 1982. Criteria for assessing the environmental behavior of chemicals: Selection and preliminary quantification. *Ecotoxicol Environ Saf* 6:283–293.

Goedkoop MJ, Hofstetter P, Müller-Wenk R, Spriensma R. 1998. The Eco-indicator '98 explained. *Int J LCA* 6(3):352–360.

Goedkoop MJ, Spriensma RS. 1999. The Eco-indicator 99: A damage oriented approach for life cycle impact assessment. The Hague, NL: Ministry of Housing, Spatial Planning and the Environment (VROM).

Heijungs R, Guinée JB, Huppes G, Lankreijer RM, Udo de Haes HA, Wegener Sleeswijk A, Ansems AMM, Eggels PG, van Duin R, de Goede HP. 1992. Environmental life cycle assessment of products: Backgrounds and guide. Leiden, NL: Centre of Environmental Science (CML).

Hertwich EG, Hammitt JK. 2001. A decision-analytic framework for impact assessment. Part I: LCA and decision analysis. *Int J LCA* 6(1):5–12.

Hofstetter P. 1998. Perspectives in life cycle impact assessment: A structured approach to combine models of the technosphere, ecosphere and valuesphere. Boston MA, USA: Kluwer Academic.

Itsubo N. 2000. Screening life cycle impact assessment with weighting methodology based on simplified damage functions. *Int J LCA* 5(5):273–280.

Keeney RL. 1992. Value-focused thinking: A path to creative decisionmaking. Cambridge MA, USA: Harvard Univ Pr.

Lindfors L-G, Christiansen K, Hoffman L, Virtanen Y, Juntilla V, Hanssen O-J, Rønning A, Ekvall T, Finnveden G. 1995. Nordic guidelines on life cycle assessment. Copenhagen, DK: Nordic Council of Ministers. Nord 1995:20.

Müller-Wenk R. 1997. Safeguard subjects and damage functions as core elements of life-cycle impact assessment. St. Gallen, CH: Universität St. Gallen. IWÖ Diskussionsbeitrag Nr. 42. ISBN 3-906502-42-2.

Spadaro JV, Rabl A. 1999. Estimates of real damage from air pollution: Site dependence and simple impact indices for LCA. *Int J LCA* 4(4):229–243.

Steen B. 1999. A systematic approach to environmental priority strategies in product development (EPS). Version 2000: General system characteristics: A systematic approach to environmental priority strategies in product development (EPS). Version 2000: Models and data of the default method. Göteborg, S: Chalmers Univ of Technology. CPM Reports 4 and 5.

Steen B, Ryding S-O. 1992. The EPS enviro-accounting method. Göteborg, S: Swedish Environmental Research Institute (IVL). Report Nr. B 1080.

Udo de Haes HA, editor. 1996. Towards a methodology for life-cycle impact assessment. Brussels, B: Society of Environmental Toxicology and Chemistry (SETAC) Europe. Report of the SETAC Europe First Working Group on Life-Cycle Impact Assessment.

Udo de Haes HA, Jolliet O, Finnveden G, Hauschild M, Krewitt W, Müller-Wenk R, editors. 1999. Best available practice regarding impact categories and category indicators in life-cycle impact assessment: Background document for the Second Working Group on Life-Cycle Impact Assessment of SETAC Europe (WIA-2). *Int J LCA* 4(2):66–74 and *Int J LCA* 4(3):167–174.

Volkwein S, Klöpffer W. 1996. The valuation step within LCA. *Int J LCA* 1(1):36–39.

Wenzel H, Hauschild M, Alting L. 1997. Environmental assessment of products. Volume 1, Methodology, tools, techniques and case studies in product development. London, GB: Chapman and Hall.

Abbreviations

ADI	acceptable (or allowable) daily intake
AEP	aquatic eutrophication potential
AFR	Avfallsforskningsrådet (Swedish Environmental Protection Agency)
AHP	Analytical Hierarchy Process
AIChE	American Institute of Chemical Engineers
AMI	Assessment of the Median Impact
AoP	area of protection
AOT	accumulated ozone over threshold
AQUIRE	Aquatic Toxicity Information Retrieval database
ASTER	Assessment Tools for the Evaluation of Risk
ASTM	American Society for Testing and Materials
ASTRAP	Advanced Statistical Trajectory Regional Air Pollution
APHEA	Air Pollution and Health: A European Approach
BASINS	Better Assessment Science Integrating Point and Nonpoint Sources
BCD	Britannica Compact Disc
Beta-ED5	slope factor based on ED5
BLP	bromine loading potential
BMBF	Bundesministerium für Forschung und Technologie (German Ministry of Education and Research)
BMD	benchmark dose
BMR	benchmark risk
BOD	biological oxygen demand
BUWAL	Bundesamt für Umwelt, Wald und Landschaft (Germany)
CLRTAP	Convention on Long-Range Transboundary Air Pollution (UNECE)
CML	Centre of Environmental Science
CNN	computational neural network
COD	chemical oxygen demand
CV	contingent valuation

CVM	Contingent Valuation Method
DALY	disability-adjusted life year
DtT	distance to target
DtcT	distance to critical (or sustainable) target
DteT	distance to equally important target
DtpT	distance to political target
DTU	Danmarks Tekniske Hojskole
EAT	ECETOC Aquatic Toxicity
EC5	effective concentration affecting 5% of the test organisms
EC50	effective concentration affecting 50% of the test organisms
ECB	European Chemical Bureau
ECETOC	European Centre for Ecotoxicology and Toxicology of Chemicals
ECOSAR	Ecotoxicity of Industrial Chemicals Based on Structure-Activity Relationships
ED5	effective dose to 5% of the test organisms
$ED5_h$	effective dose to 5% of humans
ED50	effective dose to 50% of test organisms
EDIP	Environmental Design of Industrial Products
EF	effect factor
EI	environmental impact
EIA	environmental impact assessment
EPFL	Ecole Polytechnique Fédérale de Lausanne (Swiss Federal Institute of Technology)
EPPO	European and Mediterranean Plant Protection Organization
EPS	Environmental Priority Strategies
ETH	Eidgenossichen Technischen Hochschule (Switzerland)
EU	European Union
EUSES	European Union System for the Evaluation of Substances
FAO	Food and Agriculture Organization of the United Nations
FAV	final acute value (i.e., the median HC5)
fNPP	free net primary production
GI	gastrointestinal

GIS	geographical information system
GMAV	genus mean acute value
GMV	genus mean value
GREAT-ER	Geography-Referenced Regional Exposure Assessment Tool for European Rivers
GWP	global warming potential
HC	hydrocarbon
HCFC	hydrochlorofluorocarbon
HCH	hexachlorocyclohexane
HC5	hazardous concentration affecting 5% of the species in an ecosystem
HDF	human damage factor
HEI	Health Effects Institute (U.S.)
HESI	Health and Environmental Sciences Institute
HPV	high production volume
HSDB	Hazardous Substances Data Base
HTP	human toxicity potential
ICRP	International Commission on Radiological Protection
ILSI	International Life Sciences Institute
IPCC	Intergovernmental Panel on Climate Change
IR	incremental reactivity
IRIS	Integrated Risk Information System
IRPTC	International Register of Potentially Toxic Chemicals
ISC	Industrial Source Complex model
ISO	International Organization for Standardization
IUCN	International Union for Conservation of Nature
IVL	Swedish Environmental Research Institute
IWÖ	Institut für Wirtschaft und Ökologie (Switzerland)
JECFA	Joint Food and Agriculture Organisation of the United Nations (FAO) and World Health Organization (WHO) Expert Committee on Food Additives
K_{OW}	octanol–water partitioning coefficient

LC50	lethal concentration to 50% of the test organisms
LCA	life-cycle assessment
LCI	life-cycle inventory
LCIA	life-cycle impact assessment
LD50	lethal dose to 50% of the test organisms
LOAEL	lowest observable adverse effect level
LOEC	lowest observable effect concentration
LOEL	lowest observable effect level
LSF	life support function
MADA	Multiple Attribute Decision Analysis
MAUT	Multi-Attribute Utility Theory
MAVT	Multi-Attribute Value Theory
MCDA	Multi-Criteria Decision Analysis
MIPS	material intensity per unit of service
MIR	maximum incremental reactivity
MOO	multiple objective optimization
MS	multiple species
NAPAP	National Acid Precipitation Assessment Program (U.S.)
NEX	normalised extinction of species
NMHC	nonmethane hydrocarbon
NMOG	nonmethane organic gas
NMVOC	nonmethane volatile organic compound
NOAEL	no observable adverse effect level
NOEC	no observable effect concentration
NOEL	no observable effect level
NPP	net primary production
NRC	National Research Council (U.S.)
ODP	ozone depletion potential
OECD	Organisation for Economic Co-operation and Development
OMNIITOX	Operational Models and Information Tools for Industrial Applications of Eco/toxicological Impact Assessments
OPPT	Office of Pollution Prevention and Toxics (USEPA)

PAF	potentially affected fraction of species
PAN	peroxyacetyl nitrate
PBT	persistence, bioaccumulation, and toxicity
PCB	polychlorinated biphenyl
PDF	potentially disappeared fraction of species
PM2.5	fine particles with an aerodynamic diameter < 2.5 μm
PM10	fine particles with an aerodynamic diameter < 10 μm
PNEC	predicted no-effect concentration
PNN	probabilistic neural network
POCP	photochemical ozone creation potential
PVC	polyvinylchloride
QALY	quality-adjusted life year
QSAR	quantitative structure-activity relationship
RAINS	Regional Air Pollution Information Simulation
RfC	inhalation reference concentration
RfD	oral reference dose
RIVM	National Institute of Public Health and the Environment (NL)
RTECS	Registry of Toxic Effects of Chemical Substances
SD	standard deviation
SETAC	Society of Environmental Toxicology and Chemistry
SIK	Swedish Institute for Food and Biotechnology
STG-GARLIC	SETAC Europe Scientific Task Group on Global And Regional Impact Categories
SMAV	species mean acute value
SMART	Simple Multi-Attribute Rating Technique
SMART/MOVE	Simulation Model for Acidification's Regional Trends/Multistress model voor de Vegetatie
SROM	Source–Receptor Ozone Model
SSD	species sensitivity distribution
TD50	toxic dose to 50% of test organisms
TEP	toxicity equivalency potential; terrestrial eutrophication potential (Equation 3-9)

TD5	dose with 5% tumour incidence
TIIM	Threshold Inventory Interpretation Methodology
TNO	Netherlands Organisation for Applied Scientific Research
TRI	Toxic Release Inventory
TRIM	Total Risk Integrated Methodology
TSCA	Toxic Substances Control Act (U.S.)
UNORRA	uncertainty-oriented ranking
UNECE	United Nations Economic Commission for Europe
UNEP	United Nations Environment Programme
UNEP-DTIE	UNEP Division of Technology, Industry and Environment
UNSCEAR	United Nations Scientific Committee on the Effects of Atomic Radiation
USEPA	United States Environmental Protection Agency
USES	Uniform System for the Evaluation of Substances
VOC	volatile organic compound
VROM	Ministry of Housing, Spatial Planning and the Environment (Netherlands)
WCMC	World Conservation Monitoring Centre
WHO	World Health Organization
WIA-1	SETAC Europe First Working Group on Life-Cycle Impact Assessment
WIA-2	SETAC Europe Second Working Group on Life-Cycle Impact Assessment
WMO	World Meteorological Organisation
WMPT	Waste Management Prioritization Tool (USEPA)
WRI	World Resources Institute
WTM	Windrose Trajectory Model
WtP	willingness to pay
WtPDam	WtP, based on damage costs
WtPPrev	WtP, based on prevention costs
WtPTax	WtP, based on environmental taxes
XLCA	explicit LCA

YLL	year of life lost
YOLL	year of life lost

A

Abiotic resources. *See also* Mining; Radiation, solar; Water
depletion, 16–17
environmental mechanisms of impacts, 14
extraction, 20–23, 24–25, 33
flow, 15, 17, 23, 25
fund, 15, 17, 19, 23, 26
future scarcity, 11
impact assessment, 11, 15–26
regional aspects, 34
stock dynamics, 15, 17, 18–20
stock estimates, 21
three main types, 15, 221*f*
vs. biotic resources, 17–18, 61, 221–222
Acceptable daily intake (ADI), 156
Acidification
category indicators, 84
characterisation factors, 68–69, 83, 84–86
environmental mechanism, 82
impact category description, 82
impacts on historic artefacts, 87
recommendations, 87, 96
relevant substances. *See* Acidifying substances
soil deterioration, 70
spatial aspects and site-dependence, 68–69, 83, 84–86
state-of-the-art science, 82–84
temporal aspects, 70, 86–87
Acidification potentials (AP), 84, 86
Acidifying substances, 5, 68–69, 70 n.6, 82
Acrylonitril, 151*f*
ADI. *See* Acceptable daily intake; Allowable daily intake
Advanced Statistical Trajectory Regional Air Pollution model (ASTRAP), 83, 93
Aggregated interventions
abiotic resource extractions, 20–21, 25
biotic resource extractions, 30, 34–35
Aggregation of impact categories. *See* Impact categories, grouping
Aggregation of panellists' data, 188
Agriculture
benefits of acidification, 87
benefits of climate change, 73
crop exposure, 109
crop pollination, 13, 57
draining and filling for cropland, 41
environmental interventions from, 39, 40, 41, 46
fertiliser application, 46
irrigation, 23, 40

as land occupation impact, 41, 42, 43
as land use type, 46–47
weed control, 40
AHP. *See* Analytical Hierarchy Process
Air Pollution and Health (APHEA), 138
Air quality. *See also* Emissions; Ozone depletion; Photooxidant formation
haze, 6
modelling, 111. *See also* EcoSense Model
natural purification, 13, 57
particulates, 39, 138
Albedo effect, 49
Allowable daily intake (ADI), 102, 126, 128, 129, 139*t*
Ammonia
and acidification, 82, 85*t*, 86
and aquatic eutrophication, 91, 93
secondary aerosols from, 114
and terrestrial eutrophication, 88, 89*t*
Analytical Hierarchy Process (AHP), 190, 191, 200
Anchoring effects, 189
Animals. *See also* Biodiversity; Fish
domestic, 13
existence values, 186
migration, 41
recovery time and reproduction cycle, 35–36
threatened by high extraction rates, 32*t*
wild *vs.* man-controlled culture, 27, 36, 38
Anthropogenic interventions. *See* Environmental interventions
AoP. *See* Areas of protection
Approaches to impact assessment, 4–8, 180
Aquaculture, 26, 28
Aquatic eutrophication
category indicators, 91–92
characterisation factors, 91–93
impact category description, 90–91
recommendations, 94, 96
relevant substances, 91. *See also* Nitrogen; Phosphorous
spatial aspects and site-dependence, 92–93
state-of-the-art science, 91
temporal aspects, 93
water quality models, 112
Aquatic eutrophication potential (AEP), 91–92
Aquatic fate models, 112
Aquatic toxicity databases, 158, 176
Aquatic Toxicity Information Retrieval (AQUIRE), 176
Aquifers, 23

Areas of protection (AoP)
    and biodiversity, 219
    classification of, 216–220, 221–222, 223–224, 223f
    definition, 212
    ecotoxicity impacts, 165
    environmental intervention impact, 13
    land use impacts, 48–50
    life-support functions, 220
    of the manmade environment, 13, 218–219
    of the natural environment, 13, 29, 36–39, 219–220
    of natural resources, 12–13, 29, 30, 33–36, 219
    overview, 4, 212, 220–222
    physical characteristics of, 217, 221–222, 221f
    and position of category indicators, 31
    societal values, 216t, 217, 218t, 222, 223f, 224
Assessment factors, 156, 156t, 159t
Assessment of Median Impact (AMI), 159, 161
Assessment Tools for the Evaluation of Risk (ASTER), 176
Atmospheric fate and transport, 83, 86, 92, 93

B

Background levels, 70, 86, 87, 89, 102
Behaviour, 154, 186
Benchmark dose (BMD), 130, 131f, 132, 139t
Benzaldehyde, 79
Bequest value, 186
Best available practice
    characterisation methods, 200
    in LCIA, 8, 94–96, 222
    methods development, 8, 61, 177
    need for standardisation, 8
    uncertainty, 66–67
Better Assessment Science Integrating Point and Nonpoint Sources (BASINS), 112
Bioassay studies, 142, 152, 153–157, 175
Bioavailability, 93, 153, 165, 169
Bioconcentration, 107t, 108
Biodiversity
    areas of protection, 219
    ecotoxicity impacts, 164, 165, 170
    effects of abiotic resource extraction, 23, 26
    effects of biotic resource extraction, 36, 39
    effects of consumer demand, 27
    as endpoint, 51–56
    as indicator for selection of reference state, 44
    indicators of, 57
    land use impacts, 40, 48, 49, 51–56, 57
    as a life support function, 13
    and manmade artefacts, 29–30
    in midpoint/endpoint selection, 22–23
    number of species, 26
    operationalisation characteristics, 54–56
    quantification, 51–56
Biological oxygen demand (BOD), 91, 92, 94
Biological property differentiation, 35
Biomagnification, 108, 153, 165, 169
Biomass appropriateness, 58

Biomass production, 57–58, 91
Biotic resources. *See also* Fish; Hunting; Silviculture
    classification, 221f
    environmental mechanisms of impacts, 14
    extraction, 26–39
    extraction as a benefit, 29–30
    impact assessment, 11, 26–39
    from man-controlled cultures, 11, 27, 28, 29–30, 61
    stock dynamics, 17, 18–20, 33–39
    threatened by high extraction rates, 31, 32t
    topsoil, 57–58
    vs. abiotic resources, 17–18, 61, 221–222
BLP. *See* Bromine loading potential
BMD. *See* Benchmark dose
Body burden, 107t
Bottom-up approach, 4–5, 180
Bromine loading potential (BLP), 75
Buildings and building materials, 13, 18, 23, 27, 30, 87

C

Cadmium, 114, 151f
Calibration of models, 119
CalTOX, 108f, 109, 110t, 114
Cancer, 75, 133, 137
Carbon, 91, 114
Carbon dioxide, 3, 39, 47, 71–72
Carbon monoxide, 77, 78
Carbon:nitrogen:phosphorous (Redfield ratio), 91, 92, 93, 94
Carcinogenic Potency Database, 132
Carcinogens, 128, 131–132, 137
Case-specific normalisation, 182, 183t
Category indicators (CI)
    acidification, 84
    climate change, 73
    definitions, 21, 67–68
    extraction of biotic resources, 30, 32–39
    human toxicity, 123–145
    overview, 2, 5, 17
    ozone depletion, 76
    photooxidant formation, 79–80
    position along stressor-impact chain, 31, 33
    position at midpoint *vs.* endpoint, 6, 125 n.2
    selection, 94
    terrestrial eutrophication, 88
    toxic chemical exposure, 102
    weighting, 60–61
Causality. *See* Stressor-impact chain
CemoS/Water model, 112
Centre of Environmental Science, 125, 215
Characterisation factors
    acidification, 84–86
    aquatic eutrophication, 91–93
    biotic resource extraction, 33, 34–35, 37t, 38
    climate change, 73
    human toxicity, 123–124, 124f
    and model domains, 67
    ozone depletion, 76

Index 237

photooxidant formation, 79–80
selection, 95
site-generic, 68–69, 79–80, 84, 88, 91–92, 95
site-specific, 68–69, 80–81, 84–86, 88–89, 92–93, 95
Characterisation of categories. *See* Impact categories, characterisation
ChemCAN, 109, 110*t*
Chemical contaminants
acidification, 5, 68–69, 70 n.6, 82
aquatic eutrophication, 91, 93
climate change, 71–72
databases, 126, 131*f*, 132, 133, 134, 135, 155, 157, 174, 176
double counting, 17, 46–47, 92
fate and exposure assessment, 101–119
impact assessment, 35
marginal measures, 160–162, 167
mechanism of action, 103, 127, 137, 155
metals, 109, 114, 118 n.5, 151*f*
mixtures, 83, 151, 162–164
ozone depletion, 75
photooxidant formation, 78
screening process, 7, 105, 106, 152
secondary pollutants from, 114, 138, 143
terrestrial eutrophication, 88
Chemical oxygen demand (COD), 91, 92, 94
Chlorine loading potential (CLP), 75
Chlorine oxides, 74
Chlorofluorocarbons (CFC), 74, 75, 76
Climate change
carbon dioxide as co-product, 47
category indicators, 73
characterisation factors, 73
damage modelling, 73
gases on basis of equal weight, 72
impact category description, 71
in LCIA matrix multiplications, 3
recommendations, 74, 95
relevant substances, 71–72
spatial aspects, 73
state-of-the-art science, 72
temporal aspects, 73
Climate regulation, 13, 49, 57
CLP. *See* Chlorine loading potential
CLRTAP modelling framework, 71, 71 n.8, 80, 83, 88
Coal, 17
Combi-PAF, 162–164, 169–170
Competition
for land occupation, 48–49, 50–51
for natural resources, 19, 22
for nitrogen by plants, 87–88
and property rights, 50 n.9
Comprehensiveness, 68, 113, 142–143
Computational neural network (CNN), 176
A Conceptual Framework for Life-Cycle Impact Assessment, 1
Conceptual structure of LCIA, 1, 209–224

Consistency
of criteria for weighting, 200–201
double counting issues, 17, 46–47, 92
in operationalism for biodiversity, 54*t*, 55*t*, 56
in operationalism for life support, 59, 60
Consumption
biodiversity and consumer demand, 27
and biotic resources, 33, 34, 35, 37*t*
and deposit abiotic resources, 20, 21
exposure via food. *See* Ingestion
and midpoint/endpoint selection, 22
net, of fund resources, 23, 26
of oxygen. *See* Biological oxygen demand; Chemical oxygen demand
Contingent valuation (CV), 187, 188, 197, 201
Coordination of Information on the Environment (CORINE), 51
CORINAIR, 111
Cost-benefit analysis, 135, 197
Country-average population dose, 113
Critical load functions
with acidification, 83, 85, 86
with aquatic eutrophication, 93
nutrient nitrogen, 90
Critical volume, 127
Cultural values, 49–50
CV. *See* Contingent valuation

D

Daily intake, 102, 125–126
DALY. *See* Disability-adjusted life years
Damage-oriented approaches. *See also* Top-down approach
climate change, 73
co-product issues, 47
definitions, 5, 31
ecotoxicity impacts, 166
endpoint assessment, 6, 125
in Environmental Priority Strategies, 6
method development, 6
to the natural environment, definition, 31
to natural resources, definition, 31
nitrogen and sulphur, 90
with ozone depletion potential, 77
photooxidant formation, 81
and position of category indicators, 31
selection of approach, 96
Damages, 212. *See also* Stressors
Daphnia, 175
Databases. *See* Chemical contaminants, databases
Data collection and analysis
availability of data, 167*t*, 169, 174, 193, 196
carcinogenic severity, 137
degree of information, 103, 104*f*, 104 n.1
double counting, 17, 46–47, 92
feasibility of models, 116
four types of information, 101
laboratory tests, 152

Data collection and analysis *(cont.)*
noncarcinogenic severity, 137–138
for quality, 3
role in inventory analysis, 2
uncertainty in. *See* Uncertainty
use of standard deviation, 68–69
Decision analysis, 182, 183, 188–191, 213t
Decision-making
LCIA for, 213–214
need for off-the-shelf qualities, 8
relevance of human toxicity indicators to, 127, 143
relevance of model to, 115–116
Denitrification, 84, 90
Depletion of natural resources, 16–17, 18–20
Deposition, 82, 114
Deposit resources, abiotic, 15, 20–23, 24–25
Depth of analysis, 104, 104 n.2, 113, 114
Dermal absorption, 107t, 109, 126
Detergents, 112
1,4-Dichlorobenzene, 127
Dioxin, 114
Dirty surfaces, 125 n.3
Disability-adjusted life years (DALY)
both mortality and morbidity with, 123
depth of analysis with, 105f
as endpoint, 212
overview, 135, 136–137, 140t
recommendations, 144–145
relevance to decision context, 143
Dispersion sensitivity, 22
Dissociation, 107
Distance to critical target (DtcT), 195, 196, 197–198, 199, 200
Distance to equally important target (DteT), 195, 196, 197, 200
Distance to political target (DtpT), 195, 196, 197, 199, 200, 201
Distance to target methods (DtT), 177, 191–192, 195. *See also specific methods*
Distribution, 127, 159
Diurnal variation, 81
Dobris Assessment, 39–40
Dose-fraction, 125–126
Dose-response function
acidifying substances, 70 n.6
carcinogens *vs.* noncarcinogens, 128
for ecosystems, 155
and the effect factor, 102, 106, 107t
extrapolation of curve, 132, 142
linear, 126
marginal effects, 160–162, 167
and NOAEL, 129
no-threshold, 126, 131
overview, 127
pharmacodynamics, 106
potency approach. *See* Potency-based indicators
tangential gradient, 160–162

Double counting, 17, 46–47, 92
Drought mitigation, 13, 57
DtT. *See* Distance to target methods

E
ECETOC databases, 157, 174
Ecoindicator-95, 191
Ecoindicator-99, 22, 47 n.6, 53, 61, 125, 132, 160, 195
Ecoinvent 2000, 46
Ecolabelling, 69, 69 n.4
Ecological Footprint, 21, 185–186
Economic aspects
competition for resources, 19
consumer demand and biodiversity, 27
cost-benefit analysis, 135, 197
depreciation, 45
effects of resource stocks in the system, 15, 18–20
global market, 19–20, 34
of littering, 61
methods, 6
of preservation of species, 61
prevention or abatement costs, 187, 188, 196
and product system, 214f
severity-based indicators, 135
taxes, 187, 196
weighting. *See* Valuation
willingness-to-pay. *See* Willingness-to-pay
Ecopoints, 4
ECOSAR, 175, 176
Ecoscarcity Method, 191
EcoSense Model, 86, 110–112, 114
Ecosphere, 210, 214
Ecosystem aspects. *See also* Biodiversity; Natural environment
aquatic. *See* Aquatic eutrophication
biomass productivity, 88
competition and predation, 153
co-product *vs.* damage impact, 47
critical load functions, 83
damage due to ozone depletion, 77
ecosystem diversity, 51
endpoints, 153, 160–161, 164–166. *See also* Ecotoxicity indicators
and extraction of biotic resources, 17, 18, 29
functions, *vs.* land use functions, 47 n.7
interrelations between impact categories, 70–71
nitrogen assimilation, 84, 90
phytoplankton blooms, 90
in PNEC estimation, 158–159
terrestrial, 158–159. *See also* Terrestrial eutrophication
thermal imagery, 58
Ecotax '98 method, 188
Ecotoxicity, 103
marginal effects, 160–162, 167
of mixtures, 159–160
research needs, 165
Ecotoxicity indicators
data sources, 174–176

endpoints, 160–161, 164–166
  evaluation, 166–169
  marginal measures, 160–162, 167
  mixtures, 162–164
  overview, 149–150, 167t, 168–169
  PNEC, 102, 149, 153–159
  recommendations, 169–170
  toxicity to key species, 153
Ecotoxicity of Industrial Chemicals Based on Structure-Activity Relationships (ECOSAR), 175, 176
EDIP. *See* Environmental Design of Industrial Products
Effect Category Method, 191
Effect factor, 101–102, 106, 107t, 149
Effective concentration (EC), 149, 156t, 165, 170
Effective dose (ED), 130, 131f, 132, 135
Effect score, 101–102, 149–150. *See also* Category indicators
Eidgenössischen Technischen Hochschule (ETH), 46
ELECTRE, 190, 191
Elementary flows, 211
Elicitation, 188, 189–190, 191
EMEP models, 83, 93
Emissions
  air quality modelling. *See* EcoSense Model
  background exposure, 70
  European inventory, 83, 111f, 112
  fate and exposure, causality chaining, 102f, 110
  and land use types, 46
  nitrogen, 89
  photovoltaic cell *vs.* fossil fuel, 1
  position of category indicators, 33, 38
  smog. *See* Photooxidant formation
  spatial aspects, 68–69, 78
  traffic, 78
Endangered species, 29, 31, 32t, 34, 38, 56
Endpoints
  biodiversity, for land use impacts, 51–56
  definition, 211–212
  ecosystem, 153, 160–161, 164–166
  fatal. *See* Lethal dose; Years of life lost
  indicators, 5–6
  life-support functions, 56–58, 59–60
  noncancer, 134–135
  ozone depletion, 6, 7
  and position of category indicators, 31, 33
  selection, 22–23
  severity-based, 4, 125. *See also* Disability-adjusted life years; Quality-adjusted life years; Years of life lost
  through aggregated interventions, 17
  vs. midpoints, 6, 125 n.2, 215–216, 222
Energy
  and biotic resource extraction, 36
  indicators and abiotic resources, 20, 21, 22
  usable. *See* Exergy
Energy technologies
  fossil fuel, 4, 12, 17, 19
  solar power, 2, 4, 25
Enhanced radiative forcing, 71, 73

Entandrophragma trees, 38
Entropy, 21, 22
Environment, definitions, 210, 213
Environmental degradation, 11, 17, 36
Environmental Design of Industrial Products (EDIP), 127, 215
Environmental exchanges, 211
Environmental fate and transport
  aquatic, 91, 92–93, 112
  atmospheric, 83, 86, 92, 93, 102f
  causality chain, 102f
  interface with effects, 106, 107t
Environmental interventions, 13, 31, 211. *See also* Extraction; Inventory analysis; Land occupation; Land use
Environmental mechanisms, 13–14, 17, 210–211
Environmental pathway, definition, 211
Environmental Priority Strategies (EPS), 6, 61, 125, 188, 215
Environmental quality, 40–41, 51, 56, 61
Environmental relevance, 4–5, 115–116, 180, 212
Environmental themes, 5, 6, 125 n.2
Epidemiological studies, 133, 137, 138
EPS. *See* Environmental Priority Strategies
EQC, 110t, 114
Equilibrium partitioning method, 159
Equivalency factors, 82, 127. *See also* Toxic equivalency potentials
Erosion, 41, 43, 58
ETH database, 55
Ethene, 77
Ethics, 180, 199
Ethylene, 79
European Chemical Bureau, 128t
Eutrophication. *See* Aquatic eutrophication; Nutrification; Terrestrial eutrophication
Eutrophication potential, 86
Exchange format, 211
Excretion, 127
Exergy, 21, 22, 22 n.1, 25, 36, 58
Existence values, 186
Exotic species, 29–30
Expected Value Method, 190
Explicit LCA (XLCA), 188
Exposure assessment, of toxic chemicals, 101–119
Exposure pathways, 102, 107t, 108–109, 114
External normalisation, 181–182, 183t
ExternE project, 110, 112, 125, 136, 188, 215
Extinction
  calculation of risk, 36, 38
  IUCN classification, 28–29, 31, 38
  and land use, 11
  from neglect not extraction, 27
  normalised extinction of species, 61
  potentially disappeared fraction of species, 38, 149, 160
  timeframe, 51
Extraction
  abiotic resources, 15–26, 33
  biotic resources, 26–39

Extraction *(cont.)*
    definition, 18
    impacts on areas of protection, 13
    vs. the extraction process, 13, 18
Extrapolation
    of the dose-response curve, 132, 142
    interspecific, 129
    intraspecific, 129
    overview, 129–130
    temporal, 129
    of toxicity tests, 154–157, 158
Extreme Value Method, 190

F

Fatal effects. *See* Lethal dose; Years of life lost
Fate factor, 102, 107*t*, 149
Feasibility, 116, 143–144
Feedback, 108
Fertiliser, 46
Field tests, measuring PNEC, 153, 159
Final acute value (FAV), 154
Fish
    bioconcentration, 108
    effects of phytoplankton blooms, 91
    extraction as benefit, 29
    man-controlled culture, 26, 28
    recovery of stocks, 19, 35–36
    regional issues, 20, 34
    threatened by high extraction rates, 32*t*
    toxicity tests, 175
    wild *vs.* aquaculture, 26
Fisheries industry, 13, 30, 31, 34
Floods and floodplain, 13, 23, 48, 49, 57
Flow resources, abiotic, 15, 17, 23, 25
Fluorine, 76
fNPP. *See* Free net primary production
Food and Agriculture Organization (FAO), 31, 34, 126
Food products, 30, 32*t*, 109
Forest management. *See* Silviculture
Fossil fuel, 4, 12, 17, 19
Free net primary production (fNPP), 58, 59
Functional unit
    attribution of transformation impacts, 45–46, 61
    biotic, with land occupation, 40
    in biotic resource extraction, 34, 37*t*
    overview, 1–2
    in site-specific risk assessment, 106, 150
Fund resources, abiotic, 15, 17, 19, 23, 26
Future scenarios
    availability of resources, 11, 15, 18–19, 22–23, 30
    deposit abiotic resources, 20, 21
    human needs, 17, 20

G

Generational aspects, 36, 50–51, 186
Genetic aspects, 30, 51, 137
Genus mean value (GMV), 154, 158

Geographical information system (GIS), 112
Geography-Referenced Regional Exposure Assessment Tool for European Rivers (GREAT-ER), 112, 114
GIS. *See* Geographical information system
Global aspects of natural resources, 19–20, 34
Global biodiversity indicators, 56
Global Ozone Research and Monitoring Project, 75
Global warming. *See* Climate change
Global warming potentials (GWP)
    midpoints and endpoints, 5–6
    normalisation in calculations, 72, 105 n.3
    overview, 3
    range, 72
    temporal aspects, 69, 73
GMV. *See* Genus mean value
Goals. *See* Models, comprehensiveness; Scope
GREAT-ER. *See* Geography-Referenced Regional Exposure Assessment Tool for European Rivers
Greenhouse effect. *See* Climate change
"Green taxes," 187
Ground water, 19, 23
Grouping, 2, 3, 178, 184–185
Growth dilution, 108
GWP. *See* Global warming potentials

H

Habitat degradation, 11, 36, 39. *See also* Land occupation; Land use
Half-life, 107
Hatcheries, 28
Hazard, definition, 124
Hazard index, 86, 89
Hazardous concentration (HCx), 149, 153, 154, 155, 157, 160, 161, 168
Health and Environmental Sciences Institute (HESI), 133
Health Effects Institute (HEI), 138
Hedonic pricing method, 186, 188
Henry's Law constant, 109
HESI. *See* Health and Environmental Sciences Institute
Heterogeneous mechanism of action, 103
Heuristic methods, 83
Hexachlorocyclohexane (HCH), 114
High production value substances (HPV), 128*t*, 134
Homogeneous mechanism of action, 74, 103
HPV. *See* High production value substances
Human equivalent dose, 131*f*
Human health
    areas of protection, 13
    cancer, 75, 133
    and ozone increase, 82
    quantification of impact. *See* Disability-adjusted life years
    and radiation, 75, 132–133
    wide variation in damage to, 126
Human toxicity
    additive, synergistic, or antagonistic effects, 131, 157, 164*f*
    impact chain, 104, 105*f*

indicators, 123–145
in LCIA matrix multiplications, 3
ozone, 138
particulates, 138
potency-based indicators, 125, 126, 127–133, 139t
severity-based indicators, 133–138, 141t. *See also*
Disability-adjusted life years; Quality-adjusted life
years; Years of life lost
types, 127, 134, 134t
Human toxicity potential (HTP), 104, 105f, 114
Hunting, 12
Hydrocarbons
nonmethane, 79
OH-reactive, 77, 78–79
Hydrochlorofluorocarbons (HCFC), 76
Hydrofluoroalkanes, 72
Hydrogen chloride, 68, 82, 85t
Hydrogen fluoride, 85t
Hydrogen ion, excess. *See* Acidification
Hydrogen release potentials, 83, 84
Hydrogen sulphide, 85t

I

ILSI. *See* International Life Sciences Institute
Impact assessment
beginning with. *See* Top-down approach
characteristics of methods, 14–15
extraction of abiotic resources, 15–26
extraction of biotic resources, 11, 26–39
land use, 11, 39–61
overview, 2
stressor-impact network, 13–14
two-fold, 29
uncertainty, 94
Impact categories
acidification, 82
aquatic eutrophication, 90–91
characterisation of, 2, 3, 6, 8, 30, 33, 200
extractions, 25, 30
grouping, 2, 3, 6, 17, 184–185
interrelations between, 70–71
number of subcategories. *See* Aggregated interventions
photooxidant formation, 77–78
selection, 180
spatial differentiation, 7, 19–20
terrestrial eutrophication, 87–88
weighting, 60–61
Impact factor. *See* Ecopoints
Impacts. *See* Damages; Impact categories; Stressors
Incineration, 68, 82
Incremental reactivities (IR), 79
Index of Environmental Friendliness, 189
Individual dose fraction, 107
Industrial Source Complex Model (ISC), 111, 111f
Information. *See* Data collection and analysis
Ingestion, 107t, 108, 114, 125–126, 150
Inhalation exposure, 107t, 108, 125, 138
Inhalation reference concentration (RfC), 14, 139t, 174

Input-related criteria, 193, 194t, 196
Integrated Risk Information System (IRIS), 131f, 132, 133, 134, 135
Intergovernmental Panel on Climate Change (IPCC), 72
Internal normalisation. *See* Normalisation, case-specific
International Commission on Radiological Protection (ICRP), 133, 137
International Life Sciences Institute (ILSI), 133, 140t
International Organization for Standardization (ISO), 141
methods development, 178
14001 series, 210
14040 series, 1, 2–3
14041 series, 73
14042 series, 4, 5, 16, 66, 72, 183, 210, 212
use of grouping, 184
International Union for Conservation of Nature (IUCN), 28–29, 31, 34, 38, 56
Internet. *See* Websites
Interpretation of results, 4, 178
Inventory analysis
abiotic resources, 11, 15–26
biotic resources, 11, 26–39
driven by impact analysis. *See* Top-down approach
normalisation, 2, 3, 34
overview, 2, 3
the process of resource extraction, 13
use of land use types, 46
weighting, 3
Inventory-driven approach. *See* Bottom-up approach
Iodine, 76
IPCC. *See* Intergovernmental Panel on Climate Change
IR. *See* Incremental reactivities
ISC. *See* Industrial Source Complex Model
ISO. *See* International Organization for Standardization
IVAM ER database, 51

L

Laboratory test data, 152, 153–157, 165, 175
Lakes, 23, 48, 84, 90–91, 92
Land
eutrophication on. *See* Terrestrial eutrophication
solid *vs.* intracontinental water bodies, 48
total available, 50–51
Land-cover types, 51
Land occupation
chains of, 45
competition for space, 48–49, 50–51, 61
impact, 40, 41, 43–44, 49–50
process, 40–41
Landscapes. *See* Natural landscapes
Land transformation
with a chain of occupations, 45
and competition for land, 48–49
cover. *See* Land-cover types
functional units, 45–46, 61
impacts, 40, 41–43, 45–50, 61
process, 40, 41

Land use
    allocation, 12, 18, 39
    category indicator position, 33
    chain of different uses, 45–46
    co-product issues, 47
    definitions and distinctions, 11, 40–48
    draining and filling, 41
    effects on biodiversity, 11
    elementary activities, 46–47
    environmental mechanisms, 14
    on floodplains, 23, 48
    functions, *vs.* ecosystem functions, 47 n.7
    manmade environment as extraction, 18
    and nitrogen transformations and mobility, 87
    occupation *vs.* transformation, 11, 40–43
    operationalisation for AoPs, 50–60
    relation of, to impacts on AoPs, 48–50
    renaturalisation, 12, 41, 42, 43, 44
    starting point, 39–40
    types, 46–47, 56
    weighting within impact categories, 60–61
LCA. *See* Life cycle assessment
LCIA. *See* Life cycle impact assessment
Leaching of nutrients, 82, 86, 91, 93
Lethal concentration (LC50), 139t, 154, 156t
Lethal dose (LD50), 130, 132, 139t, 157
Life cycle assessment (LCA)
    the four phases of, 1–4
    role of LCIA in, 3–4
    scope, 1–2
Life cycle impact assessment (LCIA)
    best available practice, 8, 94–96, 222
    conceptual structure, 209–224
    as decision-making tool, 8
    emissions. *See* Acidification; Climate change; Ozone depletion; Photooxidant formation
    fate and exposure assessment, 101–119
    four types of information, 101
    human toxicity in, 123–145
    impacts on ecosystems. *See* Ecotoxicity; Ecotoxicity indicators
    impacts on humans. *See* Human toxicity
    initial. *See* Screening-level assessments
    interpretation of results, 4, 178
    nitrogen and phosphorous. *See* Aquatic eutrophication; Terrestrial eutrophication
    overview, 4–8, 179–181, 210–216
    recommendations, 94–95
    scope, 7, 189, 214–215
    terminology, 210–213, 222
    vs. risk assessment, 151–152
Life-Cycle Impact Assessment: The State-of-the-Art, 1
Life expectancy, valuation, 136
Life support functions (LSF)
    areas of protection, 220, 223f
    effects of abiotic resource extraction, 23, 26
    effects of biotic resource extraction, 36, 39
    as endpoint, 11, 56–58, 59–60
    examples, 13, 57
    land use impacts, 48, 49, 50
    maintenance at regional level, 20
    in midpoint/endpoint selection, 22–23
    operationalisation characteristics, 59–60
Light, seasonal aspects, 93
Linear assumption, 126
Litter, 61
LOAEL. *See* Lowest observable adverse effect level
Local aspects, 49, 54t, 142
LOEL. *See* Lowest observable effect level
Logging. *See* Silviculture
Looping, 108
Lowest observable adverse effect level (LOAEL), 129
Lowest observable effect level (LOEL), 124
Low-level human support, 27–28
LSF. *See* Life support functions

M

MADA. *See* Multiple attribute decision analysis
Man-controlled production
    role in biotic resource impact assessment, 11, 27, 28, 29–30
    vs. biotic resource extraction, 27–28, 61
Manmade environment
    areas of protection, 13, 218–219, 223f
    degradation of cultural values and landscapes, 49–50
    ecotoxicity impacts of the, 165
    impacts on the, 58, 60, 87
    re-allocation, 18
Marginal measures, 160–162, 167, 182
Marine ecosystems, nitrogen deposition, 93
Market price, 186, 188
Mass
    aggregation of, into GWP, 73
    and biotic resources, 33, 36, 37t
    and deposit abiotic resources, 20, 21
    gas calculation on basis of equal weight, 72
    hydrogen ions per unit, 82
    loading, toxic chemicals, 102, 149
Mass balance, 108, 110t
Material intensity per unit service (MIPS), 185
MAUT/MAVT. *See* Multi-Attribute Utility/Multi-Attribute Value Theory
Maximum incremental reactivity (MIR), 79, 80, 81
MCDA. *See* Multiple criteria decision analysis
Means-end objective network, 213t
Mechanism of action, of toxins, 103, 127, 137, 155
Mechanisms, environmental, 13–14, 17, 210–211
Medium-specific models, 103, 108, 109–112, 118
Metabolism, 127
Metals, 109, 114, 118 n.5, 151f
Methane, 72, 73, 80 n.10
Method development, 6
Met-Point Method, 191
Midpoints

climate change, 72
definition, 211
overview, 5–6
ozone depletion, 6, 7, 75–76
and position of category indicators, 31, 33
selection, 22–23
vs. endpoints, 6, 125 n.2, 215–216, 222

Mining
extraction process, 18
functional units, 45
future scenarios, 21–22
impact assessment, 11, 12, 13
land use processes, 42
as land use type, 46–47
renaturalisation after, 42

MIPS. *See* Material intensity per unit service

Models
acidification, 70, 71, 71 n.7–8, 82–84, 86
air quality, 111. *See also* EcoSense Model
aquatic eutrophication, 91
aquatic fate, 112
atmospheric fate and transport, 83, 86, 111–112
climate change, 72
comprehensiveness of, 68, 113
damage, 6
domains, 67
emerging uses for, 113*t*
exposure pathway, 108–109
feasibility of, 116
medium-specific approaches, 103, 108, 109–112, 118
micro *vs.* macro, 180
modular approach for integration, 103, 108, 108*f,* 110*t,* 114
multimedia, 103, 106–109, 110*t,* 112–113, 113*t,* 114
operationalisation of biotic resources, 37*t*
operationalisation of deposit abiotic resources, 24–25
ozone concentration, 111*f,* 112
ozone depletion, 70, 71, 71 n.7–8, 75–76
ozone formation, 77, 80–81
photooxidant formation, 80–81
recommendations, 116–119
relevance to decision-making, 115–116
scientific validity and reliability, 114–115
selection, 94, 119
single-media, 113, 113*t*
spatially explicit, 101
terrestrial eutrophication, 88
transparency and reproducibility of, 115
uncertainty in, 66, 94, 115, 117, 118–119, 150
water quality, 112

Monetisation weighting, 186–188, 195–196, 203
Monte Carlo approach, 115
Montreal Protocol, 75
MOO. *See* Multiple objective optimisation
Morbidity, 125, 138
Mortality. *See* Toxicity tests; Years of life lost
Multi-Attribute Utility/Multi-Attribute Value Theory (MAUT/MAVT), 190, 191, 197
Multimedia models. *See* Models, multimedia
Multiple attribute decision analysis (MADA), 179, 184, 189–191
Multiple criteria decision analysis (MCDA), 179
Multiple objective optimisation (MOO), 179
Must criteria/candidates, 192

N

Natural environment
areas of protection, 13, 29, 36–39, 219–220, 223*f*
biodiversity as endpoint, 51–56
category indicators, 31, 36–39
damage to the, definition, 31
life-support function as endpoint, 56–58, 59–60

Natural landscapes
degradation by manmade environment, 49–50, 58, 60
human impact on, 39
as a life support function, 13
recreational function, 50 n.9

Natural resources
areas of protection, 12–13, 30, 219
category indicators, 31, 33–36
competition and depletion, 18–19
damage to, definition, 31
future scenarios, 20
regional and global approaches, 19–20, 34
role of manmade artefacts, 29–30

Nature, definition, 210
Net primary production (NPP), 58, 59, 60
NEX. *See* Normalised extinction of species
Nice-to-have criteria, 192
Nitric acid (HNO_3), 85*t,* 89*t*
Nitric oxide (NO), 78, 89*t*

Nitrogen
accumulation, 89
assimilation by ecosystems, 84, 86
benefits, 87
bioavailability, 93
emissions, 89
as limiting factor, 87, 91
ratio in phytoplankton, 91

Nitrogen dioxide (NO_2), 89*t,* 114, 138
Nitrogen oxide equivalents, 89*t*

Nitrogen oxides (NOx)
and acidification, 82, 85*t*
and aquatic eutrophication, 91, 93
climate change from, 71, 73
in formation of ground-level ozone, 77
from jets, 74, 77
ozone depletion from, 75, 78
in photooxidant formation, 79, 81
sources, 74, 77, 78
and terrestrial eutrophication, 88, 89*t*

Nitrogen:phosphorous ratio (N:P), 90
Nitrous oxide (N_2O), 72, 77, 85*t*
NOAEL. *See* No observable adverse effect level

NOEL. *See* No observable effect level
Noise, 6, 125 n.3
Noncarcinogens
  availability of data, 137–138
  vs. carcinogens, 128, 129–131, 131*f*
Nonmethane hydrocarbons (NMHC), 79
No observable adverse effect concentration (NOAEC), 154
No observable adverse effect level (NOAEL), 129, 130, 133, 135
No observable effect concentration (NOEC), 155, 156*t*, 157–158, 164–165, 168, 170
No observable effect level (NOEL), 123, 130
Normalisation
  case-specific, 177, 182, 183*t*
  definition, 2, 178, 184
  external, 181–182, 183*t*
  external *vs.* case-specific, 183–184
  and weighting, 177, 178
  when is it needed, 184
Normalised extinction of species (NEX), 61
No-threshold assumption, 126, 131
NPP. *See* Net primary production
Nuisance impacts, 6, 125, 125 n.3
Nutrients
  bioavailability, 93
  excess, impacts on plants. *See* Aquatic eutrophication; Terrestrial eutrophication
  flow from weathering processes, 23
  leached from soil, due to acidification, 82, 86
  as limiting factor for plant ecosystems, 87, 91
Nutrification, 46–47, 47 n.6, 87

O

OASIS, 176
Occupation. *See* Land occupation
Octanol-water partitioning coefficient (K_{OW}), 108, 159, 174–175
Operationalisation
  characteristics for life support, 59–60
  characteristics of biodiversity, 54–56
  in land use impact assessment, 50–60
  options for abiotic resources, 20–23, 24–25
  options for biotic resources, 33–39
Option use value, 186
Oral reference dose (RfD), 126, 129, 130, 133, 139*t*, 141, 174
Organisation for Economic Co-operation and Development (OECD), 134, 156
Output-related criteria, 194*t*, 195, 200–201
Outranking methods, 190
Oxygen consumption. *See* Biological oxygen demand; Chemical oxygen demand
Ozone
  accumulated, over threshold, 80
  formation, 70, 77, 78, 79, 96
  "hole" in the, 74, 76
  human toxicity, 138
  Source-Receptor Ozone Model, 111*f*, 112

Ozone depletion
  category indicators, 76
  characterisation factors, 76
  impact category description, 74–75
  in LCIA matrix multiplications, 3
  midpoints and endpoints, 6, 7, 75–76
  recommendations, 77, 95
  relevant substances, 75
  spatial aspects, 76
  state-of-the-art science, 75–76
  temporal aspects, 76
Ozone depletion potentials (ODP), 69, 75–76
Ozone hole, 74, 76

P

PAF. *See* Potentially affected fraction of species
PAN. *See* Peroxyacetyl nitrate
Panel weighting methods, 182, 183, 188–191
Particulates, 39, 138
Partitioning, 108, 109, 112, 116, 159, 174–175
PDF. *See* Potentially disappeared fraction of species
Peat, 15, 17
Peer review, for models, 114, 116
Perhalogenated compounds, 76
Peroxyacetyl nitrate (PAN), 77
Persistence, 112
Pesticides, 152
Pharmacodynamics, 106, 155, 175
Phosphoric acid, 85*t*
Phosphorous, 88, 90, 91, 92, 93
Photochemical ozone creation potential (POCP), 79, 80, 81, 124
Photolysis, 75, 76
Photooxidant formation
  category indicators, 79–80
  characterisation factors, 79–81
  impact category description, 77–78
  lack of integration over time, 70
  in multimedia models, 114
  recommendations, 81–82
  relevant substances, 78
  spatial aspects and site-dependence, 80–81
  state-of-the-art science, 78–79
  temporal aspects, 81
Photooxidant levels, 4, 77, 78
Photovoltaic solar cell, 4
Phytoplankton
  C:N:P ratio, 91, 92, 93, 94
  effects of blooms, 90
  N:P ratio, 90
  seasonal growth patterns, 93
Plants. *See also* Biodiversity
  algae, 90, 91, 92, 93, 94
  change in vegetation, 41
  ecological role, 57
  Entandrophragma trees, 38
  existence values, 186
  impacts of excess nutrients. *See* Aquatic eutrophication;

Terrestrial eutrophication
  man-controlled culture, 28, 30. *See also* Silviculture
  medicinal, 26, 30, 32*t*, 36
  modelling acidification-nutrification, 71 n.7
  N:P ratios, 90
  ornamental, 30, 32*t*
  primary production, 58
  threatened by high extraction rates, 32*t*
  vascular, diversity, 52–53
PNEC. *See* Predicted no-effect concentration
Political aspects, 19, 183, 187. *See also* Distance to political target
Pollination of crops, 13, 57
Pollution. *See* Chemical contaminants; Emissions
Polychlorinated biphenyls (PCB), 114
Polyvinylchloride (PVC), 68, 82
Population declines, 11, 26, 27, 34
Potency-based indicators, 125, 126, 127–133
Potentially affected fraction of species (PAF)
  as basis for PNEC, 153
  combi-, 162–164, 169–170
  definition, 38, 169–170
  in determination of hazardous concentration, 154*f*
  ecotoxicological impacts, 149
  and land use impacts, 61
  and tangential gradient, 160, 160*f*, 161–162
Potentially disappeared fraction of species (PDF), 38, 149, 160
Power plants, 111
Precautionary principle, 75, 76, 80
Precipitation, 15, 17, 82, 107
Predicted no-effect concentration (PNEC), 102, 149, 153–159
Primary production, 58
Probabilistic neural network (PNN), 176
Procedure-related criteria, 193, 194*t*, 195, 196–200
PROMETHEE, 190, 191
Proxy weighting methods, 185

Q

QALY. *See* Quality-adjusted life years
QSAR. *See* Quantitative structure-activity relationships
Quality-adjusted life years (QALY)
  decision context, 143
  as endpoint, 212
  historical background, 135
  for morbidity impacts, 136, 143
  overview, 123, 140*t*
  and weighting scheme, 144–145
Quantitative structure-activity relationships (QSAR), 156, 174–176

R

Radiation
  solar
    enhanced radiative forcing, 71, 73
    lack of competition for resource, 19, 25
    as nondepletable resource, 15, 17
    in ozone formation, 77
    power from, 2, 4, 25
    toxicity indicators, 132–133
    ultraviolet, 75, 76–77
Radiative forcing, 71, 73
Radionucleids, 107
Random Value Method, 190
Ranking of impacts, 2, 104–105, 184–185, 190–191
Rattans, 32*t*
Recommendations
  acidification, 87, 96
  aquatic eutrophication, 94, 96
  climate change, 74, 95
  ecotoxicology indicators, 169–170
  emissions LCA, 94–95
  human toxicity indicators, 144–145
  models, 116–119
  ozone depletion, 77, 95
  ozone formation, 78, 96
  photooxidant formation, 81–82
  standards, 2–3, 16
  terrestrial eutrophication, 90, 96
  weighting methods, 202–203
Recovery processes, 12, 35–36, 41, 42, 43, 44
Recreational function, 50 n.9
Recycled materials, 2, 12
Redfield ratio, 91, 92, 93, 94
Red List database, 31, 34, 56
Reference concentration. *See* Inhalation reference concentration
Reference dose. *See* Oral reference dose
Reference state, for occupation impacts, 43–44
Reference substance, for human toxicity, 127
Reference system, for normalisation, 181–183
Regional Air Pollution Information Simulation (RAINS), 71, 71 n.8, 80, 86, 89
Regional aspects
  of abiotic resources, 19–20, 68
  acidification, 80, 83–86
  assessment of land competition, 51
  atmospheric fate and transport, 83, 86
  of biotic resources, 34
  country-average population dose, 113
  land use impacts, 39, 49
  local. *See* Local aspects
  mean values for site-generic impact assessments, 69
  photooxidant formation, 80
  of stressors, 15, 17
Regulatory aspects, 114, 139*t*
Relevance-driven approach. *See* Top-down approach
Renaturalisation, 12, 41, 42, 43, 44
Replenishment, 23, 26, 28, 35, 36
Reproducibility, 115, 141–142, 167*t*, 168, 201
Risk
  acceptable, 131
  accumulated relative, 86, 88, 89

Risk *(cont.)*
    of cancer, 131–132, 133
    unit, 132
Risk assessment
    benchmark dose in, 130, 131f
    ecological, 151
    European scheme, 159
    level of sophistication, 105–106
    vs. accumulated relative risk, 86
    vs. LCIA, 151–152
Risk management, 103
Rivers, 23, 41, 48, 91, 112
Roads, 40, 45–46
Robustness, 200
Runoff, 58, 91, 93, 94, 107

S

Safeguard subjects, 4. *See also* Areas of protection
Safety factors, 129, 130, 157, 175
Scientific practice, 197
Scope, 1–2, 7, 67, 189, 214–215
Screening-level assessments, 7, 105, 106, 152
Seasonal variation, 81, 83, 93
Secondary pollutants, 114, 138, 143
Sediment, 91, 92, 158–159
Semivolatile organic compounds, 107
Sensitivity
    of biodiversity indicators, 54t, 55t, 56
    of category indicators, 17
    of criteria for weighting, 200
    ecosystem, 153
    of operationalisation for life-support, 59, 60
    of options for operationalisation, 24–25, 37t
    of species to toxin, 101, 127, 154–155, 154f, 158
Sensitivity analysis, 7, 94
SETAC. *See* Society of Environmental Toxicology and Chemistry
Severity-based indicators, 133–138, 139–144, 140t. *See also* Disability-adjusted life years; Quality-adjusted life years; Years of life lost
Silviculture
    deforestation as occupational impact, 43
    deforestation as transformation process, 41, 42
    of *Entandrophragma*, 38
    FAO monitoring, 31
    as land use type, 46–47
    planted seedlings, 28
    threatened by high extraction rates, 32t
    wood, 27, 34, 36, 38
SimpleBOX, 109, 110t
Simple Multi-Attribute Rating Technique (SMART), 190, 191
Simpson index, 56
Simulation Model for Acidification's Regional Trends/Multistress Model voor de Vegetatie (SMART/MOVE), 70
Single-media models, 113, 113t

Site-generic impact assessment
    acidification, 84
    aquatic eutrophication, 91–92
    limitations, 7
    modelling issues, 116–118
    photooxidant formation, 79–80
    regionality of mean values, 69, 69 n.4
    terrestrial eutrophication, 88
Site-specific impact assessment. *See also* Local aspects; Regional aspects
    acidification, 80–81, 84–86
    aquatic eutrophication, 92–93
    characterisation and standard deviation, 68–69
    modelling issues, 116–118
    need for, 7, 8
    photooxidant formation, 80–81
    terrestrial eutrophication, 88–89
Slope factor, 132
SMART. *See* Simple Multi-Attribute Rating Technique
Smog. *See* Photooxidant formation
Societal aspects
    areas of protection, 216t, 217, 218t, 222, 223f, 224
    competition for resources, 19
    cultural degradation, 49–50
    safeguard subjects, 215
    values, 180, 187–188, 199, 202
    willingness-to-pay, 187, 188
Society of Environmental Toxicology and Chemistry (SETAC)
    SETAC Europe First Working Group on LCIA (WIA-1), 1, 16, 40, 66, 101, 149, 178, 216
    SETAC Europe Second Working Group on LCIA (WIA-2), 1, 65–66, 117, 123, 150, 209
    UNEP-SETAC Life-Cycle Initiative, 7, 109, 141
    workshop on multimedia fate models, 109
Society of Toxicology, 129
Soil
    as combination of biotic and abiotic, 18
    deterioration with acidification, 70
    distribution of toxin, 159
    erosion, 41, 43, 58
    estimation of PNEC, 159
    leaching of nutrients, 82, 86, 93
    quality, 11, 15, 58
    regeneration, 13, 57, 58
    role of topsoil, 57–58
    sand, gravel, river clay as fund resources, 23
    temporal issues, 70
Solar power, 2, 4, 25
Solar radiation. *See* Radiation
Solvents, 75, 78
Sophistication, degree of
    human toxicity indicators, 142–143, 145
    models, 104, 105–106, 113
Source-Receptor Ozone Model (SROM), 111f, 112
Spatial aspects
    acidification, 68–69, 83, 84–86
    aquatic eutrophication, 92–93

climate change, 73
differentiation of stressors, 6–8. *See also* Regional aspects
distribution of resources, 19–20, 159
ozone depletion, 76
ozone formation, 78
photooxidant formation, 80–81
reference system for normalisation, 182
terrestrial eutrophication, 88–89
Speciation, 107
Species diversity, 51–53, 54t, 55t, 56
Species richness, 53, 56
Species sensitivity distribution (SSD), 154–155, 157, 158, 164–165
SROM. *See* Source-Receptor Ozone Model
SSD. *See* Species sensitivity distribution
Stakeholders, 189
Standards
accumulated ozone over threshold, 80
ISO. *See* International Organization for Standardization
mandatory elements/requirements, 2, 3, 16
optional elements/recommendations, 2–3, 16
performance, 190
species selection, 31, 32t
WHO guidelines for photooxidants, 78
Stepwise approach, 145
Stock resources
dynamics of abiotic, 15, 17, 18–20
dynamics of biotic, 33–39
recovery time, 35–36
stock estimates of abiotic, 21
Stoichiometric coefficient, 82
Stratospheric ozone depletion. *See* Ozone depletion
Stressor-impact chain
acidifying substances, 5
definition, 211
ecotoxicity, 151f, 164–165
environmental mechanisms, 13–14
human toxicity, 102f, 110, 113
interconnectedness, 4, 14
position for category indicators along, 31, 33
and subjectivity, 8
Stressors. *See also* Chemical contaminants; Emissions; Land occupation; Land use; *specific environmental processes*
beginning with. *See* Bottom-up approach
damages, definition, 5
definition, 211
intended *vs.* non-intended, 46 n.5
relation between land use and, 48–50
temporal and spatial differentiation, 6–8
Subjectivity, 8, 199–200, 202–203
Sulfurhexafluoride, 72
Sulphur dioxide
and acidification, 82, 84, 85t, 86
human toxicity, 138
in multimedia models, 114
in photooxidant formation, 79
Sulphur dioxide equivalent, 84, 85t
Sulphuric acid, 85t

Sulphur trioxide, 85t
Sustainability, 19
Synergism, 131, 157
Systems analysis, as function of LCIA, 213–214, 222

T

A Technical Framework for Life-Cycle Assessment, 1
Technical skills necessary, 193
Technology abatement weighting, 185–186
Technosphere, 214. *See also* Economic aspects
Tellus system, 188
Temporal aspects
acidification, 86–87
aquatic eutrophication, 93
climate change, 73
differentiation of stressors, 7–8
different land uses over time, 45–46
duration of land occupation, 45
generations, 36, 50–51, 186
ozone depletion, 76
photooxidant formation, 81
recovery time of biotic resources, 35–36
reference system for normalisation, 182
selection of time period, 69–70, 180
terrestrial eutrophication, 89
time-integrated exposure in models, 115, 118
time lag for human toxicity, 142
time lag for impacts on biodiversity, 49
Temporal extrapolation, 129
Terrestrial eutrophication
category indicators, 88
impact category description, 87–88
recommendations, 90, 96
relevant substances, 88
site-generic characterisation factors, 88
spatial aspects, 88–89
state-of-the-art science, 88
temporal aspects, 89
Terrestrial eutrophication potential (TEP), 88
Tetrachloromethane, 72
Tetrafluoromethane, 72
Thermal infrared airborne imagery, 58
Threshold Inventory Interpretation Methodology (TIIM), 84
Thresholds, 126, 131, 190
Toluene, 151f
Top-down approach, 4–5, 180
TOPKAT, 176
Total potential (EI), 185
Total Risk Integrated Methodology (TRIM), 110t
Towards a Methodology for Life-Cycle Impact Assessment, 1
Toxic chemicals. *See* Chemical contaminants
Toxic dose (TD), 132
Toxic equivalency potentials (TEP), 103, 105, 106–109, 112, 127
Toxicity assessment
interface between fate and effects, 106, 107t
level of sophistication, 103–106, 113, 142–143, 145

Toxicity assessment *(cont.)*
  ranking and scoring, 2, 104–105
  and risk assessment, 105–106
  stepwise approach, 145
Toxicity tests, 152, 153–157, 158, 175
Toxicodynamics. *See* Pharmacodynamics
Toxics Release Inventory (TRI), 155
Toxic Substances Control Act (TSCA), 174
Transfer factors, 109
Transformation impact, 40, 41–43, 43 n.2, 45–46
Transformation process, 41, 107, 109, 116. *See also* Metabolism; Photooxidant formation; Speciation
Transformation rate, 107
Transitivity, of criteria for weighting, 200–201
Transparency
  ecotoxicity indicators, 167t, 168
  human toxicity indicators, 141–142
  models, 115
  weighting criteria selection, 180–181, 196–197
Travel cost method, 186
Travel distance, 112
TRI. *See* Toxics Release Inventory
Trichlorofluoromethane, 75
Trophic level
  biomagnification, 108, 153, 165, 169
  and PNEC assessment factors, 156t

U

Ultraviolet absorption capacity, 74
Ultraviolet radiation. *See* Radiation
Umweltbundesamt ranking-method, 184–185
Uncertainty
  in best available practice, 66–67
  in data, 66, 104
  of ecotoxicity indicators, 167t, 168–169
  and extrapolation factors, 129, 155–156
  future energy consumption, 22
  in impact characterisation, 94
  in models, 66, 94, 115, 117, 118–119, 150
  Monte Carlo propagation, 115
  parameter, 161
  PNEC assessment factors, 156, 156t
  secant gradient model, 161
  tangential gradient of dose-response curve, 161–162, 161f
  value-independent, 181
Uncertainty-oriented ranking (UNORRA), 185
UNECE convention. *See* CLRTAP modelling framework
Uniform System for the Evaluation of Substances (USES), 108f, 110t
  EUSES, 114
  USES-LCA, 110t
United National Environment Programme (UNEP)
  Division of Technology, 215
  estimate on number of species, 26
  Intergovernmental Panel on Climate Change, 72
  ozone depletion activities, 75
  UNEP-SETAC Life-Cycle Initiative, 7, 8, 109, 141, 209–210, 214, 216
  UNEP-WCMC databases, 39
Unit lifetime risk, 132
UNORRA. *See* Uncertainty-oriented ranking
Uptake, 127
U.S. Environmental Protection Agency
  ecotoxicity assessment, 156
  hazard ranking system, 105
  Industrial Source Complex Model, 111
  Integrated Risk Information System, 131f, 132, 133, 134, 135
  Toxics Release Inventory, 155
  use of RfD, 126
U.S. National Acid Precipitation Assessment Program (NAPAP), 83

V

Validation
  of ecotoxicity indicators, 166–167, 167t
  of human toxicity indicators, 139–141, 139t, 140t
  of models, 114–115
  PNEC estimation methods, 157–158
Valuation. *See also* Economic aspects; Weighting
  adversity of toxic chemical, 102
  contingent, 187, 188, 197, 201
  definitions, 212
  just use, 177
  life expectancy, 136
  methods, 177
  ranking, 190–191
  and severity-based indicators, 135
  total economic, 177
  transparency in criteria selection, 180–181, 196–197
  use *vs.* non-use values, 186–187
Value of Statistical Life, 136
Value-sphere, 214
Virtually safe dose, 129
Virtual Pollution Prevention Costs 99, 188
Visibility reduction, 125 n.3
Volatile organic compounds (VOC), 78, 79, 80 n.10, 81
Volcanic outputs, 17
Vulnerable species, 31, 34, 38

W

Waste Minimization Prioritization Tool, 105
Water
  aquifers, 23
  cloud, 82
  evaporation, 49
  ground, 19, 23
  precipitation, 15, 17, 82, 107
  runoff, 58, 91, 93, 94, 107
  solid land *vs.* intracontinental water bodies, 48
  surface. *See* Lakes; Rivers
Water quality, 13, 57, 112
Websites
  IRIS database, 132

IUCN Red List database, 34
Weighting. *See also* Severity-based indicators
  of abiotic factors, 21
  criteria for evaluation, 192–201, 194*t*, 198*t*
  damage function *vs.* environmental themes, 6
  definition, 178
  economic impacts. *See* Valuation
  emissions sources, and acidification, 83, 85–86
  evaluation, 195–202, 198*t*
  intereffect, 199–200
  within land use impact categories, 60–61
  of land use types, 56
  monetisation, 186–188, 195–196, 203
  need for off-the-shelf qualities, 8
  and normalisation, timing, 178, 181
  overview, 3, 4, 177, 185–192
  panel methods, 182, 183, 188–191
  proxy, 185
  recommendations, 202–203
  subjectivity in, 199–200, 202–203
  technology abatement, 185–186
  uncertainty in, 66
  value choices. *See* Valuation
WIA-1. *See under* Society of Environmental Toxicology and Chemistry
WIA-2. *See under* Society of Environmental Toxicology and Chemistry
Willingness-to-pay (WtP), 125, 136, 140*t*, 186, 187, 188, 192
  based on damage costs (WtPDam), 195–196, 197, 198, 198*t*, 200, 201
  based on environmental taxes (WtPTax), 196, 198, 198*t*, 199, 200, 201
  based on marginal prevention costs (WtPPrev), 196, 197, 198, 198*t*, 199, 200
Windrose Trajectory Model (WTM), 111–112, 111*f*
WMO. *See* World Meteorological Organisation
Wood, 27, 34, 36, 38
World Conservation Monitoring Centre (WCMC), 26, 31, 32*t*, 35, 39, 126
World Health Organization (WHO), 78, 135, 140*t*, 141
World Meteorological Organisation (WMO), 72, 75
WTM. *See* Windrose Trajectory Model
WtP. *See* Willingness-to-pay

X

XLCA. *See* Explicit LCA

Y

Years of life lost (YOLL), 123, 135–136, 137, 140, 140*t*, 141, 144–145

Z

Zooplankton, 91

SETAC

A Professional Society for Environmental Scientists and Engineers and Related Disciplines Concerned with Environmental Quality

The Society of Environmental Toxicology and Chemistry (SETAC), with offices currently in North America and Europe, is a nonprofit, professional society established to provide a forum for individuals and institutions engaged in the study of environmental problems, management and regulation of natural resources, education, research and development, and manufacturing and distribution.

Specific goals of the society are:
- Promote research, education, and training in the environmental sciences.
- Promote the systematic application of all relevant scientific disciplines to the evaluation of chemical hazards.
- Participate in the scientific interpretation of issues concerned with hazard assessment and risk analysis.
- Support the development of ecologically acceptable practices and principles.
- Provide a forum (meetings and publications) for communication among professionals in government, business, academia, and other segments of society involved in the use, protection, and management of our environment.

These goals are pursued through the conduct of numerous activities, which include:
- Hold annual meetings with study and workshop sessions, platform and poster papers, and achievement and merit awards.
- Sponsor a monthly scientific journal, a newsletter, and special technical publications.
- Provide funds for education and training through the SETAC Scholarship/Fellowship Program.
- Organize and sponsor chapters to provide a forum for the presentation of scientific data and for the interchange and study of information about local concerns.
- Provide advice and counsel to technical and nontechnical persons through a number of standing and ad hoc committees.

SETAC membership currently is composed of more than 5,000 individuals from government, academia, business, and public-interest groups with technical backgrounds inchemistry, toxicology, biology, ecology, atmospheric sciences, health sciences, earth sciences, and engineering.

If you have training in these or related disciplines and are engaged in the study, use, or management of environmental resources, SETAC can fulfill your professional affiliation needs.

All members receive a newsletter highlighting environmental topics and SETAC activities, and reduced fees for the Annual Meeting and SETAC special publications.

All members except Students and Senior Active Members receive monthly issues of *Environmental Toxicology and Chemistry (ET&C)*, a peer-reviewed journal of the Society. Student and Senior Active Members may subscribe to the journal. Members may hold office and, with the Emeritus Members, constitute the voting membership.

If you desire further information, contact the appropriate SETAC office.

SETAC North America
1010 North 12th Avenue
Pensacola, Florida 32501-3367 USA
T 850 469 1500   F 850 469 9778
E setac@setac.org

SETAC Europe
Avenue de la Toison d'Or 67
B-1060 Brussels, Belgium
T 32 2 772 72 81   F 32 2 770 53 83
E setac@setaceu.org

www.setac.org

Environmental Quality Through Science®